# GENETICS, GENOMICS AND BREEDING OF SOYBEAN

# Genetics, Genomics and Breeding of Crop Plants

*Series Editor*
Chittaranjan Kole
Department of Genetics and Biochemistry
Clemson University
Clemson, SC
USA

## Books in this Series:

*Published or in Press*:
- Jinguo Hu, Gerald Seiler & Chittaranjan Kole: *Sunflower*
- Kristin D. Bilyeu, Milind B. Ratnaparkhe & Chittaranjan Kole: *Soybean*
- Robert Henry & Chittaranjan Kole: *Sugarcane*

*Books under preparation*:
- Jan Sadowsky & Chittaranjan Kole: *Vegetable Brassicas*
- Kevin Folta & Chittaranjan Kole: *Berries*
- C.P. Joshi, Stephen DiFazio & Chittaranjan Kole: *Poplar*
- James M. Bradeen & Chittaranjan Kole: *Potato*
- Jose Miguel Martinez Zapater, Anne-Françoise Adam Blondon & Chittaranjan Kole: *Grapes*

# GENETICS, GENOMICS AND BREEDING OF SOYBEAN

*Editors*

## Kristin Bilyeu
USDA-ARS
University of Missouri
Columbia
USA

## Milind B. Ratnaparkhe
Centre for Applied Genetic Technologies
University of Georgia
Athens, USA, 30602

## Chittaranjan Kole
Department of Genetics and Biochemistry
Clemson University
Clemson, SC
USA

CRC Press
Taylor & Francis Group
an **informa** business
www.crcpress.com

6000 Broken Sound Parkway, NW
Suite 300, Boca Raton, FL 33487
270 Madison Avenue
New York, NY 10016
2 Park Square, Milton Park
Abingdon, Oxon OX 14 4RN, UK

Science Publishers
Enfield, New Hampshire

Published by Science Publishers, P.O. Box 699, Enfield, NH 03748, USA
An imprint of Edenbridge Ltd., British Channel Islands

E-mail : _info@scipub.net_ Website : _www.scipub.net_

*Marketed and distributed by*:

CRC Press
Taylor & Francis Group
an **informa** business
www.crcpress.com

6000 Broken Sound Parkway, NW
Suite 300, Boca Raton, FL 33487
270 Madison Avenue
New York, NY 10016
2 Park Square, Milton Park
Abingdon, Oxon OX 14 4RN, UK

**Cover illustration reproduced by courtesy of Dale Rehder**

ISBN 978-1-57808-681-8

**Library of Congress Cataloging-in-Publication Data**
Genetics, genomics and breeding in soybean / editors, Kristin
Bilyeu, Milind
B. Ratnaparkhe, Chittaranjan Kole.
      p. cm.
    Includes bibliographical references and index.
    ISBN 978-1-57808-681-8 (hardcover)
  1.   Soybean--Research. 2.   Soybean--Breeding. 3.   Soybean--
Genetics.   I.
Bilyeu, Kristin D. II. Ratnaparkhe, Milind B. III. Kole,
Chittaranjan.
    SB205.S7G465 2010
    633.3'42--dc22
                                                        2009053034

*The views expressed in this book are those of the author(s) and the publisher does not assume responsibility for the authenticity of the findings/conclusions drawn by the author(s). Also no responsibility is assumed by the publishers for any damage to the property or persons as a result of operation or use of this publication and/or the information contained herein.*

Printed in the United States of America

# Preface to the Series

Genetics, genomics and breeding has emerged as three overlapping and complimentary disciplines for comprehensive and fine-scale analysis of plant genomes and their precise and rapid improvement. While genetics and plant breeding have contributed enormously towards several new concepts and strategies for elucidation of plant genes and genomes as well as development of a huge number of crop varieties with desirable traits, genomics has depicted the chemical nature of genes, gene products and genomes and also provided additional resources for crop improvement.

In today's world, teaching, research, funding, regulation and utilization of plant genetics, genomics and breeding essentially require thorough understanding of their components including classical, biochemical, cytological and molecular genetics; and traditional, molecular, transgenic and genomics-assisted breeding. There are several book volumes and reviews available that cover individually or in combination of a few of these components for the major plants or plant groups; and also on the concepts and strategies for these individual components with examples drawn mainly from the major plants. Therefore, we planned to fill an existing gap with individual book volumes dedicated to the leading crop and model plants with comprehensive deliberations on all the classical, advanced and modern concepts of depiction and improvement of genomes. The success stories and limitations in the different plant species, crop or model, must vary; however, we have tried to include a more or less general outline of the contents of the chapters of the volumes to maintain uniformity as far as possible.

Often genetics, genomics and plant breeding and particularly their complimentary and supplementary disciplines are studied and practiced by people who do not have, and reasonably so, the basic understanding of biology of the plants for which they are contributing. A general description of the plants and their botany would surely instill more interest among them on the plant species they are working for and therefore we presented lucid details on the economic and/or academic importance of the plant(s); historical information on geographical origin and distribution; botanical origin and evolution; available germplasms and gene pools, and genetic and cytogenetic stocks as genetic, genomic and breeding resources; and basic information on taxonomy, habit, habitat, morphology, karyotype, ploidy level and genome size, etc.

Classical genetics and traditional breeding have contributed enormously even by employing the phenotype-to-genotype approach. We included detailed descriptions on these classical efforts such as genetic mapping using morphological, cytological and isozyme markers; and achievements of conventional breeding for desirable and against undesirable traits. Employment of the in vitro culture techniques such as micro- and megaspore culture, and somatic mutation and hybridization, has also been enumerated. In addition, an assessment of the achievements and limitations of the basic genetics and conventional breeding efforts has been presented.

It is a hard truth that in many instances we depend too much on a few advanced technologies, we are trained in, for creating and using novel or alien genes but forget the infinite wealth of desirable genes in the indigenous cultivars and wild allied species besides the available germplasms in national and international institutes or centers. Exploring as broad as possible natural genetic diversity not only provides information on availability of target donor genes but also on genetically divergent genotypes, botanical varieties, subspecies, species and even genera to be used as potential parents in crosses to realize optimum genetic polymorphism required for mapping and breeding. Genetic divergence has been evaluated using the available tools at a particular point of time. We included discussions on phenotype-based strategies employing morphological markers, genotype-based strategies employing molecular markers; the statistical procedures utilized; their utilities for evaluation of genetic divergence among genotypes, local landraces, species and genera; and also on the effects of breeding pedigrees and geographical locations on the degree of genetic diversity.

Association mapping using molecular markers is a recent strategy to utilize the natural genetic variability to detect marker-trait association and to validate the genomic locations of genes, particularly those controlling the quantitative traits. Association mapping has been employed effectively in genetic studies in human and other animal models and those have inspired the plant scientists to take advantage of this tool. We included examples of its use and implication in some of the volumes that devote to the plants for which this technique has been successfully employed for assessment of the degree of linkage disequilibrium related to a particular gene or genome, and for germplasm enhancement.

Genetic linkage mapping using molecular markers have been discussed in many books, reviews and book series. However, in this series, genetic mapping has been discussed at length with more elaborations and examples on diverse markers including the anonymous type 2 markers such as RFLPs, RAPDs, AFLPs, etc. and the gene-specific type 1 markers such as EST-SSRs,

SNPs, etc.; various mapping populations including $F_2$, backcross, recombinant inbred, doubled haploid, near-isogenic and pseudotestcross; computer software including MapMaker, JoinMap, etc. used; and different types of genetic maps including preliminary, high-resolution, high-density, saturated, reference, consensus and integrated developed so far.

Mapping of simply inherited traits and quantitative traits controlled by oligogenes and polygenes, respectively has been deliberated in the earlier literature crop-wise or crop group-wise. However, more detailed information on mapping or tagging oligogenes by linkage mapping or bulked segregant analysis, mapping polygenes by QTL analysis, and different computer software employed such as MapMaker, JoinMap, QTL Cartographer, Map Manager, etc. for these purposes have been discussed at more depth in the present volumes.

The strategies and achievements of marker-assisted or molecular breeding have been discussed in a few books and reviews earlier. However, those mostly deliberated on the general aspects with examples drawn mainly from major plants. In this series, we included comprehensive descriptions on the use of molecular markers for germplasm characterization, detection and maintenance of distinctiveness, uniformity and stability of genotypes, introgression and pyramiding of genes. We have also included elucidations on the strategies and achievements of transgenic breeding for developing genotypes particularly with resistance to herbicide, biotic and abiotic stresses; for biofuel production, biopharming, phytoremediation; and also for producing resources for functional genomics.

A number of desirable genes and QTLs have been cloned in plants since 1992 and 2000, respectively using different strategies, mainly positional cloning and transposon tagging. We included enumeration of these and other strategies for isolation of genes and QTLs, testing of their expression and their effective utilization in the relevant volumes.

Physical maps and integrated physical-genetic maps are now available in most of the leading crop and model plants owing mainly to the BAC, YAC, EST and cDNA libraries. Similar libraries and other required genomic resources have also been developed for the remaining crops. We have devoted a section on the library development and sequencing of these resources; detection, validation and utilization of gene-based molecular markers; and impact of new generation sequencing technologies on structural genomics.

As mentioned earlier, whole genome sequencing has been completed in one model plant (Arabidopsis) and seven economic plants (rice, poplar, peach, papaya, grapes, soybean and sorghum) and is progressing in an array of model and economic plants. Advent of massively parallel DNA sequencing using 454-pyrosequencing, Solexa Genome Analyzer, SOLiD system, Heliscope and SMRT have facilitated whole genome sequencing in many other plants more rapidly, cheaply and precisely. We have included extensive coverage on the level (national or international) of collaboration and the strategies and status of whole genome sequencing in plants for which sequencing efforts have been completed or are progressing currently. We have also included critical assessment of the impact of these genome initiatives in the respective volumes.

Comparative genome mapping based on molecular markers and map positions of genes and QTLs practiced during the last two decades of the last century provided answers to many basic questions related to evolution, origin and phylogenetic relationship of close plant taxa. Enrichment of genomic resources has reinforced the study of genome homology and synteny of genes among plants not only in the same family but also of taxonomically distant families. Comparative genomics is not only delivering answers to the questions of academic interest but also providing many candidate genes for plant genetic improvement.

The 'central dogma' enunciated in 1958 provided a simple picture of gene function—gene to mRNA to transcripts to proteins (enzymes) to metabolites. The enormous amount of information generated on characterization of transcripts, proteins and metabolites now have led to the emergence of individual disciplines including functional genomics, transcriptomics, proteomics and metabolomics. Although all of them ultimately strengthen the analysis and improvement of a genome, they deserve individual deliberations for each plant species. For example, microarrays, SAGE, MPSS for transcriptome analysis; and 2D gel electrophoresis, MALDI, NMR, MS for proteomics and metabolomics studies require elaboration. Besides transcriptome, proteome or metabolome QTL mapping and application of transcriptomics, proteomics and metabolomics in genomics-assisted breeding are frontier fields now. We included discussions on them in the relevant volumes.

The databases for storage, search and utilization on the genomes, genes, gene products and their sequences are growing enormously in each second and they require robust bioinformatics tools plant-wise and purpose-wise. We included a section on databases on the gene and genomes, gene expression, comparative genomes, molecular marker and genetic maps, protein and metabolomes, and their integration.

Notwithstanding the progress made so far, each crop or model plant species requires more pragmatic retrospect. For the model plants we need to answer how much they have been utilized to answer the basic questions of genetics and genomics as compared to other wild and domesticated species. For the economic plants we need to answer as to whether they have been genetically tailored perfectly for expanded geographical regions and current requirements for green fuel, plant-based bioproducts and for improvements of ecology and environment. These futuristic explanations have been addressed finally in the volumes.

We are aware of exclusions of some plants for which we have comprehensive compilations on genetics, genomics and breeding in hard copy or digital format and also some other plants which will have enough achievements to claim for individual book volume only in distant future. However, we feel satisfied that we could present comprehensive deliberations on genetics, genomics and breeding of 30 model and economic plants, and their groups in a few cases, in this series. I personally feel also happy that I could work with many internationally celebrated scientists who edited the book volumes on the leading plants and plant groups and included chapters authored by many scientists reputed globally for their contributions on the concerned plant or plant group.

We paid serious attention to reviewing, revising and updating of the manuscripts of all the chapters of this book series, but some technical and formatting mistakes will remain for sure. As the series editor, I take complete responsibility for all these mistakes and will look forward to the readers for corrections of these mistakes and also for their suggestions for further improvement of the volumes and the series so that future editions can serve better the purposes of the students, scientists, industries, and the society of this and future generations.

Science publishers, Inc. has been serving the requirements of science and society for a long time with publications of books devoted to advanced concepts, strategies, tools, methodologies and achievements of various science disciplines. Myself as the editor and also on behalf of the volume editors, chapter authors and the ultimate beneficiaries of the volumes take this opportunity to acknowledge the publisher for presenting these books that could be useful for teaching, research and extension of genetics, genomics and breeding.

Chittaranjan Kole

# Preface to the Volume

The soybean is an economically very important leguminous seed crop for feed and food products that is rich in seed protein (about 40 %) and oil (about 20%); soybean enriches the soil by fixing nitrogen in symbiosis with bacteria. Soybean was domesticated in northeastern China about 2500 BC and subsequently spread to other countries. The enormous economic value of soybean was realized in the first two decades of the 20th century. In the international world trade markets, soybean is ranked number one in the world among the major oil crops. In addition to human consumption, it is a major protein source in animal feeds and is also becoming a major crop for biodiesel production. For many decades, plant breeders have used conventional breeding techniques to improve soybeans. World production of soybean has tripled in the last 20 years. Soybean production continues to expand as demand for soybeans and soybean products increases.

In the past decade or so, there has been a virtual explosion in the field of soybean research. This volume deals with the recent advances in soybean genome mapping, molecular breeding, genomics, sequencing and bioinformatics and is intended to bridge traditional research with modern molecular investigations on soybean. There are 15 chapters in all, each of which is relatively independent. When information from one chapter is needed for understanding the other, cross references are provided. This book begins with basic information about the soybean plant and germplasm diversity (Chapter 1), followed by classical genetics and traditional breeding (Chapter 2). Two chapters review activities on mapping single-gene traits (Chapter 3) and linkage map construction (Chapter 4). The construction of genetic linkage maps has enabled many soybean geneticists and breeders to dissect the genetic loci of interest into their genetic contributions to trait variation, additive or dominant effect and their interactions. Chapter 5 gives a concise overview on QTL mapping in soybean. Progress in mapping and identifying molecular markers associated with many agriculturally important traits has provided the foundation for marker-assisted selection in soybean. Traits that improve the value and functionality of soybean to give greater utility in food, health and industry uses have been emphasized by soybean breeders. Three chapters, molecular breeding (Chapter 6), gene cloning (Chapter 7) and candidate gene analysis of mutant soybean germplasm (Chapter 8) provide comprehensive reviews in the respective areas.

In recent years an understanding of the composition and organization of the soybean genome has developed. This has been the result of numerous advances in functional and structural genomics. In addition to development of an integrated genetic-physical map and DNA sequencing, large scale comparative and functional genomics studies are in progress. Recently, the US Department of Energy Joint Genome Institute (DOE JGI) has released a 7x sequence coverage of the soybean genome, making it widely available to the research community to advance new breeding strategies. Two chapters describe the recent advances in soybean functional genomics (Chapter 9) and whole genome sequencing (Chapter 10).

Soybean is also an attractive choice for comparative genomics and genome evolution studies as it is a major food crop, a legume (a large and diverse plant family that is both ecologically and economically important), and is an ancient polyploid. Soybean is known to have undergone two rounds of whole genome duplication since its divergence from the Rosid clade. The soybean genome is highly duplicated which complicates genome mapping and sequencing. Recent advances in soybean comparative genomics are highlighted in Chapter 11. Chapter 12, discusses common types of data that are currently available to be used in bioinformatic analysis including specific databases that exist and numerous tools that can be utilized and can aid the soybean community. Insights into soybean proteomics and metabolomics are accelerating at an impressive rate and are reviewed in chapters on proteomics (Chapter 13) and metabolomics (Chapter 14). The book ends with a chapter on future prospects of the soybean crop (Chapter 15). In the 15 chapters, reputed specialists provide concise and comprehensive reviews on the current status of soybean genome research. Each chapter has been written by one or more experts who have worked diligently in compiling information about their respective areas of expertise. We greatly appreciate their effort and time devoted to this book. We hope that this book is useful to soybean researchers as well as to people working with other crop species.

Kristin Bilyeu
Milind B. Ratnaparkhe
Chittaranjan Kole

# Contents

# List of Contributors

**Noureddine Benkeblia**
Department of Life Sciences, The University of West Indies, Jamaica.
Phone: +1-1-876-927-1202
Fax: +1-876-977-1075
Email: *noureddine.benkeblia@uwimona.edu.jm*

**Madan Bhattacharyya**
Department of Agronomy and Interdepartmental Genetics, Program, G303
Agronomy Hall, Iowa State University, Ames, IA 50011-1010, USA.
Phone: +1-515-294-2505
Fax: +1-515-294-2299
Email: *mbhattac@iastate.edu*

**Steven B. Cannon**
USDA-ARS, Corn Insects and Crop Genetics Research Unit, Ames, IA 50011,
USA.
Phone: +1-515-294-6971
Fax: +1-515-294-9359
Email: *steven.cannon@ars.usda.gov*

**Thomas J. Caperna**
USDA-ARS, Animal Biosciences and Biotechnology Laboratory, Beltsville,
MD 20705, USA.
Phone: +1-301-504-8506
Fax: +1-301-504-8623
Email: *thomas.caperna@ars.usda.gov*

**Pengyin Chen**
Department of Crop, Soil, and Environmental Sciences, University of
Arkansas , Fayetteville, AR 72701, USA.
Phone: +1-479-575-7564
Fax: +1-479-575-7465
Email: *pchen@uark.edu*

**Kei Chin C. Cheng**
Department of Plant Science, McGill University, 21111 Lakeshore Road,
Sainte Anne de Bellevue, Quebec H9W 5B8 Canada.
Phone: +1-514-398-8627
Fax: +1-514-398-7897
Email: *kei.cheng@elf.mcgill.ca*

**Kerry M. Clark**
Soybean Breeding, University of Missouri, 3600 New Haven Road, Columbia,
MO 65201, USA.
Phone: +1-573-882-0198
Fax: +1-573-884-5911
Email: *clarkk@missouri.edu*

**Ralph E. Dewey**
Department of Crop Science, North Carolina State University, Raleigh, NC
27695-8009, USA.
Phone: +1-919-515-2705
Email: *Ralph_dewey@ncsu.edu*

**Sangeeta Dhaubhadel**
Southern Crop Protection and Food Research Center, Agriculture and Agri-
Food Canada, London, Ontario, Canada N5V 4T3.
Phone: +1-519-457-1470 x 670
Fax: +1-519- 457-3997
Email: *dhaubhadels@agr.gc.ca*

**Wesley M. Garrett**
USDA-ARS, Animal Biosciences and Biotechnology Laboratory, Beltsville,
MD 20705, USA.
Phone: +1-301-504-7413
Fax: +1-301-504-8623
Email: *wesley.garrett@ars.usda.gov*

**Mark Gijzen**
Southern Crop Protection and Food Research Center, Agriculture and Agri-
Food Canada, London, Ontario, Canada N5V 4T3.
Phone: +1-519-457-1470 (280)
Fax: +1-519-457-3997
Email: *mark.gijzen@agr.gc.ca*

**David Goodstein**
Joint Genome Institute, 2800 Mitchell Drive, Walnut Creek, CA 94598 CA
94598, USA.
Phone: +1-510-643-9943
Fax: +1-925-296-5693
Email: *dmgoodstein@lbl.gov*

**David M. Grant**
USDA/ARS, Department of Agronomy, Iowa State University, Ames, IA
50011, USA.
Phone: +1-515-294-1205
Fax: +1-515-294-9359
Email: *david.grant@ars.usda.gov*

**Sachiko Isobe**
Kazusa DNA Research Institute, 2-6-7 Kazusa-kamatari, Kisarazu, Chiba
292-0818, Japan.
Phone: +81-438-52-3928
Fax: +81-438-52-3934
Email: *sisobe@kazusa.or.jp*

**Scott A Jackson**
Department Of Agronomy, Purdue University, West Lafayette, IN 47907, USA.
Phone: +1-765-496-3621
Fax: +1-765-496-7255
Email: *sjackson@purdue.edu*

**Setsuko Komatsu**
Soybean Physiology Research Team, National Institute of Crop Science,
Tsukuba, Japan.
Phone: +81-29-838-8693
Fax: +81-29-838-8693
Email: *skomatsu@affrc.go.jp*

**David A. Lightfoot**
Plant Genomics and Biotechnology, Public Policy Institute, 113, Department
of Plant, Soil and General Agriculture, Southern Illinois University—
Carbondale Carbondale, IL 62901-4415, USA.
Phone: +1-618-453-1797
Fax: +1-618-453-7457
Email: *ga4082@siu.edu*

**Julie M. Livingstone**
Department of Plant Science, McGill University, 21111 Lakeshore Road,
Sainte Anne de Bellevue, Quebec H9W 5B8 Canada.
Phone: +1-514-398-8627
Fax: +1-514-398-7897
Email: *julie.livingstone@mail.mcgill.ca*

**Devanand L Luthria**
USDA-ARS, Food Composition and Methods Development Laboratory, Beltsville, MD 20705, USA.
Phone: +1-301-504-7247 x 266
Fax: +1-301-504-8314
Email: *D.Luthria@ars.usda.gov*

**Jianxin Ma**
Department Of Agronomy, Purdue University, West Lafayette, IN 47907, USA.
Phone: +1-765-496-3662
Fax: +1-765-496-7255
Email: *maj@purdue.edu*

**Frédéric Marsolais**
Southern Crop Protection and Food Research Center, Agriculture and Agri-Food Canada, London, Ontario, Canada N5V 4T3.
Phone: +1-519-457-1470 (311)
Fax: +1-519-457-3997
Email: *Frederic.Marsolais@agr.gc.ca*

**Hideyuki Matsuura**
Graduate School of Agriculture, Hokkaido University, Sapporo, Japan.
Phone: +-81-11-706-2495
Fax: +81-11-706-2495
Email: *matsuura@chem.agr.hokudai.ac.jp*

**Therese Mitros**
Center for Integrative Genomics and
Department of Molecular and Cell Biology, University of California at Berkeley, Berkeley, CA 94720, USA.
Phone: +1-510-643-9943
Fax: +1-925-296-5693
Email : *tmitros@berkeley.edu*

**Maria J. Monteros**
Samuel Roberts Noble Foundation, 2510 Sam Noble Parkway, Ardmore, OK 73401, USA.
Phone: +1-580-224-6810
Fax: +1-580-224-6802
Email: *mjmonteros@noble.org*

**Takuji Nakamura**
Soybean Physiology Research Team, National Institute of Crop Science,
Tsukuba, Japan.
Phone: +81-29-838-8392
Fax: +81-29-838-8392
Email: *takuwan@naro.affrc.go.jp*

**Savithiry S. Natarajan**
USDA-ARS, Soybean Genomics and Improvement Laboratory, PSI-10300,
Baltimore Avenue, Beltsville, MD 20705, USA.
Phone: +1-301-504-5258
Fax: +1-301-504-5728
Email: *savi.natarajan@ars.usda.gov*

**Will Nelson**
University of Arizona, Bio5 Institute, 1657 E. Helen Street, Tucson, AZ 85721,
USA.
Phone: +1-520-621-1945
Fax: +1-520-621-7186
Email: *will@agcol.arizona.edu*

**Keiki Okazaki**
Rhizosphere Environment Research Team, National Agricultural Research
Center for Hokkaido Region, Sapporo, Japan.
Phone: +81-11-857-9243
Fax: +81-11-857-9243
Email: *okazakik@affrc.go.jp*

**James Orf**
Department of Agronomy and Plant Genetics, University of Minnesota, 411
Borlaug Hall, 1991 Upper Buford Circle, St. Paul, MN 55108-6026, USA.
Phone: +1-612-625-8275
Fax: +1-612-625-1268
Email: *orfxx001@umn.edu*

**Ed Ready**
United Soybean Board 16305 Swingley Ridge Rd., Suite 120, Chesterfield,
MO 63017, USA.
Phone: +1-314-579-1580
Fax: +1-314-579-1599
Email: *eready@smithbucklin.com*

**Dan Rokhsar**
Joint Genome Institute, 2800 Mitchell Drive, Walnut Creek, CA 94598 CA 94598, USA.
Phone: +1-510-642-8314
Fax: +1-925-296-5693
Email: *dsrokhsar@lbl.gov*

**Andrew M. Scaboo**
Department of Crop, Soil, and Environmental Sciences, University of Arkansas, 115 Plant Science Building, Fayetteville, AR 72701, USA.
Phone: +1-479-575-2109
Fax: +1-479-575-7465
Email: *ascaboo@uark.edu*

**Jeremy Schmutz**
HudsonAlpha Genome Sequencing Center, 601 Genome Way, Huntsville, AL 35806, USA.
Phone: +1-256-327-5213
Fax: +1-256-327-0964
Email: *jschmutz@hudsonalpha.org*

**Takuro Shinano**
Rhizosphere Environment Research Team, National Agricultural Research Center for Hokkaido Region, Sapporo, Japan.
Phone: +81-11-857-9243
Fax: +81-11-857-9243
Email: *shinano@affrc.go.jp*

**Randy C Shoemaker**
USDA-ARS, Corn Insects and Crop Genetics Research Unit, Ames, IA 50011, USA.
Phone: +1-515-294-6233
Fax: +1-515-294-2299
Email: *randy.shoemaker@ars.usda.gov*

**Shengqiang Shu**
Joint Genome Institute, 2800 Mitchell Drive, Walnut Creek, CA 94598, USA.
Phone: +1-510-643-9943
Fax: +1-925-296-5693
Email: *SQShu@lbl.gov*

**David A. Sleper**
National Center for Soybean Biotechnology, Division of Plant Sciences, University of Missouri, Columbia, MO 65211-7310, USA.
Phone: +1-573-882-7320
Fax: +1-573-884-9676
Email: *sleperd@missouri.edu*

**Martina V. Strömvik**
Department of Plant Science, McGill University, 21111 Lakeshore Road, Sainte Anne de Bellevue, Quebec H9W 5B8 Canada.
Phone: +1-514-398-8627
Fax: +1-514-398-7897
Email: *martina.stromvik@mcgill.ca*

**Satoshi Tabata**
Kazusa DNA Research Institute, 2-6-7 Kazusa-kamatari, Kisarazu, Chiba 292-0818, Japan.
Phone: +81-438-52-3933
Fax: +81-438-52-3934
Email : *tabata@kazusa.or.jp*

**Jennifer Tedman-Jones**
Southern Crop Protection and Food Research Center, Agriculture and Agri-Food Canada, London, Ontario, Canada N5V 4T3.
Phone : +1-800-667-2547 (2816)
Fax : +1-450-686-7012
Email: *jennifer.tedman-jones@roche.com*

**Hirofumi Uchimiya**
Institute of Molecular and Cellular Biosciences, The University of Tokyo, Tokyo, Japan.
Phone: +81-3-5841-7845
Fax: 81-3-5841-8466
Email: *uchimiya@iam.u-tokyo.ac.jp*

**David R. Walker**
USDA-ARS, Soybean/Maize Germplasm, Pathology and Genetics Research Unit, Urbana, IL 61801, USA.
Phone: +1-217-244-1274
Fax: +1-217-244-7703
Email: *david.walker@ars.usda.gov*

**Dechun Wang**
Department of Crop and Soil Sciences, Michigan State University, A384-E
Plant and Soil Sciences Building, East Lansing, MI 48824-1325, USA.
Phone: +1-517-355-0271 x 1188
Fax: +1-517-353-3955
Email: *wangdech@msu.edu*

**Jun Wasaki**
Graduate School of Biosphere Science, Hiroshima University, Higashi
Hiroshima, Japan.
Phone: +81-82-424-4370
Fax: +81-82-424-4370
Email: *junw@hiroshima-u.ac.jp*

**Toshihiro Watanabe**
Graduate School of Agriculture, Hokkaido University, Sapporo, Japan.
Phone: +81-11-706-2498
Fax: +81-11-706-2498
Email: *nabe@chem.agr.hokudai.ac.jp*

**Jennifer L. Yates**
Monsanto Company, 32545 Galena-Sassafras Road, Galena, MD 21635,
USA.
Phone: +1-410-648-5093 x 35
Fax: +1-410-648-5715
Email: *jennifer.yates@monsanto.com*

**P. Zhang**
Department of Crop Science, North Carolina State University, Raleigh, NC
27695-8009, USA.
Phone: +1-919-515-2705
Fax: +1-919-515-7959
Email: *pzhang3@unity.ncsu.edu*

# Abbreviations

| | |
|---|---|
| 1YT | First year trial |
| 2D-PAGE | Two-dimensional polyacrylamide gel electrophoresis |
| 2YT | Second year trial |
| 3YT | Third year trial |
| 4DTv | Transversion rate at four-fold synonymous codon positions |
| 4YT | Fourth year trial |
| 5′ RATE | Robust analysis of 5′-transcript ends |
| ACP | Acyl carrier protein |
| AFLP | Amplified fragment length polymorphism |
| ANOVA | Analysis of variance |
| ARF | Auxin response transcription factor |
| BAC | Bacterial artificial chromosome |
| BBI | Bowman Birk Inhibitor |
| BC | Backcross |
| BES | BAC-end sequence |
| BLAST | Basic local alignment search tool |
| BLOSUM | Block substitution matrix |
| BPMV | Bean pod mottle virus |
| CAMTA | Calmodulin-binding transcription activator |
| CAPS | Cleaved amplified polymorphic sequences |
| CCP | Cysteine cluster protein |
| cDNA | Complementary-DNA |
| CE-MS | Capillary electrophoresis-Mass spectrometry |
| CGH | Comparative genomic hybridization |
| CGI | Common gate interface |
| CHCA | $\alpha$-Cyano-4-hydroxycinnamic acid |
| CHS | Chalcone synthase |
| CIG | Center for Integrative Genomics |
| cM | Centi-Morgan |
| CNV | Copy number variation |
| CPMV | Cowpea chlorotic mottle virus |
| DAG | Directed acyclic graph |
| DBI | Database interface |

| DDBJ | DNA Databank of Japan |
|---|---|
| DOE | Department of Energy |
| DP | Donor parent |
| EBI | European Bioinformatics Institute |
| EC | Enzyme Commission |
| EMBL | European Molecular Biological Laboratory |
| EMS | Ethylmethane sulfonate |
| ER | Endoplasmic reticulum |
| ESI | Electrospray ionization |
| EST | Expressed sequence tag |
| FAT | Fatty acid thioesterase |
| FDA | Food and Drug Administration |
| FPC | Fingerprinted contigs |
| FT-ICR-MS | Fourier transform-Ion cyclotron resonance-Mass spectrometry |
| GABA | $\gamma$-Aminobutyric acid |
| GC-MS | Gas chromatography-Mass spectrometry |
| GCOS | GeneChip operating software |
| GEO | Gene expression omnibus |
| GMO | Genetically modified organism |
| GO | Gene ontology |
| HI | Harvest index |
| HPLC | High-pressure liquid chromatography |
| HR | Hypersensitive response |
| ID | Identifier |
| INDEL | Insertion-deletion |
| IPG | Immobilized pH gradient |
| IRGSP | International Rice Genome Sequencing Project |
| JA | Jasmonic acid |
| JGI | Joint Genome Institute |
| KEGG | Kyoto Encyclopedia of Genes and Genomes |
| KTI | Kunitz trypsin inhibitor |
| LCM | Laser capture microdissection |
| LC-MS | Liquid chromatography-mass spectrometry |
| LC-MS(/MS) | High-performance liquid chromatography with tandem mass spectrometry |
| LD | Linkage disequilibrium |
| LG | Linkage group |
| LIS | Legume Information System |
| LRR | Leucine-rich repeat |
| LTR | Long terminal repeat |
| MABC | Marker-assisted backcrossing |
| MALDI | Matrix-assisted laser desorption ionization |

| | |
|---|---|
| MAS | Marker-assisted selection |
| MIAME | Minimum information about a microarray experiment |
| MIPS | Myo-inositol 1-phosphate synthase |
| miRNA | micro-RNA |
| MPSS | Massively parallel signature sequencing |
| mRNA | Messenger-RNA |
| MS | Mass spectrometry |
| MSTFA | N-Methyl-N-(trimethylsilyl)trifluoroacetamide |
| MYA | Million years ago |
| NAQF | Non-aqueous fractionation |
| NBD | Nucleotide-binding domain |
| NCBI | National Center for Biotechnology Information |
| NIL | Near-isogenic lines |
| NMR | Nuclear magnetic resonance |
| NSF | National Science Foundation |
| NUE | Nitrogen use efficiency |
| O&O | Order and orientation |
| ORF | Open-reading frame |
| PAM | Point-accepted mutation |
| PASA | Program to assemble spliced alignments |
| PCA | Principal component analysis |
| PCR | Polymerase chain reaction |
| PFGE | Pulse-field gel electrophoresis |
| Pi | Inorganic phosphorous |
| PI | Plant introduction |
| PMF | Peptide mass fingerprinting |
| PSSM | Position-specific scoring matrices |
| QIT MS | Quadrupole ion trap mass spectrometer |
| QTL | Quantitative trait locus |
| RAPD | Random(ly) amplified polymorphic DNA |
| rDNA | Ribosomal-DNA |
| RFLP | Restriction fragment length polymorphism |
| RGA | Resistance gene analog |
| *R*-gene | Resistance gene |
| RIL | Recombinant inbred line |
| RNAi | RNA interference |
| RP | Recurrent parent |
| SACPD | Stearoyl-ACP-desaturase enzyme |
| SAGE | Serial analysis of gene expression |
| SCN | Soybean cyst nematode |
| SDS | Sudden death syndrome |
| SGMD | Soybean Genomics and Microarray Database |
| SIB | Swiss Institute for Bioinformatics |

| | |
|---|---|
| SMV | Soybean mosaic virus |
| SNI | Single nucleotide insertion |
| SNP | Single nucleotide polymorphism |
| SoyGD | Soybean Genome Database |
| SPD | Single-pod decent |
| SPT | Single-plant-threshed |
| SSD | Single-seed descent |
| SSR | Simple sequence repeat |
| STS | Sequence tagged site |
| TAIR | The Arabidopsis Information Resource |
| TC | Tentative contig |
| TCA | Trichloroacetic acid |
| TE | Transposable element |
| TIGR | The Institute for Genome Research |
| TILLING | Targeted induced local lesions in genomes |
| TMCS | Trichlormethylchlorsilane |
| TOF | Time of flight |
| UCD | University of California, Davis |
| UCLA | University of California, Los Angeles |
| UPLC-MS/MS | Ultra Performance LC-MS/MS |
| URL | Uniform resource identifier |
| USDA | United States Department of Agriculture |
| VIGS | Virus-induced gene silencing |
| WGD | Whole-genome duplication |
| WGS | Whole-genome shotgun |
| WUE | Water use efficiency |
| YAC | Yeast artificial chromosome |

# Introduction

## James Orf

### ABSTRACT

Soybean, *Glycine max* (L.) Merr., is the world's most important oil seed crop. In the US during 2008, soybean was grown on over 30 million hectares, producing over 81 million metric tons valued at over US$27 billion. Soybean is processed to produce soybean oil, which is mainly used as human food or to produce biodiesel, and high-protein soybean meal that is used as animal feed. Soybean was domesticated in China, but only in the last 50 years or so has it become an important crop worldwide. The cultivated soybean has one wild annual relative, *G. soja*, and 23 wild perennial relatives. Soybean spread to many Asian countries two to three thousand years ago, but was not known in the West until the 18th century. Soybean has had less cytogenetic research than many other important crop species because of its small chromosomes and relatively large number of chromosomes ($2n = 40$). Soybean as a species has considerable genetic diversity, much of which remains to be explored. This presents breeders with opportunities to further improve soybean in the future.

**Keywords:** soybean production; history; taxonomy; genetic diversity; gene pools; cytogenetics

## 1.1 Soybean

The soybean, *Glycine max* (L.) Merr., is one of numerous domesticated plants used as human food. It is a major crop in the United States, Brazil, China and Argentina and important in many other countries. Currently world production of soybean is greater than any other oilseed crop. It has only

Department of Agronomy and Plant Genetics, University of Minnesota, 411 Borlaug Hall, 1991 Upper Buford Circle, St. Paul, MN 55108-6026, USA; e-mail: *orfxx001@umn.edu* been since the 1960s that soybean emerged as the dominant oilseed crop produced (Smith and Huyser 1987). Demand for soybean continued and still is continuing to grow on a worldwide basis (Wilcox 2004; Goldsmith 2008; Orf 2008).

## 1.2 Soybean Production

Soybean production continues to expand as demand for soybeans and soybean products increase. The demand comes from the increasing use of its oil for human consumption and for biodiesel, and an increasing demand for high-protein meal for animal feed in both developed and developing countries. Although soybean is classified as an oilseed, the soybean seed contains about 38–44% protein and 18–23% oil on a moisture-free basis. Traditional foods such as tofu, miso, natto, tempeh and soy sauce are made from soybean. The two main products derived from the soybean seed after processing are seed oil and the protein-containing meal. In the US, soybean oil is used as cooking oil as well as a main ingredient in margarine, salad oils, salad dressings, mayonnaise and shortening. The meal is used primarily as a protein source for swine, poultry, beef, dairy and fish. Soybean meal can also be used to make protein concentrate, texturized protein and protein isolates that are used in food products for human consumption.

Currently the US is the largest producer of soybean followed by Brazil, Argentina and China (USDA-FAS 2008). Table 1-1 shows the most recent data on world supply of soybean. Soybean is the oilseed with the greatest production on a worldwide basis accounting for approximately 56% of total oilseed production (USDA-FAS 2009). Soybean production occupies approximately 6% of the world's available land (Goldsmith 2008). Soybean hectarage, production and yield in the US from 1924 through 2008 are shown in Table 1-2. During that time harvested area ranged from 168,000 hectares in 1925 to 30,214,000 in 2006, yields from 0.74 tons per hectare in 1924 to 2.89 tons per hectare in 2005, and total production from 132,000 metric tons in 1925 to 86,848,000 metric tons in 2006. Over this time period the trend for hectarage, yield and production has been upward.

In 2008, soybean hectarage and production was reported from 31 states, all in the eastern part of the US (Table 1-3). The leading states in terms of production were Iowa (15%), Illinois (14%), Minnesota (9%), Indiana (8%), Nebraska (8%), Missouri (6%), Ohio (5%), South Dakota (5%), Kansas (4%), and Arkansas (4%). In the last 12 years, production has shifted from the southern and eastern parts of the soybean growing area more to the northern and western areas. This is evidenced from the fact that in 1969 the north-central states of Iowa, Illinois, Minnesota, Indiana, Nebraska, Ohio, Missouri, South Dakota, North Dakota, Michigan and Wisconsin produced 69% of the total soybeans in the US, while the southern states of Arkansas,

Table 1-1 Soybeans: World supply and distribution (000' metric tons).

| | 2004/05 | 2005/06 | 2006/07 | 2007/08 | 2008/09 (Jan) |
|---|---|---|---|---|---|
| **Production** | | | | | |
| US | 85,019 | 83,507 | 87,001 | 72,859 | 80,536 |
| Brazil | 53,000 | 57,000 | 59,000 | 61,000 | 59,000 |
| Argentina | 39,000 | 40,500 | 48,800 | 46,200 | 49,500 |
| China, Peoples Republic | 17,400 | 16,350 | 15,967 | 14,000 | 16,800 |
| India | 5,850 | 7,000 | 7,690 | 9,300 | 9,700 |
| Paraguay | 4,040 | 3,640 | 6,200 | 6,800 | 5,600 |
| Canada | 3,042 | 3,161 | 3,460 | 2,700 | 3,300 |
| Other | 8,422 | 9,513 | 9,428 | 8,028 | 8,765 |
| **Total** | 215,773 | 220,671 | 237,546 | 220,887 | 233,201 |
| **Imports** | | | | | |
| China, Peoples Republic | 25,802 | 28,317 | 28,726 | 37,816 | 36,000 |
| EU-27 | 14,539 | 13,937 | 15,291 | 15,148 | 14,150 |
| Japan | 4,295 | 3,962 | 4,094 | 4,014 | 4,000 |
| Mexico | 3,640 | 3,667 | 3,844 | 3,650 | 3,585 |
| Argentina | 692 | 584 | 1,986 | 2,954 | 2,535 |
| Taiwan | 2,256 | 2,498 | 2,436 | 2,149 | 2,350 |
| Thailand | 1,517 | 1,473 | 1,532 | 1,733 | 1,650 |
| Indonesia | 1,112 | 1,187 | 1,309 | 1,200 | 1,300 |
| Korea, Republic of | 1,240 | 1,190 | 1,231 | 1,231 | 1,260 |
| Egypt | 762 | 776 | 1,325 | 1,100 | 1,200 |
| Other | 7,629 | 6,489 | 7,280 | 7,640 | 7,907 |
| **Total** | 63,484 | 64,080 | 69,054 | 78,635 | 75,937 |
| **Exports** | | | | | |
| United States | 29,860 | 25,579 | 30,386 | 31,598 | 29,937 |
| Brazil | 20,137 | 25,911 | 23,485 | 25,364 | 25,250 |
| Argentina | 9,568 | 7,249 | 9,559 | 13,830 | 14,400 |
| Paraguay | 2,888 | 2,315 | 4,500 | 5,080 | 4,000 |
| Canada | 1,124 | 1,318 | 1,683 | 1,775 | 1,830 |
| Other | 1,210 | 1,408 | 1,889 | 1,830 | 1,771 |
| **Total** | 64,787 | 63,780 | 71,502 | 79,477 | 77,188 |

Most countries are on an October/September Marketing Year (MY). The US, Mexico, and Thailand are on a September/August MY. Canada is on an August/July MY. Paraguay is on a March/February MY and Turkey is on a March/February MY. Foreign Agricultural Service/USDA Office of Global Analysis January 2009.

Mississippi, Louisiana, South Carolina, Georgia and Alabama produced 19%. In 2008, the production in the north-central States was 84% and in the south only 9% of the total production. A number of reasons have been suggested for this shift including greater yield potential (and thus greater breeding efforts) in the north-central area, more diseases, insects and other challenges in the south and more available hectarage to shift to soybeans in the north-central states.

Table 1-2 Soybeans: Hectare, yield, and production, US 1924 to 2008.

| Year | Hectares harvested (000) | Yield per harvested hectare ton/ha | Production (000) MT |
|------|------|------|------|
| 1924 | 181 | 0.74 | 134 |
| 1925 | 168 | 0.79 | 132 |
| 1926 | 189 | 0.75 | 142 |
| 1927 | 230 | 0.82 | 189 |
| 1928 | 235 | 0.91 | 214 |
| 1929 | 287 | 0.89 | 257 |
| 1930 | 435 | 0.87 | 379 |
| 1931 | 462 | 1.01 | 470 |
| 1932 | 405 | 1.01 | 412 |
| 1933 | 423 | 0.87 | 368 |
| 1934 | 630 | 1.00 | 630 |
| 1935 | 1,181 | 1.13 | 1,332 |
| 1936 | 955 | 0.96 | 918 |
| 1937 | 1,047 | 1.20 | 1,257 |
| 1938 | 1,229 | 1.37 | 1,686 |
| 1939 | 1,748 | 1.40 | 2,455 |
| 1940 | 1,947 | 1.09 | 2,125 |
| 1941 | 2,385 | 1.22 | 2,920 |
| 1942 | 4,007 | 1.28 | 5,108 |
| 1943 | 4,211 | 1.23 | 5,179 |
| 1944 | 4,149 | 1.26 | 5,252 |
| 1945 | 4,350 | 1.21 | 5,261 |
| 1946 | 4,022 | 1.34 | 5,340 |
| 1947 | 4,621 | 1.10 | 5,078 |
| 1948 | 4,326 | 1.43 | 6,189 |
| 1949 | 4,245 | 1.50 | 6,379 |
| 1950 | 5,592 | 1.46 | 8,151 |
| 1951 | 5,514 | 1.40 | 7,730 |
| 1952 | 5,846 | 1.39 | 8,140 |
| 1953 | 6,006 | 1.22 | 7,332 |
| 1954 | 6,904 | 1.34 | 9,290 |
| 1955 | 7,541 | 1.35 | 10,179 |
| 1956 | 8,351 | 1.46 | 12,237 |
| 1957 | 8,447 | 1.56 | 13,168 |
| 1958 | 9,717 | 1.63 | 15,806 |
| 1959 | 9,166 | 1.58 | 14,516 |
| 1960 | 9,580 | 1.58 | 15,120 |
| 1961 | 10,936 | 1.69 | 18,484 |
| 1962 | 11,181 | 1.63 | 18,229 |
| 1963 | 11,589 | 1.64 | 18,228 |
| 1964 | 12,471 | 1.53 | 19,093 |
| 1965 | 13,952 | 1.65 | 23,034 |
| 1966 | 14,801 | 1.71 | 25,291 |
| 1967 | 16,121 | 1.65 | 26,598 |
| 1968 | 16,763 | 1.79 | 30,154 |
| 1969 | 16,741 | 1.84 | 30,866 |
| 1970 | 17,111 | 1.79 | 30,702 |
| 1971 | 17,296 | 1.85 | 32,037 |

| Year | Hectares harvested (000) | Yield per harvested hectare ton/ha | Production (000) MT |
|---|---|---|---|
| 1972 | 18,502 | 1.87 | 34,611 |
| 1973 | 22,545 | 1.87 | 42,155 |
| 1974 | 20,793 | 1.59 | 33,132 |
| 1975 | 21,715 | 1.94 | 42,177 |
| 1976 | 20,007 | 1.75 | 35,102 |
| 1977 | 23,421 | 2.06 | 48,140 |
| 1978 | 25,784 | 1.98 | 50,905 |
| 1979 | 28,489 | 2.16 | 61,581 |
| 1980 | 27,464 | 1.78 | 48,965 |
| 1981 | 26,796 | 2.02 | 54,183 |
| 1982 | 28,124 | 2.12 | 59,664 |
| 1983 | 25,323 | 1.76 | 44,558 |
| 1984 | 26,776 | 1.89 | 50,690 |
| 1985 | 24,948 | 2.29 | 57,178 |
| 1986 | 23,616 | 2.24 | 52,915 |
| 1987 | 23,155 | 2.28 | 52,784 |
| 1988 | 23,236 | 1.81 | 42,190 |
| 1989 | 24,113 | 2.17 | 52,401 |
| 1990 | 22,887 | 2.29 | 52,463 |
| 1991 | 23,494 | 2.30 | 54,173 |
| 1992 | 23,584 | 2.53 | 59,665 |
| 1993 | 23,209 | 2.19 | 50,931 |
| 1994 | 24,628 | 2.78 | 68,505 |
| 1995 | 24,925 | 2.37 | 59,227 |
| 1996 | 25,656 | 2.53 | 64,839 |
| 1997 | 27,990 | 2.61 | 73,242 |
| 1998 | 28,529 | 2.61 | 74,665 |
| 1999 | 29,341 | 2.46 | 72,288 |
| 2000 | 29,325 | 2.56 | 75,123 |
| 2001 | 29,555 | 2.66 | 78,742 |
| 2002 | 29,361 | 2.55 | 75,077 |
| 2003 | 29,353 | 2.28 | 66,838 |
| 2004 | 29,953 | 2.84 | 85,089 |
| 2005 | 28,857 | 2.89 | 83,419 |
| 2006 | 30,214 | 2.87 | 86,848 |
| 2007 | 25,890 | 2.81 | 73,539 |
| 2008 | 30,280 | 2.67 | 81,287 |

Brazil is the second largest soybean producer in the world (Table 1-1). Soybean production has been increasing steadily in the last five years. Reports from within and outside Brazil indicate that there are large areas in the Cerrados ecological zone, especially in the states of Mato Grosso, Mato Grosso do Sul, Goias and Bahia, and perhaps even in the tropical rainforest zone for expansion of soybean production. There are a number of challenges for soybean production and export in Brazil including poor transportation infrastructure, diseases and insects and input costs. Nevertheless, it is expected that Brazil will surpass the US as the largest soybean producer in the world in the not too distant future. Brazil has well developed research organizations and has been able to consistently produce high yields with rapid adoption of modern technologies (Wilcox 2004).

Table 1-3 Soybean production by states of US in 2008 (million metric tons).

| | | | |
|---|---|---|---|
| Alabama | 0.34 | Nebraska | 6.20 |
| Arkansas | 3.29 | New Jersey | 0.07 |
| Delaware | 0.15 | New York | 0.29 |
| Florida | 0.03 | North Carolina | 1.51 |
| Georgia | 0.34 | North Dakota | 2.89 |
| Illinois | 11.75 | Ohio | 4.43 |
| Indiana | 6.71 | Oklahoma | 0.25 |
| Iowa | 12.21 | Pennsylvania | 0.47 |
| Kansas | 3.30 | South Carolina | 0.47 |
| Kentucky | 0.74 | South Dakota | 3.79 |
| Louisiana | 0.86 | Tennessee | 1.36 |
| Maryland | 0.40 | Texas | 0.14 |
| Michigan | 1.92 | Virginia | 0.50 |
| Minnesota | 7.25 | West Virginia | 0.02 |
| Mississippi | 2.15 | Wisconsin | 1.53 |
| Missouri | 5.25 | | |

Argentina is the third largest soybean producer in the world and second largest in South America following Brazil (Table 1-1). There has been some expansion of soybean production in Argentina in the last few years but there is not nearly as much opportunity for additional expansion of soybean production as in Brazil. The expansion of soybean production in Argentina can be attributed at least in part to more favorable economic policies by the government, although that has decreased in the last few years, the use of minimum and no-tillage production systems, adoption of double-cropping soybean after wheat and improvement in storage and transportation infrastructure (Wilcox 2004).

China continues to be a major producer of soybeans. The production of soybeans has been about the same in the last five years (Table 1-1). Wilcox (2004) indicates that the provinces of Heilongjiang, Liaoning and Inner Mongolia produce about 45% of the total in China. Most of the soybeans in these areas are seeded in the spring. He indicates that about 30% of the production is double-cropped following wheat in Henan, Shandong, Hebei and Anhui provinces. The remaining production is in the south and frequently follows rice. Although China was the largest soybean producer in the past, some soybean production areas have been occupied by other crops. In the last few years, soybean use has increased dramatically in China as the standard of living has increased and there is a greater demand for soybean oil and meat products from animals that consume soybean meal. In the last few years, the government has adopted policies to encourage more soybean production. Recently, China has become the largest importer of soybeans (Table 1-1).

India ranks fifth in soybean production on a worldwide basis (Table 1-1). India has recently expanded its soybean production (Wilcox 2004). Major soybean growing states include Madhya Pradesh, Maharashtra and Rajasthan. As population and living standards increase, it is expected that soybean production will increase in India.

Paraguay has become a significant producer of soybean because of fertile land for growing soybeans and favorable transportation for export (Table 1-1). Cultivars developed for Argentina or Brazil can be planted in Paraguay. Recently, there have been demands from some groups to reduce soybean plantings.

Canada has a limited area where soybean can be grown. Recently, Canada has ranked eighth in soybean production in the world. Cultivars and production practices are similar to Midwest US.

Soybean imports for the last five marketing years are shown in Table 1-1. The Peoples Republic of China has become the largest importer of soybeans. Until about 30 years ago, China was an exporter of soybeans (Wilcox 2004). The European Union continues to be a major importer of soybeans since there is a large demand for the protein in soybean meal and most countries do not have large areas that are favorable for soybean production. Japan has also been a major soybean importer for many years. Japan imports soybeans for human food use as well as for crushing. Other important importing countries include Mexico, Taiwan, Thailand, The Republic of Korea, Indonesia, and Egypt (Table 1-1).

About 29% of the world's production of soybeans was exported in 2008/09 (Table 1-1). Although Brazil has been increasing exports in recent years the United States has remained the number one soybean and soybean product exporter. Argentina is also a major exporter of soybeans. Paraguay exports over half of the soybeans they produce. Many of the soybeans exported from Canada are used for human food in Asian countries.

Production practices vary considerably around the world. The size of fields varies from a few square meters to thousands of hectares. The work of growing soybeans may be mainly done by hand or almost totally mechanized. While most soybeans are grown under rainfed conditions, irrigation is used for at least some production in many countries. Inputs also vary from little, if any to large amounts of fertilizer and pesticides throughout the growing season.

Since soybean is a leguminous crop, it fixes its own nitrogen in association with *Bradyrhizobium japonicum*. If soybeans have not been grown on the field or it has been many years since soybeans were raised, inoculant should be applied at planting to establish the bacteria in the soil (Hoeft et al. 2000). The specific amount of other major and minor nutrients to apply to the soil depends on the results of soil tests and the yield level anticipated. Heatherly and Elmore (2004) have described lime and fertility needs for soybeans in greater detail.

Cultivar selection is a very important step in achieving maximum soybean production. Improved cultivars are available for all soybean producing areas. Selecting a cultivar should be done on an individual field basis. Important aspects of a cultivar include yield potential in its area of adaptation, resistance to diseases, nematodes and insects, tolerance to various abiotic stresses (including soil pH, drought and salt), levels of protein and oil, and tolerance to herbicides. In the US a relative maturity system is used to indicate where cultivars are considered full season. Other parts of the world use a number of other systems to classify when and where cultivars should be planted. In most soybean growing areas of the world, there are many cultivars (both publicly and/or privately developed) available with different characteristics that will be suitable for almost any given environmental situation. In most cases, cultivar tests by public or private organizations are conducted to aid growers in selecting the best cultivar(s) for their fields. Use of high quality seed helps assure good results.

The value of the soybean crop in the US for the 2007/08 marketing year was approximately US\$27 billion based on the price farmers received (USDA-ERS 2009). Of course the value of production in other countries is more difficult to estimate given local markets and currency fluctuations. However, the value of the soybean crop in other countries is approximately proportional to the production figures shown in Table 1-1.

### 1.3 Taxonomy

The genus *Glycine* Willd. is a member of the family Fabaceae/Leguminosae, subfamily Papilionoideae, and the tribe Phaseoleae. Within the tribe Phaseoleae, there are 16 genera in the subtribe Glycininae (Lackey 1977a; Polhill 1994; Hymowitz 2008). Lackey (1977a) originally recognized 16 genera in the subtribe and subdivided them into two groups, Glycine and Shutaria. Polhill (1994) added two genera and rearranged the subtribe. Hymowitz (2008) has additional details and a table of the current subtribe Glycininae. The genus *Glycine* is divided into two subgenera, *Glycine* (perennials) and *Soja* (Moench) F.J. Herm. (annuals).

*Glycine* as a genus has a somewhat confused taxonomic history. Linnaeus originally introduced the name *Glycine* in his first edition of *Genera Plantaram* (Linneaus 1737). This was based on *Apios* of Boerhaave (Linnaeus 1754). *Glycine* is derived from the Greek *glykus* (sweet). The name probably refers to the sweetness of the edibile tubers produced by *G. apios* L. (Henderson 1881), now known as *Apios americana* Medik. Linnaeus in his 1753 publication *Species Plantarum* listed eight *Glycine* species. All these were subsequently moved to other genera, thus the Greek word *glykys*, currently does not refer to any *Glycine* species (Hymowitz and Singh 1987). In his 1753 publication, Linnaeus described the cultivated soybean as both

*Phaseolus max*, based on specimens he saw and *Dolichos soja* based on descriptions of other writers. Hymowitz (2008) provides further details and references regarding the current Latin name of the cultivated soybean.

Since the time of Linnaeus scholars have discussed the correct nomenclature for the cultivated soybean (Piper and Morse 1923; Ricker and Morse 1948; Lawrence 1949; Paclt 1949). The combination *Glycine max* as proposed by Merrill in 1917 is now generally accepted as the valid designation of the cultivated soybean. The genus *Glycine* has had many species added and removed over the years including removal of the original lectotype (Bentham 1864, 1865; Hitchcock and Green 1947; Hermann 1962; Verdcourt 1970; Lackey 1977a, b). Further discussions about the evolution of *Glycine* nomenclature including tables are presented by Hymowitz and Singh (1987) and Hymowitz (2004, 2008).

The genus *Glycine* Willd. is currently divided into two subgenera, *Glycine* and *Soja* (Monech) F.J. Herm. The subgenus *Glycine* currently contains 23 wild perennial species and the subgenus *Soja* contains the cultigen *G. max* (L.) Merr. and its wild annual purported ancestor *G. soja* Sieb and Zucc. Table 1.4 shows the current information on the genus *Glycine*.

The geographical origin of the genus *Glycine* was in South-East Asia. Hymowitz (2004, 2008) indicates that recent taxonomic, cytological and molecular systematic research and publications on the genus *Glycine* and related genera suggest the following: a putative ancestor of the current genus *Glycine* originated in South-East Asia with $2n = 2x = 20$ (Kumar and Hymowitz 1989; Singh and Hymowitz 1999; Lee and Hymowitz 2001; Singh et al. 2001) (Fig. 1-1). From this ancestral area Singh et al. (2001) assume the northward migration to China of a wild perennial ($2n = 4x = 40$, unknown or extinct) with subsequent evolution to a wild annual ($2n = 4x = 40$; *G. soja*) and finally to the cultivated (domesticated) soybean ($2n = 4x = 40$; *G. max*, cultigen) (Fig. 1-1). Also, as shown in Figure 1-1, they assume the wild perennial species found in Australia and Pacific islands today evolved from the putative ancestor in South-East Asia. Additional details are described by Hymowitz (2008).

## 1.4 Brief History of the Crop

Although most accounts suggest the soybean was domesticated in the northeast area of modern day China, much of the literature that describes the historical development relates longstanding errors and misconceptions (Hymowitz and Shurtleff 2005). Hymowitz (2008) suggests that historical records were in Chinese, a language not known by almost all western scientists and the records were not, until recently, available to research scholars and scientists, and thus studies on soybean domestication were extremely difficult. This summary relies on the most current information available on the history of soybean.

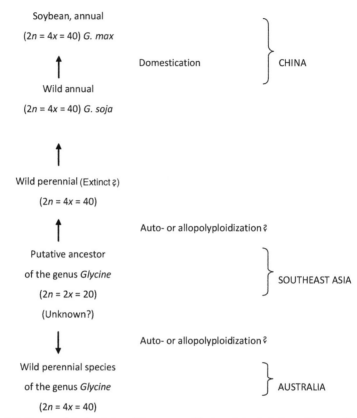

**Figure 1-1** Geographical origin of the genus *Glycine*. Adapted from Hymowitz (2004, 2008).

The farmers of China domesticated the soybean. Since domestication occurs over a period of time (many decades or centuries), the period of soybean domestication is uncertain. Hymowitz (2008) suggests that the domestication process of soybean probably took place during the Shang Dynasty (ca. 1766-1125 BCE; Bray 1984; Ho 1969, 1975; Hymowitz 1970; Hymowitz and Newell 1980). He further states that linguistic, geographical and historical evidences suggest that cultivated soybean emerged as domesticated during the Zhou Dynasty (ca. 1125-256 BCE) in the eastern part of northern China. Ho (1969) indicates that the movement of soybean (mainly landraces) in China was generally north to south, as this is how consolidation of territories and degeneration of dynasties in China occurred. The literature contains many factual errors about soybean domestication. Hymowitz (2008) discusses some of these and provides documentation of the correct evidence.

The history of how the domesticated soybean was disseminated is not fully known. Since China is the primary gene center, as people in Asia moved on land or sea trade routes or as certain groups of people moved from China from about the first century AD until the Age of Discovery (15th to 17th century AD), soybeans were most likely carried, to many Asian countries (Hymowitz 2008). In many of the places soybeans were carried, additional landraces developed (secondary gene centers) in the modern countries of Indonesia, Japan, Malaysia, Myanmar, Nepal, North India, the Philippines, Thailand, and Vietnam. Based on the literature and research on seed protein extracts and published morphological and physiological data, Hymowitz (2008) summarized the suggested paths of dissemination as:

1. The soybeans grown in the former USSR (Asia) came from Northeast China.
2. The soybeans grown in Korea were derived from two or three possible sources—Northeast China, North China, and the introduction of soybeans from Japan especially in the southern part of Korea.
3. The soybeans grown in Japan were derived from the intermingling of two possible sources of germplasm—Korea and Central China. The first points of contact were probably in Kyushu, and from there the soybean moved slowly northward to Hokkaido. In addition, the soybean moved southward from Kyushu to the Ryukyu Islands, where they came in contact with the soybeans moving northward from Taiwan. The earliest Japanese reference to the soybean is in KoJiKi or "Records of Ancient Matters", which was published in 712 CE (Chamberlain 1906).
4. The soybeans originally grown in Taiwan came from coastal China.
5. The germplasm source for the soybeans grown in South-East Asia is Central and South China.
6. The soybeans grown in the northern half of the Indo-Pakistan subcontinent came from Central China.
7. The soybeans grown in Central India were introduced from Japan, South China and South-East Asia.

There is also misinformation regarding the introduction of the soybean and/or soybean products into the western world. Hymowitz (2008) provides a detailed account of the dissemination of soybean to the West. He notes that there are some descriptions of possible soybean products as early as the 13th century up to the 17th century. However, he notes—"The soybean reached Europe quite late. It must have reached the Netherlands before 1737 as Linnaeus described the soybean in *Hortus Cliffortianus*". Hymowitz also notes that soybean seeds were planted in the Jardin des Plantes in 1740 in Paris, France and in 1790 in the Royal Botanic Garden in Kew, England. The first report of soybeans being grown in the current US was in 1865 by

Samuel Bowen in the Colony of Georgia (Hymowitz and Harlan 1983). The next earliest documentation is in 1770 when Benjamin Franklin sent soybean seed to John Bertram in Philadelphia (Smyth 1907). The first person believed to use the word *soybean* in American literature was Dr. James Mease in 1804 (Mease 1804; Hymowitz 2008). Further details about soybean and its uses are given by Hymowitz (2008).

## 1.5 Chromosome Number and Genomic Relationships

The soybean has a chromosome number of $2n = 40$. Goldblatt (1981) reports— "The base number for Phaseoleae is almost certainly $x = 11$, which is also probably basic in all tribes". He has also suggested that aneuploid reduction ($x = 10$) is prevalent throughout the Papilionoideae. These reports and more recent taxonomic, cytological, and molecular systematic research on the genus *Glycine* and allied genomes lead to the conclusion that a putative ancestor of the genus *Glycine* with $2n = 2x = 20$ likely arose in South-East Asia and then through auto- or allopolyploidization and adaption to an annual life cycle the wild and cultivated soybean is $2n = 4x = 40$ (Darlington and Wylie 1955; Kumar and Hymowitz 1989; Lee and Hymowitz 2001; Singh and Hymowitz 1999; Singh et al. 2001). All the *Glycine* species studied by Singh and Hymowitz (1985) showed diploid like meiosis, thus soybean can be considered a diploidized tetraploid.

The subgenus *Soja*, which contains the wild annual soybean, *G. soja* and the cultivated soybean *G. max* is considered one genome, since hybrids made between accessions of the two species are almost always successful, viable and produce fertile $F_1$ plants (Table 1-4). Although all the 23 perennial species of the subgenus *Glycine* could be considered potential sources of useful genes, so far only backcrossed-derived fertile progeny between *G. max* and *G. tomentella* has been reported (Singh et al. 1990, 1993).

The genomic relationships among most of the diploid ($2n = 40$) perennial species in the subgenus *Glycine* was reported by Hymowitz (2004, 2008) (Table 1-4). Hymowitz (2004) reported that these relationships were established using cytogenetic analysis, biochemical techniques and molecular tools by many authors in numerous publications. Thus, species with the same genome designation are expected to be able to be crossed and produce viable, vigorous and fertile $F_1$ plants (Table 1-4). For *Glycine* species with dissimilar genome designations the crossability is extremely low, the pods and/or seeds may abort, any seedling that develop are weak and sterile (Singh and Hymowitz 1988, Singh et al. 1992; Kollipara et al. 1993). A much more detailed discussion of the species and species relationships in the subgenus *Glycine* is presented by Hymowitz (2004). The polyploid and aneuploid members of the perennial species have been studied less and appear to have more complex relationships (Hymowitz 2004).

**Table 1-4** The genus *Glycine*, 2*n* number, genome, and distribution.*

| Subgenus *Glycine* | 2n | Genome | Geographic distribution |
|---|---|---|---|
| 1. *G. albicans* Tind. and Craven | 40 | I | Australia |
| 2. *G. aphyonota* B. Pfeil | 40 | ? | Australia |
| 3. *G. arenarea* Tind. | 40 | H | Australia |
| 4. *G. argyria* Tind. | 40 | A | Australia |
| 5. *G. canescens* F.J. Herman | 40 | A | Australia |
| 6. *G. clandestine* Wendl. | 40 | A | Australia |
| 7. *G. curvata* Tind. | 40 | C | Australia |
| 8. *G. cyrtoloba* Tind. | 40 | C | Australia |
| 9. *G. falcata* Benth. | 40 | F | Australia |
| 10. *G. gracei* B.E. Pfeil and Craven | 40 | A | Australia |
| 11. *G. hirticaulis* Tind. and Craven | 40 | H | Australia |
| | 80 | ? | Australia |
| 12. *G. lactovirens* Tind. and Craven | 40 | I | Australia |
| 13. *G. latifolia* (Benth.) Newell and Hymowitz | 40 | B | Australia |
| 14 *G. latrobeana* (Meissn.) Benth. | 40 | A | Australia |
| 15. *G. microphylla* (Benth.) Tind. | 40 | B | Australia |
| 16. *G. montis-douglas* B.E. Pfeil and Craven | 40 | ? | Australia |
| 17. *G. peratosa* B. Pfeil and Tind. | 40 | A | Australia |
| 18. *G. pindanica* Tind. and Craven | 40 | H | Australia |
| 19. *G. rubiginosa* Tind. and B. Pfeil | 40 | A | Australia |
| 20. *G. stenophita* B. Pfeil and Tind. | 40 | B | Australia |
| 21. *G. syndetika* B.E. Pfeil and Craven | 40 | A | Australia |
| 22. *G. tabacina* (Labill.) Benth. | 40 | B | Australia Australia, W.C. and |
| | 80 | Complex | S. Pacific Islands |
| 23. *G. tomentella* Hayata | 38 | E | Australia |
| | 40 | D | Australia, PNG |
| | 78 | Complex | Australia, PNG |
| | 80 | Complex | Australia, PNG Indonesia Philippines, Taiwan |
| **Subgenus *Soja* (Moench) F.J. Herm.** | | | |
| 24. *G. soja* Sieb. and Zucc. | 40 | G | China, Japan, Korea, Russia, Taiwan (Wild Soybean) |
| 25. *G. max* (L.) Merr. | 40 | G | Cultigen (Soybean) |

*Adapted from Hymowitz (2004, 2008) and Pfeil et al. (2006).

Even though soybean is one of the world's major crops, basic information on cytological and cytogenetic aspects lag behind other major crops such as rice (*Oryza sativa* L.), wheat (*Triticum aestivum* L.) and maize (*Zea mays* L.). Soybean chromosomes are smaller than those of other crops, so individual chromosome identification is difficult (Hymowitz 2004). He reports that all the 20 chromosomes can be differentiated using pachytene analysis. In addition, marker and cytological stocks have been developed including primary trisomics, tetrasomics, monosomics, translocations, inversions and monosomic alien addition lines (Hymowitz 2004). Additional research is needed on soybean in the areas of cytology and cytogenetics.

## 1.6 Gene Pools

The concept of three gene pools: primary (GP-1), secondary (GP-2) and tertiary (GP-3) as proposed by Harlan and deWet (1971) can be applied well to the genus *Glycine*. The primary gene pool (GP-1) for soybean would include cultivars, landraces and *Glycine soja* genotypes. GP-1 is defined as biological species that can easily be crossed within the gene pool and produce $F_1$ hybrids that are vigorous, exhibit normal meiotic chromosome pairing and possess total seed fertility, such that segregation is normal and gene exchange is basically easy. GP-2 as defined by Harlan and deWet (1971) consists of species that can be crossed with GP-1 and produce $F_1$ hybrids that have some fertility. By this definition there are no currently described *Glycine* species in GP-2. GP-3 is the extreme limit of potential genetic resources classically defined (this does not include transgenes). Harlan and deWet (1971) suggest gene transfer is almost not possible or requires rescue techniques that result in sterile (or lethal) hybrids. In *Glycine* the 23 wild perennial species would be considered GP-3.

Although the wild perennial species carry resistance to several diseases, nematodes, and have tolerance to salt and certain herbicides and lack some biologically active seed components (see Hymowitz 2008 for a detailed listing), the transfer of useful genes into soybean has not been accomplished. Thus, at least for the time being, breeders/geneticists really only have access to the primary gene pool for expanding the germplasm base.

Soybean germplasm collections are listed and/or maintained by several countries (FAO 1996). The US germplasm collection has over 21,000 strains of soybeans, wild soybeans and wild perennial *Glycine* species. Dr. R.L. Nelson, USDA-ARS, Urbana, IL is the curator of the collection. Information on the accessions in the collection can be found on the GRIN system (*http:/ /www.ars.grin.gov*) maintained by USDA-ARS. The collection is the primary source for new genetic traits for cultivar improvement and for basic studies of soybean in the US. The germplasm collection also contains about 200 genetic types (T-lines) and over 600 genetic isolines.

## 1.7 Diversity

As noted earlier, soybean seed contains approximately 20% oil and 40% protein on a dry matter basis. An estimated 98% of soybeans produced are processed into soybean oil and high-protein meal. The other 2% are used directly for human consumption. The number of products that use soybean, soybean oil, soybean protein, soybean carbohydrates, phytochemicals found in soybeans and other constituents numbers at least in the hundreds, if not thousands (Deak et al. 2008). Space limitations preclude a discussion of these products. Details about soybean products and uses can be found in Johnson et al. (2008).

Soybean (*G. max*) and its wild annual relative *G. soja* contain a great deal of diversity (Carter et al. 2004). This includes diversity for many obvious morphological traits like flower, pubescence, seed and hilum color, disease and insect resistance traits, physiological and biochemical traits as well as content of protein, oil and carbohydrates and their constituents (Boerma and Specht 2004). Genetic diversity in soybean is covered in great detail by Carter et al. (2004). In summary, they cover four main aspects of soybean genetic diversity: formation, collection, evaluation and utilization. They note that germplasm collections exist in many countries that contain landraces as well as current cultivars. It is unlikely that many additional landraces will be collected since modern cultivars have replaced most landraces, however there are likely genotypes of *G. soja* as well as perennial species from the subgenus *Glycine* that remain to be collected. Evaluation of accessions in collections is an ongoing process especially as new traits of interest for diseases and insects, biochemical processes or molecular and gene expression patterns are identified and studied. Carter et al. (2004) state—"Genetic diversity has no impact unless it is utilized". In order to utilize the genetic diversity in soybeans, interactions between collections and breeding/genetics programs in both the public and private sectors is necessary. Since the soybean genome has now been sequenced there is even more opportunity for scientists to utilize the genetic diversity in soybean for the good of human kind.

# References

Bentham G (1864) Flora Australiensis, vol 2. L. Reeve, London, UK.

Bentham G (1865) On the genera *Sweetia* Sprengel and *Glycine* Linn., simultaneously published under the name of Leptolobium. J Linn Soc Bot 8: 59–267.

Boerma HR, Specht JE (eds) (2004) Soybeans: Improvement, Production and Uses. 3rd edn. Agron Monogr 16, Am Soc Agron, Madison, WI, USA, pp 303-416, 949–1118.

Bray F (1984) Joseph Needham, Science and Civilization in China. Biology and Biological Technology, part II: Agriculture, vol 6. Cambridge Univ Press, Cambridge, UK.

Carter TE Jr, Nelson RL, Sneller CH, Cai Z (2004) Genetic diversity in soybean. In: HR Boerma JE Specht (eds) Soybeans: Improvement, Production and Uses. Am Soc Agron, Madison, WI, USA, pp 303–416.

Chamberlain BH (1906) Ko-Ji-Ki, records of ancient matters. Trans Asiat Soc Japan (translation) 10 (Suppl).

Darlington CD, Wylie AP (1955) Chromosome Atlas of Flowering Plants. Allen and Unwin, London, UK.

Deak NA, Johnson LA, Lusas EW, Rhee KC (2008) Soy protein products, processing and utilization. In: LH Johnson, PA White , R Galloway (eds) Soybeans Chemistry, Production, Processing, and Utilization. AOCS Press, Urbana, IL, USA, pp 661–724.

FAO (1996): *ftp://ftp.fao.org/docrep/FAO/Meeting/015/aj633e.pdf*

Goldblatt P (1981) Cytology and the phylogeny of Legumminosae. Advances in legume systematics, part II In: RM Polhill , PH Raven (eds) Royal Botanic Garden: Kew, UK, pp 427–463.

Goldsmith PD (2008) Economics of soybean production, marketing and utilization. In: LP Johnson, PA White, R Galloway (eds) Soybeans Chemistry, Production, Processing, and Utilization. AOCS Press, Urbana, IL, UK, pp 117–150.

Harlan JR, deWet JMJ (1971) Toward a rational classification of cultivated plants. Taxon 20: 509–517.

Heatherly LG, Elmore RW (2004) Managing inputs for peak production. In: HR Boerman, JE Specht (eds) Soybeans: Improvement, Production and Uses. Am Soc Agron, Madison, WI, USA, pp 451–536.

Henderson P (1881) Henderson's Handbook of Plants. Henderson, New York, USA.

Herman FJ (1962) A revision of the genus *Glycine* and its immediate allies. USDA Tech Bull 1268, pp 1–79.

Hitchcock AS, Green ML (1947) Species lectotypical generum Linnaei. Brittonia 16: 114–118.

Ho P-T (1969) The loess and the origin of Chinese agriculture. Am Hist Rev 75: 1–36.

Ho P-T (1975) The Cradle of the Eastern University of Hong Kong, Hong Kong.

Hoeft RG, Nafziger ED, Johnson RR, Aldrich SR (2000) Modern corn and soybean production. MCSP Publ, Champaign, IL, USA.

Hymowitz T (1970) On the domestication of the soybean. Econ Bot 24: 408–421.

Hymowitz T (2004) Speciation and cytogenetics. In: HR Boerman, JE Specht (eds) Soybeans: Improvement, Production and Uses. Am Soc Agron, Madison, WI, USA, pp 97–136.

Hymowitz T (2008) The history of the soybean. In: LA Johnson, PJ White, RGalloway (eds) Soybeans Chemistry, Production, Processing, and Utilization. AOCS Press, Urbana, IL, USA, pp 1–31.

Hymowitz T, Newell, CA (1980) Taxonomy, speciation, domestication, dissemination, germplasm resources and variation in the genus *Glycine*. In: RJ Summerfield, AH Bunting (eds) Advances in Legume Science. Royal Botanic Gardens, Kew, UK, pp 251–264.

Hymowitz T, Harlan JR (1983) Introduction of soybean to North America by Samuel Bowen in 1765. Econ Bot 37: 371–379.

Hymowitz T, Singh RJ (1987) Taxonomy and speciation. In: JR Wilcox (ed) Soybeans: Improvement, Production and Uses. 2nd edn. Agron Monogr 16. Am Soc Agron, Madison, WI, USA, pp 23–48.

Hymowitz T, Shurtleff WR (2005) Debunking soybean myths and legends in the historical and population literature. Crop Sci 45: 473–476.

Johnson LA, White PA, Galloway R (eds) (2008) Soybeans Chemistry, Production, Processing, and Utilization. AOCS Press, Urbana, IL, USA, pp 193–772.

Kollipara KP, Singh RJ, Hymowitz T (1993) Genomic diversity in aneuploid (2*n* = 38) and diploid (2*n* = 40) *Glycine tomentella* revealed by cytogenetic and biochemical methods. Genome 36: 391–396.

Kumar PS, Hymowitz T (1989) Where are the diploid (2n = 2x = 20) genome donors of *Glycine* Wild. (Leguminosae, Papilionoideae)? Euphytica 40: 221–226.

Lackey JA (1977a) A Synopsis of the Phaseoleae (Leguminosae, Papilionaideae). PhD Dissert, Iowa State Univ, Ames, IA, USA.

Lackey JA (1977b) Neonotonia, a new generic name to include *Glycine wighti* (Arnott) Verdecourt (Leguminosae, Papilionoideae). Phytologia 37: 209–212.

Lawrence GHM (1949) Name of the soybean. Science 110: 566–567.

Lee J, Hymowitz T (2001) A molecular phylogenetic study of the subtribe Glycinnae (Leguminosae) derived from the chloroplast DNA rps16 intron sequence. Am J Bot 88: 2064–2073.

Linnaeus C (1737) Genera Plantarum. 1st edn (In Latin). Wishoff, Leiden, Netherlands.

Linnaeus C (1754) Genera Plantarum. 5th edn (In Latin). Lars Salvius, Stockholm, Spain.

Linnaeus C (1968) Hortus Cliffortianus. Historiae Naturalis Classica. In: J.Cramer, HK Swan (eds) Stechert-Hafner, New York, USA (in Latin) vol 63, p 1737.

Mease J (1804) Willich's Domestic Encyclopedia. 1st Am edn, vol 5. Murray and Highle, London, UK, p 13.

Orf JH (2008) Breeding, genetics and production of soybeans. In: LH Johnson, PA White, R Galloway (eds) Soybeans Chemistry, Production, Processing, and Utilization. AOCS Press, Urbana, IL, USA, pp 33–65.

Paclt D (1949) Nomenclature of the soybean. Science 109: 339.

Piper CV, Morse WJ (1923) The Soybean. McGraw-Hill, New York, USA.

Polhill RM (1994) Classification of the Leguminosae, vols 25–27. In: FA Bisby, J Buckingham, JB Harborne (eds) Phytochemical Dictionary of the Leguminosae. Chapman and Hall, New York, USA.

Ricker PL, Morse WJ (1948) The correct botanical name for the soybean. J Am Soc Agron 40: 190–191.

Singh RJ, Hymowitz T (1985) Diploid-like meiotic behavior in synthesized amiphiploids of the genus *Glycine* Willd. subgenus Glycine. Can J Genet Cytol 27: 655–660.

Singh RJ, Hymowitz T (1988) The genomic relationships between *Glycine max* (L.) Merr. and *G. soja* Sieb. and Zucc. as revealed by pachytene chromosome analysis. Theor Appl Genet 76: 705–711.

Singh RJ, Hymowitz T (1999) Soybean genetic resources and crop improvement. Genome 42: 605–616.

Singh RJ, Kollipara KP, Hymowitz T (1990) Backcross-derived progeny from soybean and *Glycine tomentella* Hayata intersubgeneric hybrids. Crop Sci 30: 871–874.

Singh RJ, Kollipara KP, Hymowitz T (1992) Genomic relationships among diploid wild perennial species of the genus *Glycine* Willd. Subgenus Glycine revealed by crossability, meiotic chromosome pairing and seed protein electrophoresis. Theor Appl Genet 85: 276–282.

Singh RJ, Kollipara KP, Hymowitz T (1993) Backcross (BC$_2$–BC$_4$)- derived fertile plants from *Glycine max* and *G. tomentella* intersubgeneric hybrids. Crop Sci 33: 1002–1007.

Singh RJ, Kim HH, Hymowitz T (2001) Distribution of rDNA loci in the genus *Glycine* Willd. Theor Appl Genet 103: 212–218.

Smith K, Huyser W (1987) World distribution and significance of soybean. In: JR Wilcox (ed) Soybeans Chemistry, Production, Processing and Utilization, 2nd edn. Agron Monogr 16, Am Soc Agron, Madison, WI, USA, pp 1–22.

Smyth AH (1907) Writings of Benjamin Franklin, vol 5. Macmillan, New York, USA.

USDA-ERS (2009): *http://usda.mannlib.cornell.edu/ers/89002/2009/index.html*

USDA-FAS (2008): *http://usda.mannlib.cornell.edu/MannUsdaviewDocumentinfo.do? document ID=1194*

USDA-FAS(2009):*http://usda.mannlib.cornell.edu/MannUsdaviewDocumentInfo.do? document ID =1194*

Verdcourt B (1970) Studies in the Leguminosae-Papilionoideae for the Flora of Tropical East Africa. II Kew Bulletin 24: 235–307.

Wilcox JR (2004) World distribution and trade of soybean. In: HR Boerma, JE Specht (eds) Soybeans Chemistry, Production, Processing, and Utilization, 3rd edn. Agron Monogr 16, Am Soc Agron, Madison, WI, USA, pp 1–14.

# Classical Breeding and Genetics of Soybean

*Andrew M. Scaboo,[1] Pengyin Chen,[1*] David A. Sleper[2] and Kerry M. Clark[3]*

## ABSTRACT

This chapter addresses two major technical aspects, classical genetics and traditional breeding, which are directly concerned with soybean cultivar development and germplasm enhancement. Today, most of the soybean cultivar development occurs in the private sector, while public sector breeders focus on germplasm enhancement, breeding methodology and molecular technology development, and education of students who become professional plant breeders. Continued enhancement of soybean cultivars relies on identification and genetic manipulation of novel desirable genes for adaptation to new environments, new management practices, and new end uses. Selection efficiency is largely dependent on specific traits of interest that are either qualitative or quantitative. While many agronomic traits, such as disease resistance, are simply inherited and easy to select for, yield is a polygenic and complex trait to manipulate. Several traits may be linked or correlated, which may be advantageous or deleterious to breeders in terms of selection. The traditional breeding scheme can be summarized in three basic steps: 1) selecting parents with desired characteristics and intercrossing them, 2) growing hybrid

[1]Department of Crop, Soil, and Environmental Sciences, University of Arkansas, Fayetteville, AR 72701, USA.
[2]Division of Plant Sciences, 271-F Life Sciences Center, University of Missouri, Columbia, MO 65211, USA.
[3]Soybean Breeding, University of Missouri, 3600 New Haven Road, Columbia, MO 65201, USA.
*Corresponding author: *pchen@uark.edu*

populations for four to five generations to allow genetic segregation and recombination while reaching allelic homozygosity (true-breeding), and 3) selecting and evaluating pure lines carried forward from each cross. This is a continuous, simultaneous, and cyclic process in which there is much variation in methodology and strategies for handling each component. After all, plant breeding is a number's game, long-term by nature, time sensitive, and resource dependent. The success of a breeding program rests on proper use of genetics, methodologies, time, and resources.

**Keywords:** soybean breeding; soybean genetics; cultivar development; variety selection; breeding methodology; variety protection; germplasm enhancement

## 2.1 Introduction

Soybean [*Glycine max* (L.) Merrill] was domesticated in northeastern China about 2500 BC and subsequently spread to southern China, Korea, Japan, and other countries in South-Eastern Asia. Soybean was introduced into the US during the 1700s and was grown initially as a forage crop (Hymowitz 1990, 2004). It was only in the 1920s and 1930s that it was used as a grain crop. Early US plant breeders, mostly from the Agricultural Experiment Stations of the states and the United States Department of Agriculture (USDA), developed lodging- and shattering-resistant varieties, which were responsible for changing soybean from a forage crop to an oilseed crop. Variety development remained largely with the USDA and Agricultural Experiment Stations until 1970 when the Plant Variety Protection Act was passed. With the passage of this act, commercial variety development began as private breeders could protect their intellectual property (proprietary varieties), which provided them an opportunity to capture additional financial resources to conduct large, comprehensive, and expensive soybean breeding programs. Today, most of the variety development of soybean occurs in the private sector; however, public sector breeders still have an important role to play in variety development. In addition to variety development, public sector breeders place emphasis on germplasm enhancement, breeding methodology and molecular technology development, and education of students who become professional plant breeders. In the future, the productivity of modern agriculture will depend largely on the ability of breeders to constantly adapt new varieties to changing environmental conditions and management strategies.

Success of soybean breeding is dependent on germplasm availability, genetic variation, selection strategies, and resource management. Various crosses between different varieties or germplasm lines are often attempted by breeders to generate increased genetic variation through gene recombination and change of allele frequency in a breeding population in which selection is exercised. Breeding soybean, like other crops, is a number's game, long term and continuous process, and involves manipulation of genetics of an array of important and complex traits. The strategy to handle a mating scheme and a breeding population structure becomes critical in providing increased potential for genetic superiority, while proper selection and resource management help in improving plant breeding efficiency and success rate. This chapter addresses two major technical aspects, classical genetics and traditional breeding, which are directly concerned with a variety development program.

## 2.2 Classical Genetics of Soybean

### 2.2.1 *Taxonomy and Cytogenetics*

Soybean is a self-pollinated diploid and has a chromosome number of $2n = 4x = 40$. Taxonomically, soybean is classified in the legume family, Leguminosae, subfamily Papilionoideae, tribe Phaseoleae, and genus *Glycine*. The genus *Glycine* contains two subgenera *Soja* and *Glycine*. The wild soybean species, *Glycine soja* (L.) Sieb. and Zucc., is native to the Far East and has a viny, prostrate growth habit and a great tendency for its seed to shatter (Hymowitz 2004). Regular soybean is cross-compatible with the wild species *Glycine soja*, but undesirable growth characteristics of *Glycine soja* are apparent in the progeny. Wild and perennial tetraploid *G. tomentella* ($2n = 78, 80$) accessions are also used by breeders to introgress specific traits of interest. However, crossability between *G. max* and *G. tomentella* is extremely low due to early pod abortion (Hymowitz 2004). Although tissue culture techniques can be successfully used to overcome interspecific hybridization difficulties and fertility barriers, genomic elimination of wild relatives from progeny still present a challenge for breeders to introgress exotic genes of interest and traits of value.

### 2.2.2 *Qualitative and Quantitative Traits*

Inheritance of phenotypic traits in soybean is largely controlled by the environment in which the plant is grown, the genes which the plant inherited from its parents, and the interaction or response of these genes in a respective environment. Qualitative traits are defined as those traits which have discrete phenotypic categories. For example, the soybean flower is either white or purple in color; there are no intermediate shades such as light purple. Qualitative traits such as flower and pubescence color, disease and insect resistance, and herbicide resistance are useful selectable markers used by soybean breeders for improvement and population development, such as distinguishing between a hybrid population and a self-pollinated population. These traits are largely controlled by one or a few genes, and are usually affected very little by changes in environmental conditions. Qualitative traits are usually found by determining phenotypic ratios or patterns of inheritance in segregating populations.

In contrast, the quantitative traits are controlled by many genes, each having a small effect and characterized by a non-discrete continuum of phenotypic classes. The environment has a much greater effect on quantitative traits than qualitative, making the search for genes controlling quantitative traits much more difficult, because it is difficult to measure a gene which has an infinitely small effect, but when these gene effects are combined they have a greater influence on the variation of the trait. The phenotypic expression of genes controlling a quantitative trait is measured and analyzed by determining tendencies, population distribution, or mean values, and genetic, phenotypic, and environmental variances, instead of phenotypic ratios. The most infamous and important quantitative trait would be seed yield in soybean.

### 2.2.3 Major Traits and Associated Genes

There are four basic categories of research in which scientists have discovered major genes associated with particular phenotypic traits. These include, but are not limited to morphological, physiological/biochemical, seed composition and quality, and biotic and abiotic stresses. Examples of major genes that have been found for these four classes of phenotypic traits in soybean are listed in Table 2-1. Gene symbols are regulated and assigned by the Soybean Genetics Committee.

Seed yield is one the most important economic traits to farmers. The ability of farmers to grow a new and improved cultivar that has significant yield advantages over existing cultivars may be the best way to improve farmer profitability because it usually requires no additional resource inputs from the farmer, and in some cases, such as novel disease resistance, it would require less input. The most recent studies indicate that soybean yields are improving at a rate of 23 kg ha$^{-1}$yr$^{-1}$ due to improved genetic gain, improved cultural practices, and the rise in atmospheric $CO_2$ concentrations (Orf et al. 2004). Developing higher yielding conventional and glyphosate resistant soybean cultivars, which are adapted to the most recent environmental and cultural practices, is a key to ensure competitive and profitable rates of soybean yield improvement with regard to other row crops grown nationally. Much genetic gain has been achieved in soybean seed yield as a result of conventional breeding techniques such as recurrent selection (Piper and Fehr 1987; Guimaraes and Fehr 1989; Burton et al. 1990; Werner and Wilcox 1990; Upholf et al. 1997). Genes that have been

**Table 2-1** Major soybean genes, their symbols, and associated phenotypic traits.

| Trait–Phenotype | Gene(s) | References |
|---|---|---|
| **Morphological Traits** | | |
| Maturity (late/early) | $E1–E7/e1–e7$ | Owen (1927); Bernard (1971); Buzzell (1971); Kilen and Hartwig (1971); Buzzell and Voldeng (1980); McBlain and Bernard (1987); Bonato and Vello (1999); Cober and Voldeng (2001) |
| Indeterminate/Determinate | $Dt1/dt1$ | Woodworth (1932, 1933); Bernard (1972); Thompson et al. (1997) |
| Dwarfness | $df2–df7, df8$ | Porter and Weiss (1948); Byth and Weber (1969); Fehr (1972); Palmer (1984); Werner et al. (1987); Soybean Genetics Committee (1995) |
| **Physiological Traits** | | |
| Flower pigmentation (purple/white) | $W1/w1$ | Takahashi and Fukuyama (1919); Woodworth (1923); Zabala et al. (2007) |
| Pubescence pigmentation (tawny/gray) | $T/t$ | Piper and Morse (1910); Nagai (1921); Woodworth (1921); Williams (1950); Buttery and Buzzell (1973); Zabala et al. (2003) |
| β-(1-6)-glucoside (present/absent) | $Fg1/fg1$ | Buttery and Buzzell (1975) |
| **Seed Composition** | | |
| Palmitate (normal/low) | $Fap1–Fap7/$ $fap1–fap7$ | Erickson et al. (1988); Wilcox and Cavins (1990); Rahamn et al. (1999); Fehr et al. (1991a); Fehr et al. (1991b); Schnebly et al. (1994); Stoltzfus et al. (2000a); Stoltzfus et al. (2000b); Narvel et al. (2000) |
| Stearate (normal/low) | $Fas/fas; St1–St2$ $/st1–st2$ | Graef et al. (1985); Hammond and Fehr (1983); Rahman et al. (1997) |
| Linoleneate (normal/low) | $Fan1–Fan3/$ $fan1–fan3$ | Rennie and Tanner (1989); Hammond and Fehr (1983); Wilcox and Cavins (1985, 1987); Rennie et al. (1988); Fehr et al. (1992); Fehr and Hammond (1996); Ross (1999); Ross et al. (2000); Bilyeu et al. (2003); Anai et al. (2005); Bilyeu et al. (2005); Bilyeu et al. (2006); Chappell and Bilyeu (2006); Chappell and Bilyeu (2007) |
| Phytate (normal/low) | $Pha1–Pha2/$ $pha1–pha2$ | Oltmans et al. (2004); Walker et al.(2006) |
| **Biotic and Abiotic Stresses** | | |
| Bacterial Blight (resistant/susceptible) | $Rpg1–Rpg4/$ $rpg1–rpg4$ | Mukherjee et al. (1966); Keen and Buzzell (1991) |

*Table 2-1 contd....*

*Table 2-1 contd....*

| Trait–Phenotype | Gene(s) | References |
|---|---|---|
| Frogeye Leafspot (resistant/susceptible) | *Rcs1–Rcs3/ rcs1 – rsc3* | Athow and Probst (1952); Probst (1965); Boerma and Phillips (1983) |
| Sudden Death Syndrome (resistant/susceptible) | *Rfs/rfs* | Stephens et al. (1993); Mclean and Byth (1980); Hartwig and Bromfield (1983) |
| Soybean Mosaic Virus (resistant/susceptible) | *Rsv1/rsv1, Rsv3/rsv3, Rsv4/rsv4* | Kiihl and Hartwig (1979); Chen et al. (1991); Buzzell and Tu (1989); Ma et al. (1995); Buss et al. (1999); Gunduz (2000) |
| Drought and Salt Tolerance | *GmDREBa-GMDREBc* | Li et al. (2005) |
| Salt Stress Tolerance | *GmPAP3, GmCAX1* | Liao et al. (2003); Luo et al. (2005) |

identified to be associated with seed yield have been highly population- and environment-specific, so breeders must recognize elite gene pools and incorporate novel germplasm into their programs to maintain genetic diversity and improve genetic gain. Wang et al. (2003) successfully identified quantitative trait loci (QTL) controlling yield in five soybean populations and two environments using lines derived from a recurrent backcross between IA2008 (*G. max*) and PI 468916 (*G. soja*). This research demonstrated the ability to acquire novel gene combinations with positive effects on seed yield from wild soybean relatives to create improved *G. max* cultivars.

Important morphological traits to soybean researchers include flowering and maturity, stem and petiole growth, plant height (dwarfness), leaf form, determinacy, pod wall color, and pubescence type. A total of seven gene pairs have been identified in soybean as being significantly associated with flowering and maturity, denoted as *E1e1–E7e7* (Owen 1927; Bernard 1971; Buzzell 1971; Kilen and Hartwig 1971; Buzzell and Voldeng 1980; McBlain and Bernard 1987; Bonato and Vello 1999; Cober and Voldeng 2001). These genes are responsible for a variety of responses including late and early flowering and maturity and are important in developing cultivars adapted to target regions. Genes affecting stem determinacy or termination in soybean have been denoted as *Dt1* and *dt1* and *Dt2* and *dt2*, and are used as selectable markers in segregating populations (Woodworth 1932, 1933; Bernard 1972; Thompson et al. 1997). Several genes associated with plant height, specifically dwarfness, have been reported and denoted as *df2-df8* (Porter and Weiss 1948; Byth and Weber 1969; Fehr 1972; Palmer 1984; Werner et al. 1987; Soybean Genetics Committee 1995). These genes confer a phenotype, which results in a relatively short stature plant.

Important physiological traits in soybean include, but are not limited to, nutrient assimilation, flavanols, isoflavones, chlorophyll deficiency, pigmentation, sterility, water and radiation use efficiency, nodulation, nitrogen fixation, and many more. Many genes for pigmentation have been found and described in detail, such as the *W1w1* gene pair, which is a pleiotropic locus responsible for pigmentation of the flower and the seedling hypocotyl in soybean (Takahashi and Fukuyama 1919; Woodworth 1923; Zabala and Vodkin 2007). The *Tt* gene pair has been found to be associated with the color of pubescence in soybean, resulting in either gray (lack of pigmentation) or tawny (brown) pubescence (Piper and Morse 1910; Nagai 1921; Woodworth 1921; Williams 1950; Buttery and Buzzell 1973; Zabala and Vodkin 2003). The gene *T* is also a single dominant gene, and both flower and pubescence colors are useful selectable markers for determining hybrids versus self-pollinated populations. Other genes that confer traits, which can be used as selectable markers for developing hybrid populations, include *L1l1* and *L2l2* for pod wall color (black, brown, or tan), *Lf1lf1* and *Lf2lf2* for leaf foliate number (3-, 5-, or 7-foliate), and *Lnln* for leaf shape (ovate and narrow) (Takahashi and Fukuyama 1919; Woodworth 1932, 1933; Takahashi 1934, Domingo 1945; Bernard 1967; Fehr 1972).

Protein and oil concentration are arguably the most important seed composition traits in soybean. Soybean seed oil concentration is important for a variety of industrial processes and for the nutritional value for animal and human consumption. Normal soybean oil is composed of five fatty acids including palmitic (16:0), stearic (18:0), oleic (18:1), linoleic (18:2), and linolenic (18:3) acid with an average composition of 10%, 4%, 22%, 54%, and 10%, respectively. Development of alternative fatty acid profiles in soybean oil has been targeted at three major phenotypes aimed at enhancing value for their respective markets. These include oils used for frying (and bio-diesel), baking, and industrial processes. The general alternative fatty acid profiles are as follows: for frying/bio-diesel—7% saturated (palmitic and stearic), 60% oleic, 31% linoleic, and 2% linolenic; for baking—42% saturated, 19% oleic, 37% linoleic, and 2% linolenic; for industrial—11% saturated, 12% oleic, 55% linoleic, and 22% linolenic (Wilson 2004). Development of high yielding cultivars with modified fatty acid profiles, and higher than normal seed oil concentration, could give farmers a profitable alternative if processors pay a premium for the modified seed.

Dry soybean seeds typically contain ~ 400 g kg$^{-1}$ protein on a dry weight basis, which is one of the highest protein fractions for grain crops, and soybean meal is a valuable source of protein for humans and animals throughout the world. Improving soybean cultivars for higher protein content and lower phytate content is a goal in many soybean breeding programs. The most important objectives for enhancing soybean seed protein are: to improve amino acid balance and composition, to increase digestibility of meal, and to help reduce the environmental impact of animal production (Wilson 2004). Soybean protein has a good balance of amino acid composition for poultry and swine dietary needs, so soybean farmers and poultry farmers could benefit from this type of value added cultivar.

The reported range of seed protein and oil concentrations in the USDA Soybean Germplasm Collection are 34.1 to 56.8% for protein (mean = 42.1%) of seed dry mass, and 8.3 to 27.9% for oil (mean = 19.5%) (Wilson 2004). As reported by Hurburgh et al. (1990), there is a strong negative correlation between seed protein and seed oil, indicating that as protein increases oil decreases. A negative correlation between seed yield and seed protein, and a positive correlation between seed yield and seed oil concentrations has also been reported; elite southern soybean varieties generally have a higher mean seed protein concentration than elite northern cultivars (Wilson 2004). Evidently, there are both environmental and genetic factors that influence protein and oil concentrations; developing a soybean cultivar which is high yielding with high protein and oil concentrations is extremely difficult. Although these correlations have been deleterious in obtaining both high protein and oil, recurrent selection has enabled breeders to develop several useful high yielding, high protein lines, such as Prolina and Osage (Burton et al. 1999; Chen et al. 2007)

Another important seed compositional trait which has received attention of late is seed phytate concentration. Most phosphorus in dry soybean seeds is found as phytate, a mixed cation salt of phytic acid (*myo*-inositol 1,2,3,4,5,6-hexa*kis*phosphate), and is mostly unavailable to monogastric animals such as human, poultry, swine, and fish. Manure from swine and poultry production, when land applied, provides non-point source Pi contamination of ground water and can cause eutrophication in fresh water ecosystems, which promotes algal bloom production and reduces available oxygen for macroorganisms. Therefore, development of soybean cultivars, which are lower in the allocation of phosphorus into phytic acid, may help provide more mineral nutrients to growing monogastric animals without the addition of phytase or phosphate supplements, while at the same time saving valuable ecosystems inundated with non-point source Pi pollution. Wilcox et al. (2000) successfully developed mutant soybean lines with ~1.9 g kg$^{-1}$ of phytate P and ~3.1 g kg$^{-1}$ of Pi, compared to normal conventional cultivars with ~4.3 g kg$^{-1}$ of phytate P and ~0.7 g kg$^{-1}$ of Pi. Oltmans et al. (2004) found that this trait was controlled by a duplicate dominant epistasis, and denoted the alleles involved as *pha1* and *pha2*, where both recessive alleles are needed to produce a low-phytate phenotype. These mutant lines have been widely used by breeders across the country to develop agronomical high quality soybean seed with low seed phytate concentration. The low phytate cultivars could play a crucial role in providing the solutions to environmental problems associated with current agricultural practices around the world.

Biotic and abiotic stresses on plants are also a major concern in soybean breeding. Insect and pathogen resistance is of major concern for breeders and farmers because genetic resistance in a cultivar can lead to a reduction of pesticide application and better management without additional inputs.

Genetic resistance has been found in soybean for a variety of pathogens, including bacterial blight, brown stem rot, frogeye leaf spot, downy mildew, powdery mildew, Phytophthora root rot, sudden death syndrome, soybean mosaic virus, peanut mottle virus, cyst nematode, reniform nematode, and root-knot nematode. Resistance to most of them was found in novel germplasm and is generally conferred by one or a few genes. For example, three major gene pairs have been found to be associated with soybean mosaic virus resistance including *Rsv1rsv1, Rsv3rsv3, and Rsv4rsv4*. Genetic resistance has been successfully used by breeders in developing resistant cultivars for many major soybean diseases.

The major abiotic stresses for soybeans grown in commercial production include water stresses such as drought and flooding and soil fertility issues such as aluminum and salt tolerance. Improvement of soybean cultivars for areas prone to flooding or drought has largely relied on selecting for traits associated with water use efficiency (WUE) and harvest index (HI). WUE refers to the relative amount of shoot biomass produced per unit of water transpired and HI refers to the ratio between grain mass and total shoot mass (Purcell and Specht 2004). Breeders can select tolerant varieties by measuring traits associated with WUE and HI such as transpiration rate, wilting, rooting depth and morphology, leaf area, heat tolerance, nitrogen fixation rates, transpiration efficiency and many more. Other abiotic stresses such as toxic aluminum concentration in soil have been found to be beneficial to breeders when selecting for drought tolerant lines. Carter and Rufty (1993) established that lines, which showed tolerance to high aluminum concentration, were at least partially related to maintaining leaf vigor during prolonged drought. New and improved soybean germplasm have been released that have drought tolerance and prolonged nitrogen fixation under drought stress (Chen et al. 2007; Sinclair et al. 2007).

## 2.3 Traditional Breeding

### 2.3.1 Breeding Objectives

Plant breeding is defined as the art and science of improving the heredity of plants in relation to their economic use (Fehr 1991a; Sleper and Poehlman 2006; Acquaah 2007). Art is the practical part of the definition of plant breeding. A breeder can practice plant breeding strictly as an art and expect to make progress in certain instances. For example, prehistoric plant breeders were gatherers of food from plants in the wild. In order to gather seed for planting next year's crop, these early selectors would harvest non-shattered seed from the largest heads, ears, or other plant parts containing more seeds (Gai et al. 1997). Doing so changed the architecture of the plant so that in the modern age, many of our crops could not survive in the wild, including soybean, because they have been selected by man over many centuries for traits that suit man's use rather than helping them to survive in nature. Seed shattering, the premature release of seeds from the plant pod, is an example of a plant trait that helps plants to survive in nature (Funatsuki et al. 2006). Plant breeders select vigorously against seed shattering for obvious reasons. Today, plant breeding is still practiced as an art but much less so. The science of plant breeding began in the early part of the last century as the result of Mendel's work with the garden pea. The plant breeder has to have a vision as to what a new crop variety should look like and this is where the art comes in to play. Today plant breeding is practiced largely as a science depending heavily on such disciplines as genetics, cytogenetics, molecular genetics, genetic engineering, statistics, plant pathology, entomology, plant physiology, and etc. In a sense, engineering could be part of the modern definition of plant breeding. The overall aim of plant breeding is to develop a variety that is superior in one or more traits as compared to the best available varieties currently grown, and hopefully improve upon the livelihood of producers and consumers. It is not possible to develop a variety that is superior in all traits. Most plant breeders choose their objectives carefully and try to improve one or several traits at a time. In the following segment, we will discuss several important agronomic and quality traits for soybean improvement via cultivar development.

### 2.3.1.1 Yield

The number one objective in soybean breeding is improvement in yield (Fehr 1991b). The bottom line for profitability of the soybean producer is high seed yield. If the breeder develops an improved soybean variety, for example, with a superior disease resistance package, but is poor yielding, producers will not bother to grow it because it is likely not to be economically feasible to do so. Yield potential in soybean is expressed phenotypically through complex plant morphological features and physiological functions, and genetically expressed as a complex quantitative (controlled by many genes) character that interacts with the environment in which the soybean variety is grown (Sleper and Poehlman 2006). The soybean breeder measures yield potential by the mass or weight of the seed produced per unit area of land, through extensive yield trials, in which the harvested seed yield is compared with that of standard varieties or commercial checks. After evaluation of yield potential over many locations for several years, the breeder will identify those new soybean varieties with high yield potential over a wide spectrum of environments and ultimately release those new varieties that have high yield potential and stability under producers' variable growing conditions. In 2007, Kip Cullers of Purdy, MO reported a world record soybean yield of 154.7 bushels per acre (Pioneer online news release 2007). The national average soybean yield for 2005 was estimated at 43.0 bushels per acre. This record breaking growing season shows that most growers and breeders have yet to maximize the yield potential of soybean through genetic improvement and cultural management (soystats.com; verified July 24, 2009).

### 2.3.1.2 Maturity

Maturity must be given serious consideration when establishing plant breeding objectives. There are 13 major soybean maturity groups established in North America, designated as 000, 00, 0, and I–X (Fehr 1991b). Maturity has an influence on where the new variety will be grown and best adapted in what production system or cropping sequence, and has an effect on yield and seed quality. Full-season varieties are generally more productive than early varieties for a given location. In the South, early-maturing varieties are getting more popular as they may offer flexibility for farm operations and potential for drought avoidance and escape from pests. However, seed quality and germination become an issue in some years and under extreme environmental conditions such as drought and heat stress. Late-maturing varieties usually perform well in a double cropping system after a wheat crop as they tend to have prolonged vegetative growth stage. Soybean flowers in response to shortening days and is photoperiod sensitive. As a result, a variety that is planted in the North will flower later, and when the same variety is planted further South will flower earlier. Soybean breeders have attempted to develop cultivars that are daylength neutral and potentially adapted to a wide growing area in latitude (Polson 1972).

### 2.3.1.3 Pest Resistance

Soybean varieties developed with genetic resistance to destructive disease pathogens are among the foremost contributions of soybean breeding. In breeding for resistance to a particular pest, each pest is treated as a separate breeding objective. Because the soybean plant is attacked by many pests, the breeder must establish priorities and concentrate available resources on developing varieties that are resistant to the most destructive pests prevalent in the area where the variety is to be grown. Pest resistant soybean varieties are developed by crossing parents with desired pest resistance to high yielding parents and selecting among segregating progenies those individuals with both high yield potential and resistance to the targeted pest. Examples of pests influencing production of soybean include the soybean cyst nematode (*Heterodera glycines*), Phytophthora rot (*Phytophthora soja*), sudden death syndrome (*Fusarium solani*), frogeye leaf spot (*Cercospora sojina*), brown stem rot (*Phialophora gregata*), stem canker (*Diaporthe phaseolorum*), charcoal rot (*Macrophomina phaseolina*), Asian soybean rust

(*Phakopsora meibomiae* and *P. pachyrhizi*), and many others (Sinclair 1982). In addition, weeds are also considered pests and breeding programs are actively incorporating herbicide resistance genes, such as resistance to glyphosate, into improved varieties. Many genetic mutants and transgenic events have been incorporated into breeding programs across the US to develop varieties with herbicide resistance genes. However, there is increasing evidence that certain weed species have developed resistance to specific herbicides. This may serve as a friendly warning for the breeders to use diverse germplasm and breeding strategies in future breeding processes to avoid genetic uniformity and vulnerability.

### 2.3.1.4 Lodging Resistance

Plant lodging is a measure of the bending and breaking over of plants before harvest. Lodging causes difficulty in harvesting and hence, reduces seed yield. Lodging, if occurring early, may also affect seed quality. Resistance to lodging is a quantitative trait with an approximate heritability of 55% (Brim 1973). Resistance to lodging may be improved by selecting for sturdy stems, vigorous and strong root systems, and in many cases shorter plant height. Soybean breeders often evaluate breeding lines for lodging resistance at multiple locations with different soil types and over years to be sure that selected lines will hold up well under adverse environmental conditions.

### 2.3.1.5 Shattering Resistance

Shattering refers to the dispersion of soybean seeds from the pods before harvest or during the harvest operation. Resistance to shattering is important to maintain high yield potential of soybean. Shattering resistance has been introduced to most elite soybean cultivars in North America (Bailey et al. 1997), yet in other regions of the world such as Japan and China breeding for resistance to shattering is still of major concern (Funatsuki et al. 2006). Seed shattering is more a concern and challenge for breeders working with exotic germplasm in their breeding programs. Selection in different environments and for extended period of time after optimum harvest time may be helpful in identification of true genetic resistance to seed shattering in the field.

### 2.3.1.6 Seed Quality and Composition

Seed quality can be influenced by environment and pathogens such as *Phomopsis longicolla* (Hobbs et al 1985), which is responsible for Phomopsis seed decay. Seed infection by *P. longicolla* increases between growth stages

R7 and R8. This disease is most prevalent in moist and warm conditions, and is more of a problem with early maturing varieties. Phomopsis seed decay can severely affect germination and if present necessitates some sort of seed treatment to improve germination the following spring. Seed quality becomes a major concern when the cost for planting seed is high for soybean producers. Another important pathogen affecting soybean seed quality is *Cercospora kikuchii* which causes purple seed stain. Genetic resistance to both *Phomopsis longicolla* and *Cercospora kikuchii* has been identified in soybean germplasm and can be used in developing resistant cultivars (Jackson et al. 2005, 2006).

Seed quality can also refer to modifying such seed traits as the content of protein, amino acids, oil, fatty acids, carbohydrates, phytate, and isoflavones. One of the goals of many soybean breeders is to select for higher levels of seed protein. Selecting for higher protein level often results in lowering the percentage of oil with an associated decrease in seed yield potential (Wilcox 2001). An example of another seed quality breeding objective is to reduce the levels of linolenic acid. Soybean oil quality is largely a function of its fatty acid composition. Oxidative instability of soybean oil occurs in the presence of high levels of linolenic acid and reduces the shelf life of the oil. Inheritance of low linolenic acid is relatively simple as single genes have been identified which lower this fatty acid (Bilyeu et al. 2003; Anai et al. 2005; Bilyeu et al. 2005, 2006; Patil et al. 2007). Reducing the amount of saturated fat is another goal of some soybean breeding programs. This can be accomplished by reducing the level of palmitic fatty acid. The Food and Drug Administration (FDA) requires that an oil must contain less than 7% of total saturated fats to be classified as a "low saturate". Most soybean varieties are between 13 and 14% saturated fat. Single alleles have been identified that will lower palmitic acid in soybean (Erickson et al. 1988; Wilcox and Cavins 1990; Fehr et al. 1991a, b; Schnebly et al. 1994; Rahamn et al. 1999; Stoltzfus et al. 2000a, b). In addition, soybean breeders try to increase the levels of oleic acid in attempts to improve the quality of soybean oil. Many breeding programs also aim to increase certain amino acids (such as methionine and lysine), isoflavones, digestible sugars (such as sucrose) while decreasing phytate and indigestible sugars (such as raffinose and stachyose) for the improvement of nutritional value and functionality of soybean meal. Adjustments can be made via plant breeding to modify these seed traits, however, the challenge is to maintain or simultaneously improve seed yield potential.

## 2.3.1.7 Niche Market

Other objectives may be important to fill a certain niche, for example, breeding for traits which are deemed desirable by consumers, processors, and producers of food grade soybeans. Food grade soybeans include varieties that are used in the making of tofu, soymilk, natto, edamame, soy sauce and many others. The traits which are desired in soybean cultivars for these types of soy foods vary according to specific consumer and processor needs.

For instance, the size of dry soybean seeds is important for a variety of specialty soybean cultivars generally used for consumption by humans. Small-seeded cultivars ($\leq$ 80 mg seed$^{-1}$) are used to make natto or bean sprouts. Large-seeded cultivars (> 220 mg seed$^{-1}$) are used to make tofu, miso, soymilk, or harvested at R6 as green beans for edamame. These values are relative to a normal soybean seed size of conventional cultivars ranging from 120 to 160 mg seed$^{-1}$. In addition to the seed size requirement, uniform round seed with yellow hilum is desired, and high protein and sugar content and high water absorption capacity are preferred. These types of specialty soybean cultivars could enhance the premium paid to farmers for their crop, and become a valuable source of income to small farmers. There is also much room for genetic gain to be made relative to the yield potential of specialty soybean cultivars. Other important traits in breeding food grade soybean cultivars include isoflavone contents, calcium content, seed texture, taste and flavor, protein subunit fractionation, and seed-soluble sugar concentrations.

### 2.3.1.8 New and Future Objectives

The future of plant breeding depends on growers' and consumers' needs and growth of respective markets. Seed yield will always be a major concern for breeders. One of the major hurdles for plant breeders is introgression of traits into high yielding lines without affecting yield. Drought tolerance, genetic diversity, herbicide and pest resistance, high oil for bio-diesel and edible oil consumption, and environmental issues will all be important traits for breeders in the future. Currently, more than 90% of the soybeans grown in the US go to feed animals such as livestock, poultry, swine, and fish. Breeding for more efficient, functional, and environmental sensitive traits will be important in dealing with future health and ecosystem concerns. An example of this would be seed phytate content in soybean.

### 2.3.2 Crossing Technique

Equipment used to make crosses in soybean is minimal. High quality forceps are necessary to manipulate the flower and make pollination. Plastic tags are also a necessary item and are fastened to the soybean plant to identify the location of the pollinated flowers after the cross has been made. The person making the cross will write on the tag such information as the date the cross was made, identification of the male and female parents, and any other information deemed important. It is imperative that the tags be weather proof, including information written on the tags. This is why plastic tags are used with the information written with a lead pencil and tags fastened to the plants with copper wire. Tags are distinctly visible at harvest and the pods resulting from crossing can be individually hand harvested.

The soybean flower is a typical legume flower in that it has a calyx with five sepals, a corolla consisting of five petals which enclose the pistil or female part of the flower, and ten stamens or the male portion. Of the ten stamens, nine surround the pistil in a tube-like fashion and the remaining one stands free. At the ends of the stamens are the anthers, which contain pollen grains or the male sex cells. Pollen from the anthers sheds directly on the stigma resulting in self-pollination. Soybean does have a small amount of outcrossing ranging from 0.5 to possibly 2.0%. The flower will usually open in the early morning unless there are cool, damp, or cloudy conditions which will delay this occurrence. Shedding of pollen usually occurs shortly before the flower opens. An open soybean flower is only about 6 mm wide across the standard.

The initial step in making a cross is to prepare the female flower. The soybean flower is bisexual in that it contains both sexes, male and female. Flowers that are expected to open the next day are chosen on the plant that is going to be the female. Flowers selected are those where the floral buds are swollen and the corolla is visible through the calyx or just emerging from it. The soybean typically has from 3 to 15 floral buds in the axil of a leaf branch. Usually one to possibly three flowers in a given leaf axil are chosen as females. Care must be taken to insure removal, through the use of forceps, of all other floral buds, including immature buds that are hidden under the stipules in the leaf axil. These immature buds could develop into flowers at a later date and make identification of the pollinated flower difficult, if not impossible.

The flower with the corolla, visible through the calyx, is grasped between the thumb and index finger. The calyx is removed by pulling it down and around the flower. This is repeated until all five sepals are removed. After the calyx is removed, the next step is to remove the corolla. This is done by carefully placing the forceps just above the calyx scar and gently wiggling the corolla upwards until it comes free. After the corolla is removed, the stigma surrounded by the anthers will be visible. If done correctly, the anthers can be removed if the corolla is grabbed in the proper place with the forceps. If the anthers are removed, the flower is emasculated. Emasculation is not necessary as the stigma is receptive for one day before the anthers start to shed pollen.

After the female flower is prepared, it should be pollinated as soon as possible. The exposed stigma will remain viable for several hours. Pollen is initially shed in the early morning hours and up until early to mid-day depending on environmental conditions. Flowers about to open or recently opened are chosen from the male parent. Forceps are used to remove the stamens from these flowers and the pollen-containing anthers are gently brushed over the exposed stigma of the female plant. After pollination is complete, the tag as previously described, is tied to the female plant to identify location of the pollinated flower.

One of the challenges in making crosses is being able to identify those seeds that came from inadvertent self-pollination. If the flowers were not chosen carefully, self-pollination can result. To eliminate plants resulting from self-pollinated seed, genetic markers are used. For example, flower color is determined by a single gene with purple flower color being genetically dominant to white. If a purple male is crossed to a white female, the resulting $F_1$ plants should all be purple. If they turn out to be white, this would indicate that they were not $F_1$s and should be discarded. Other genetic markers can be used including resistance to herbicides. If resistance to a particular herbicide is genetically dominant, the breeder could cross a herbicide resistant male to a susceptible female. True $F_1$ hybrids will not be harmed after spraying with the herbicide, but self-pollinated plants will be destroyed. Additional markers that can be used to distinguish true crosses from selfes include pubescence color, hilum color, seed and seedcoat color, pod wall color, seed size, growth habit, leaf shape, and maturity (Fehr 1980).

### 2.3.3 Selection of Parents

One of the most important aspects of improving upon the probability of success in any plant breeding program is to carefully choose the objectives and parents used for crossing. The breeder must identify changes in the improved soybean variety which, if made will increase, for example, yield, stabilize production, or improve seed quality. The next step is to search for those parents that possess the desired traits to be improved which will hopefully lead to a superior variety and accomplish the stated objectives. There are many means in obtaining novel soybean lines for use as parents in a breeding program, such as: national and international germplasm collections, material transfer agreements with public and private breeders, and the use of experimental lines within the current breeding programs. The choice of which parents to use is made much clearer once the objective(s) have been firmly established. For example, if the objective is to develop an improved variety that is both high yielding and resistant to the soybean cyst-nematode, parents should be chosen that have resistance to the soybean cyst-nematode and have high yield potential. If the breeder chooses to use two parents to accomplish the above breeding objective, a soybean cyst-nematode resistant line could be crossed to a susceptible high yielding variety followed by selection for both high yield and resistance to the soybean cyst-nematode.

Construction of the initial population can happen by using two or more parents, depending on the plant breeding objectives and the perceived worthiness of the parents. If a two-parent cross, also referred to as single cross, is made, $P_1 \times P_2$, 50% of the genes in the segregating population will be from each parent. This would be the type of cross to make if the breeder believes that both parents have equal value and adequate genetic potential. The biparental cross is simple and widely used by many breeders.

A three-parent cross, also referred to as top cross or three-way cross, is made by crossing the two-parent population to a third parent, $(P_1 \times P_2) \times P_3$. These results in a segregating population where on an average 25% of the genes come from $P_1$, 25% from $P_2$ and 50% of the genes from $P_3$. Construction of the segregating population in this manner is desirable if the breeder wants the largest influence from $P_3$. If one of the parents has only one desirable trait $(P_1)$ that the breeder wishes to use, it may not be desirable to use only a two-parent cross involving $P_1$. By using a three-parent cross, genes from a highly desirable parent $(P_3)$ will have considerable influence on deficiencies, for example, found in $P_1$. Parents $P_2$ and $P_3$ are desirable parents and need to be mated with $P_1$ to pick up a desirable trait lacking in $P_2$ and $P_3$. $P_3$ may have exceptional high yield potential and that would be the reason the breeder may want most of the genes from this parent.

A breeder may construct a segregating population by using a four-parent cross. This is referred to as a double cross or a four-way cross. It involves the mating of two single crosses or two two-parent crosses, $(P_1 \times P_2) \times (P_3 \times P_4)$. If the segregating population is formed in this manner, on an average, 25% of the genes will be contributed from each of the four parents. The breeder would use this mating if it was perceived that all the four parents had equal value or complementary traits. Four parents could be mated as $[(P_1 \times P_2) \times P_3] \times P_4$. $P_1$ and $P_2$ are involved in the first two-parent cross and contribute on average 12.5% of their genes each to the segregating population. $P_3$ contributes 25% and $P_4$ 50% of the genes on average to the segregating population.

The breeder has a tremendous amount of flexibility in developing the segregating populations from which to commence selection. How the parents are to be mated can have a large influence on genetic recombination and the outcome of a particular soybean variety. In some instances, the breeder may choose to use more than four parents. This would be termed a complex population and is very seldom done because of the time taken to develop such a population. It should be pointed out that breeding success is not necessarily dependent on the number of parents involved in the crossing scheme and that the duration of the breeding cycle is an important consideration for all breeders.

## 2.3.4 Inbreeding and Selection Methodology

Because soybean is a self-pollinated crop, the breeders use the breeding procedures developed to improve self-pollinated plant species. Examples of breeding procedures used to improve self-pollinated crops include bulking populations, bulk with mass selection, single-seed-descent, pedigree, and backcross breeding.

Bulk breeding, as described by Fehr (1991a), Newman (1912) and Sleper and Poehlman (2006), is the plant breeding method in which seed used to grow each inbreeding generation is a sample of that harvested from all plants of the previous generation. The major advantages of bulk breeding include: the ease of maintaining breeding populations, an increased frequency of desirable genotypes due to natural selection, and the ease of using mass selection during bulk breeding. The disadvantages of bulk breeding include: not all plants are represented in the next generation of progeny, genotypic frequencies and genetic variability is not easily defined, and unsuitable environments may lead to natural selection to favor undesirable genotypes (Fehr 1991a).

For self-pollinated species, bulk breeding accompanied by mass selection is one of the oldest breeding techniques used for crop improvement. Mass selection was used by the earliest farmers by selecting desirable plants and seeds from heterogeneous native populations, based on crop use and personal preferences, to develop cultivars commonly referred to as landraces (Fehr 1991a). Mass selection, in principle, is based on a wide selection of individuals in a population for a desirable trait or against any undesirable traits, in which only a sample of those selected individuals is carried forward in the breeding program (Allard 1960; Fehr 1991a). This selection ultimately results in the elimination of undesirable genotypes from the population, and increase in the frequency of desirable genotypes greater than that if no selection was applied. The main advantage of mass selection, during the inbreeding of self-pollinated crop species, is the ability to easily and inexpensively increase the frequency of desirable genotypes for cultivar development. The disadvantages of using mass selection include: selection can only be applied in environments were the desirable trait is readily expressed, and the effectiveness is largely dependent on the heritability of the trait in the respective population (Fehr 1991a).

Unlike the bulk breeding method, single-seed descent (SSD) is more suited to use environments which may not be in the targeted region or environment associated with cultivar release, for generational advancement. The SSD method was first described by Johnson and Bernard (1962) for soybean, although Jones and Singleton (1934) and Goulden (1941) also described similar methods for rapid inbreeding of populations before evaluation of individual lines or families, and Brim (1966) later described the methodology as a modified pedigree method. The most basic strategy for the SSD method is to harvest a single seed from each plant within the segregating population for bulk planting in the following generation (Fehr 1991a, Sleper and Poehlman 2006). This procedure allows for each $F_2$ plant to be represented in subsequent generations during inbreeding. In practice, if only one seed is taken from each plant, the germination rate of seeds in that population will determine the number of available progeny, and if the germination or resulting seed viability is not 100%, then the population size will decrease and gene frequency will change during inbreeding. Breeders may use several modifications of SSD to preserve genotypic variation and population size acquired at the $F_2$ stage during inbreeding. One method, called single hill method, uses hill plots to plant a few seeds from each individual in the $F_2$ population and subsequent generations. When the desired level of homozygosity is achieved in the population, a single plant is harvested from within each hill plot (Fehr 1991a).

Another modification to the single-seed decent described by Fehr (1991a) and Allard (1960) is the multiple-seed procedure or modified SSD, which is also described by many soybean breeders as the single-pod decent (SPD) or the bulk-pod method, depending on the methodology. This procedure ensures genotypic variation and adequate population size is not influenced by seed viability. A small number of seeds, usually two to four, is taken from each individual in the $F_2$ population and bulked, and a similar number of seeds is taken in subsequent generations of inbreeding, and individual plants within the population are selected at the desired inbreeding stage (Fehr 1991a)

The advantage of using SSD, or a modification of SSD, is that one can maintain an adequate population size for selection in advanced inbreeding stages, as well as maintain the genetic diversity within a segregating population. Normally, natural selection based on environmental conditions will not influence the integrity of genetic variation using SSD, so the procedure is ideal for use in non-targeted environments, i.e. greenhouses and winter nurseries (Fehr 1991a). A major disadvantage in using the SSD method is the time and resources used during planting and harvesting, which are required for accurate depiction of the methodology.

The pedigree breeding and selection methodology can be used for inbreeding both self- and cross-pollinated crops (Sleper and Poehlman 2006). The method utilizes the ability to select and evaluate single plants to create uniform homozygous plant families and lines. Selection normally begins with plants in the $F_2$ generation, where individuals are evaluated and only desirable plants are selected and carried forward to plant independent progeny rows in the $F_3$ generation. During selection in the $F_3$ generation, breeders may select the best plant families, plant rows, or individual plants depending on the desired outcome and resources available to the program. This continues for two or three more generations until homozygous recombinant inbred lines (RILs) are selected for yield trials. The advantages of the pedigree method include the ability to discard inferior lines during early generations and to maximize genetic variability among lines during selection. The major disadvantage of pedigree breeding is the amount of time and maintenance of the accuracy of record keeping during generational advancement.

The backcross breeding procedure is used if the breeder already has a good variety, but it is deficient in one important trait. For example, a line may be high yielding, but susceptible to a particular disease. Backcross breeding can be used to overcome this deficiency and is used when transferring traits that are controlled by one or a few genes. Backcross breeding is also used to get a genetically engineered trait into an improved line or variety such as resistance to the glyphosate herbicide. Genetically engineered traits of soybean today are controlled by a few genes, and backcross breeding can be used initially to get these traits into the proper genetic background (Sleper and Poehlman 2006).

The backcross breeding procedure is a type of recurrent hybridization by which a desirable gene is substituted for an inferior one in an otherwise desirable variety. Two parents are used in the backcross breeding procedure. One of the parents contains the desirable gene to improve an already existing variety. This parent is referred to as the donor parent and enters into only the first cross. The other parent is the variety that is to be improved by receiving the desirable gene from the donor parent. This parent is called the recurrent parent and is part of every cross throughout the entire procedure. The backcross breeding procedure is a stepwise procedure whereby recovery of the genes from the recurrent parent is predictable, and where the final product includes the new gene from the donor parent and almost complete recovery of the genes from the recurrent parent. Therefore, backcrossing is a very efficient method for developing an improved, if not new, variety in a short period of time.

A practical example of using backcrossing to develop improved varieties involves introgression of the glyphosate resistance gene into a highly productive conventional cultivar. Resistance to glyphosate is controlled by a single dominant gene denoted as $R$. The recurrent parent A is completely susceptible to glyphosate and has the genetic constitution of $rr$. The donor parent B has the genotype $RR$ and is crossed to recurrent parent A only once. The $F_1$ is heterozygous ($Rr$) and is backcrossed to recurrent parent A. The first backcross generation ($BC_1F_1$) will produce plants in a one to one ratio of resistant to susceptible. These $BC_1F_1$ plants are sprayed with glyphosate and only the resistant ($Rr$) ones will remain and these will in turn be backcrossed again to recurrent parent A to produce the $BC_2F_1$ generation. The process is repeated until near-recovery of the entire genetic constitution of recurrent parent A is achieved. In $BC_1F_1$, 75% of the genes will be recovered from parent A. In $BC_2F_1$ 87.5% of the genes from recurrent parent A will be recovered. This process is repeated until the breeder is satisfied that enough of the genes from recurrent parent A are recovered. In the last backcross generation, the *Rr* plants are selfed and only the *RR* plants are saved and released as a new, improved variety with resistance to the glyphosate herbicide.

Some of the most relevant and recent research comparing breeding methodologies is summarized by Orf et al. (2004) in the American Society of Agronomy Publication: Soybeans: Improvement, Production, and Uses. The efficiency of the pedigree, SSD, and SSD with concurrent selection (maturity) methodologies on developing early maturing cultivars was investigated by Byron and Orf (1991) in four populations. They found no significant differences in methodologies for selection on maturity, yield, plant height, lodging, and seed weight. They concluded that SSD with concurrent selection used the fewest resources and was most suitable for developing early maturing soybean cultivars. Cooper (1990) also indicated a novel methodology to reduce resources used by plant breeders for population development. The author used a modified early generation testing procedure to evaluate single replication and location data for $F_{2:3}$ through $F_{4:6}$ lines. This procedure allowed for a 10-fold decrease in the number of yield plots evaluated as compared to the early-generation yield testing, pedigree selection, and single-seed descent selection methodologies described by Boerma and Cooper (1975).

Degago and Caviness (1987) compared bulk populations grown from 10 to 18 years at 3 locations in Arkansas. One of the locations was annually infested with Phytophthora root rot and stem rot infections, while the others had very slight infestations. Results indicated that populations grown at infested sites had significantly higher seed yields than those grown at non-infested sites. These results indicate that bulk breeding can be useful in developing high yielding, disease resistant cultivars for targeted environments. A breeder may choose a specific method or use a combination of methods in the process of inbreeding and selection. It appears that SSD and bulk pod methods combined with mass selection are most popular among the soybean breeders today.

## 2.3.5 Variety Development and Release

The following text depicts a typical season by season account of variety development from cross to release by using the single-pod descent methodology as an example. Plant breeding procedures can vary a lot depending on the individual breeder and the resources available. The following depicts what can happen in a breeding program using single-pod descent, but one must realize that this serves as an example as each breeder will modify the single-pod descent method to fit their programs with their specific objectives and resources available. There may also be variability in breeding programs in relation to the number of generations grown per season using winter nurseries, as well as the specific generation in which a winter nursery is used to advance plant material and populations.

## Season 1

Crosses are made during the summer in the field to produce $F_1$ (first filial generation) seed. The example shown depicts a two-parent cross (Fig. 2-1). Hybrid seeds are harvested in the fall and planted in an off-season nursery in the tropics prior to mid-December. Because the soybean plant is a daylength sensitive species, artificial lights are often used to ensure that near normal plant growth occurs for maximum seed production. In certain instances it is possible to obtain two or more generations in a tropical off-season nursery.

## Season 2

Seeds from $F_1$ plants grown in a tropical nursery are single-plant-threshed (SPT) and placed in separate bags. Seeds ($F_2$) are returned to the breeder and the seed from each SPT are planted in an individual plot to allow for rouging off-type plants and identifying the segregating populations. When the plants are mature, single-pods are picked from a random group of $F_2$ plants. The number of pods picked will depend on the wishes of the breeder and resources available. Pods are threshed and the seed ($F_3$) planted in the tropical nursery prior to mid-December.

## Season 3

Pods are harvested in the tropical nursery from $F_3$ plants during April through May depending on maturity and planting date in the previous fall. $F_3$ plants are not grown with the assistance of artificial lights because only one to several pods is harvested from each plant. Seed is threshed in bulk for each population, which originated from the cross produced in season 1, and planted as $F_4$ bulk populations by the breeder. Bulk populations are monitored very carefully during the growing season to note incidences of disease, lodging, shattering and other undesirable characteristics. During the fall, individual plants are selected and harvested from each population and SPT. At the time of harvest, the breeder can select for such agronomic traits as resistance to lodging and proper maturity. The number of plants selected from each population is variable and depends upon the wishes of the breeder and the availability of resources. A high degree of homozygosity

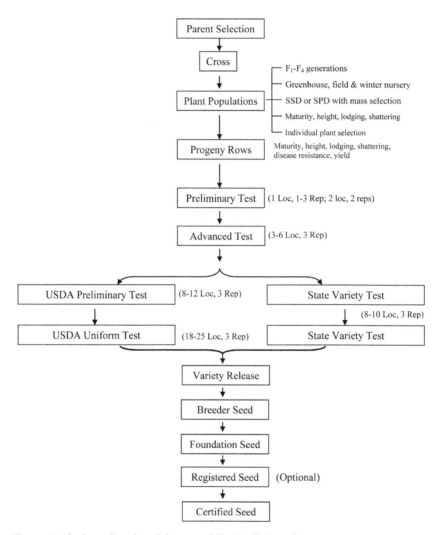

**Figure 2-1** Soybean Breeding Scheme and Variety Release Process.

has been achieved by this time because the seed produced on these $F_4$ plants has reached the $F_5$ generation. An alternative approach to handling the early stage of the breeding process from crossing to $F_4$ would be: making crosses in the greenhouse in the spring, growing $F_1$ plants in the field in summer at the home station, advancing the $F_2$ and $F_3$ plant populations using SSD or bulk pod for two generations in a winter nursery, and growing $F_4$ population at home station for plant selection. This strategy would potentially save a year for developing pure lines.

## Season 4

During the winter, individual SPT lines can be screened by marker-assisted selection (Pathan and Sleper 2008) and/or phenotypically in the greenhouse for resistance to diseases such as soybean cyst-nematode and Phytophthora rot. Those not possessing the desired level of resistance can be discarded and need not be advanced in the breeding program. Seeds from each SPT harvested in season 3 are planted to an individual progeny row at the home station. Each individual row is given a number. For example if the breeder has 20,000 progeny rows, they will be numbered from 1 to 20,000. This is the first time in the breeding program that selections are referred to as lines. Line implies breeding true, and because a high degree of homozygosity has been achieved by the $F_5$, breeders refer to these selections as lines. The breeder observes these lines for desirable agronomic traits such as proper maturity, resistance to foliar diseases, resistance to lodging, proper plant height, and yield potential. At harvest, selected rows are noted and the seeds ($F_6$) threshed and bulked from an individual progeny row. A breeder may employ a selection intensity of 5 to 15%, or possibly higher depending upon the breeding objectives. At this time, maturity is recorded for the selected lines based on relative maturity of several check varieties planted in the progeny row nursery. The selected lines receive an experimental designation, based on their row number in the nursery. For example, if the breeder chose an entry in row number 789 and selected this in the year 2008, it would receive the designation VA08-789. This number will stay with the new experimental selection until it is renamed as a released variety or discarded because of poor performance. VA represents the location code, as each breeder has a unique one. When other breeders view this experimental number, it is obvious that the experimental line originated from Virginia, was selected in the year 2008, and came from progeny row number 789. In this way, the complete history of how the variety came about can be traced.

## Season 5

Yield testing is initiated in season 5 or approximately around this season depending on how many generations were advanced over the previous seasons. Yield testing is the most expensive endeavor of any soybean breeding program because multiple replications and locations are involved and expensive specialized field plot equipments are needed. The breeder usually has very little seed of each selection during this phase of the breeding program so decisions have to be made on how many locations and replications are possible for initial testing, or as we refer to it as the Preliminary Test which may be called as First Year Trial (1YT) by breeders in private industry. Many entries will be evaluated in the Preliminary Test with most of the lines being discarded after harvest because their yield did not compare to high performing check varieties. Average selection intensity at this stage might be 10%, so the breeder will discard 90% of the poorest performing lines after harvest. Selection will occur for yield and other important agronomic traits such as resistance to lodging, proper maturity, and resistance to diseases.

## Season 6

The next stage of testing will incur in year 2, referred to as Advanced Trial by public breeders or Second Year Trial (2YT) by private breeders. At this stage, the breeder is likely to have enough seed to plant replicated four-row plots at each location. The number of replications and locations (perhaps three to six locations with three replications) will depend upon the resources in the breeding program. The center two rows of each plot will be harvested for yield. Again, selection will occur for yield and other desirable agronomic traits with a discard of 90 to 95% of the poorest performing entries based on performance compared to high performing check varieties.

## Season 7

The final stage of testing will likely be in year 3 with as many replications and locations as deemed necessary. This test is usually conducted concurrently with the State Variety Trials and the USDA Uniform Tests. Each test will include new experimental selections that have performed superiorly over a series of years and locations. If the experimental line is superior in one or more characteristics compared to the best check varieties, it is released. This is a key difference between a public and private breeding program. In the private sector, elite lines (or year 3 and 4 lines) are not tested in the USDA Uniform Yield Trials, rather they are privately yield tested in advanced trials (3YT and 4YT) within the company.

In addition to the breeders own yield tests, public breeders also participate in regional tests, such as the USDA Uniform Soybean Tests (Uniform and Preliminary). The purpose of the two tests are to evaluate the best experimental soybean lines developed by federal and state experiment station breeders in the USA and lines developed by Canadian public breeders. The tests are organized according to maturity and each test has a uniform set of checks, so meaningful comparisons can be made across a number of locations. It is not uncommon for both tests to involve more than 20 locations with over 10 states participating each year. The preliminary tests include experimental strains and are conducted for only one year at multiple locations with the highest performing lines advanced for more extensive regional testing in the Uniform Tests. Data collected for both tests include yield, lodging, height, seed quality, seed size, seed composition (percentage of oil and protein), shattering, emergence, major disease reactions, and other pertinent data.

After considerable testing, an experimental strain of soybean may be considered for release if it has been shown to be superior in one or more traits compared to high performing check varieties. Public breeders will consult with a Varietal Release Committee and together they will make a recommendation on release to parties as deemed appropriate. If a decision is made to release the improved variety, seed increases need to take place to ensure adequate seed supply for interested growers. The seed increases will go through a certification process to assure that the seed has high purity and quality. There are four classes of certified seed, namely, breeder's seed, foundation seed, registered seed, and certified seed. Figure 2-1 gives a schematic demonstration of the entire breeding and variety release process. However, breeders often skip the registered class of seed and go directly to certified seed to save time in getting varieties to the farmer. Private companies often have their own purity unit to handle seed increase, purification, and production from breeder's seed to certified seed.

Breeder's seed is the seed produced by the breeder and that which the breeder has direct control of. Breeder's seed will be in limited supply and may range from a few pounds to several bushels. The breeder has taken care to rogue off-type plants to assure genetic purity of the new variety. The breeder's seed is identified with a white tag affixed to the bag containing the seed.

The next class of seed in the certification process is foundation seed. Foundation seed results directly from breeder's seed. Usually, foundation seed program is part of the Agricultural Experiment Station and it produces foundation seed of new soybean varieties for sale to seed dealers to propagate. Foundation seed production is carefully monitored and genetic identity and purity of the new variety is maintained. Bags containing foundation seed are identified with white tags.

Registered seed is the first-generation increase of breeder or foundation seed. Again as before, genetic identity and purity is maintained. Registered seed is identified with a purple tag and is used to produce certified seed. Oftentimes, the registered class of seed is skipped to hasten the release process.

Certified seed is the first-generation increase of breeder, foundation, or registered seed. Highest seed production standards are utilized to assure genetic identity and purity. Bags containing certified seed have blue tags attached to them. The tags contain such information as who labeled them, name of soybean variety, percentage of pure seed, percentage of inert material, percent germination, and other pertinent information.

Varieties may be released as branded varieties where the exact name of the variety is not stated. This allows more than one company to market the same variety and apply their own name to it. Branded varieties can go through the green tag program, which is not part of the national seed certification program but seed is increased under the same strict seed quality standards as the certification procedure. Today, many public varieties are released exclusively to companies who produce their own certified seed and donot rely upon a state-certified seed certification service.

## 2.3.6 Variety Protection and Patent

While a soybean is being released, there are several factors in protecting the integrity of the new variety. The two most common methods of protecting new soybean varieties include protection through the Plant Variety Protection Act and through a plant utility patent. The Plant Variety Protection Act came into being in 1970 and was amended in 1994. It provides legal protection to sexually (by seed) produced varieties such as soybean for a period of 20 years. If a variety is protected through a plant utility patent, protection is good for 17 years.

Protection of soybean varieties is often warranted because it allows owners of varieties to maintain control over their purity and marketing. Without these types of controls, soybean varieties would essentially be pirated and the owner would not receive a fair return, which may jeopardize further research and development of new soybean varieties. The purpose of protecting soybean varieties is to encourage development of novel varieties and make them available to the public. The public certainly benefits directly as the recipient from such intellectual property protection because it receives the use of improved soybean varieties and the multitude of uses and products developed from soybean. A soybean variety is eligible for Plant Variety Protection if it meets the following criteria: Distinctiveness—Distinctiveness means a soybean variety differs by one or more identifiable morphological, physiological, or other characteristics, Uniformity—Uniformity means a soybean variety must demonstrate that variations within the variety are describable, predictable, and commercially acceptable, Stability—Stability means that the soybean variety must remain essentially unchanged with regard to essential and distinctive characteristics with a reasonable degree of reliability commensurate with that of soybean varieties of the same type.

Plant utility patents have been allowed for use on sexually propagated plants such as soybean since 1985. The USA Government has certain requirements before it will issue a plant patent. These include: Novelty—Novelty means that the person who is the first inventor or developer of a soybean variety must show that it is original in some manner, Utility—Utility means something that is capable of being used beneficially for the purpose for which it was developed, Non-obviousness—Non-obviousness means that the new soybean variety is something that goes beyond what people who have ordinary skill in the art would know. To determine if it is non-obvious the question needs to be asked if it would be obvious to the soybean breeder or inventor guided by previous patents and any other information that might be available in soybean breeding, such as publications. This is likely the most difficult of the three to comprehend. In addition to public release, a breeder may choose to release a new variety as an exclusive proprietary product through private channels such as exclusive release, licensing, sub- licensing, contract, subcontract, and special agreement; in all cases specific legal protection is implemented.

### 2.3.7 Resource Management and Breeding Program Efficiency

Plant breeding is a number's game, long-term by nature, time sensitive, and resource dependent. The success of a breeding program is largely dependent on when and how to do what and how many. Breeders require specialized planting and harvest equipment as well as powerful computers to track and record the thousands of experimental selections that are screened and tested.

Computers keep track of every experimental selection from its conception as a new cross to its release as an improved variety approximately 6 to 10 years later. They are used to design tests, to print field books so that notes on new lines may be taken, and to analyze the statistics of yield. Some of the other data that computers store and analyze includes information on parentage, disease resistance, maturity, lodging, height, flower and pubescence color, emergence, seed size, and protein and oil content. Without the incredible tracking abilities and speed of today's computers, breeders would be able to test only a small percentage of the new lines they now grow.

Research combines may also house small computers that have software developed specifically for reading the weight and moisture of a harvested sample. Many of these small computers can also be used to take field notes on developing soybean lines. Because these computers can transfer their data directly into a larger office computer, they eliminate the need for manual data entry. This reduces typographical errors and allows the researcher to quickly access data and make decisions about the value of the thousands of experimental lines the breeding project tests each year. Computers used on equipment and in the field have contributed to the increased size and efficiency of modern soybean breeding projects.

Research fields for soybean breeders are usually prepared and cared for in much the same way as a neighboring farmer would prepare a field. Fields are tilled with traditional equipment and treated with an appropriate herbicide for the prevailing conditions. Many public soybean breeders grow research plots on both university-owned farms and on private land, where they work with the farmer in land preparation and weed control.

Breeding projects often use specialized planters that have a platform with a seat where a person can sit and feed the different seed needed for each plot into the planter. Most attach to the tractor with a three-point hitch so that they can be easily loaded onto a trailer and moved around to different testing sites. Research planters are usually 4-rows wide and have a spinning cone that evenly distributes seed through a plot, which may be 10 to 20 ft. long. A divider that sits above the cone allows for one package of seed to be dropped in by the operator at the beginning of each plot and to be dispersed to multiple row units. The operator presses a button on the platform to trip a solenoid that opens the bottom of the cone and drops the seed into the planter row units. Many breeding projects also save time during planting by using a long cable spread the length of the field with metal buttons set at a prescribed length. The buttons force open a gate on the planter tool bar which trips the solenoid, freeing up the operator's hands to do other jobs.

During the first few years of developing a new soybean variety, several types of small harvest equipment are used. In the single-pod descent breeding method, breeding populations are quickly advanced to homozygosity by picking one pod from every plant and pooling this seed to produce the next generation. After several years of this, single plants are harvested and threshed separately. A drum or a belt-driven single-plant thresher is normally used. This small machine threshes one plant or a few pods at a time and keeps each sample pure of contamination from other plants. The season after the single plants are harvested, the seed from each plant is grown in a 5 to 10 ft long row. A productive breeding program may have from 10,000 to 70,000 of these rows. A one-row combine with a pneumatic seed delivery system is utilized to harvest these one-row plots.

The research combine used to harvest yield plots also utilizes an air delivery (pneumatic) system to get seed from the sickle bar to the grain bin or to the weigh bucket. The use of air delivery instead of augers in a combine is very important to a breeding program because an auger can damage seed and because the air delivery system has very quick and thorough clean-out, ensuring against contamination from different varieties. A good research combine will be set up with a weigh bucket and moisture probe connected to a computer, which eliminates the need for manual weighing and moisture testing. As plots are harvested, they are bagged and tagged with an identifying tag so that they can be identified later for further disease screening and yield testing the following year. Most research combines harvest two rows at a time and are small enough to be transported on a trailer.

One of the most recent resources used by breeders is the molecular marker for marker-assisted selection (Pathan and Sleper 2008). The development of fast and easy protocols to genotype large numbers of plants or lines has enabled public and private breeders to more efficiently select lines with desirable traits. This will be discussed further in a later chapter.

The overall goal of a breeding program is to provide a steady flow of new and improved varieties. There is a great deal of flexibility involved in every step and all aspects of the breeding process. Although there is no uniform protocol to run a breeding program, the efficiency and success of a breeding program depend largely on resources available and decisions made by the breeder. While breeders use every resource to make the breeding process less time consuming and less labor intensive, they attempt to make best decisions based on science, experience, and reality. The two most important factors impacting the efficiency of a breeding program are time and number; breeders tend to move a large number of materials as quickly as possible. In this process, a proper decision is crucial as to how many and what crosses to make, how many and what size of populations to maintain or advance, how many plants to select, and how many lines to select and evaluate, how many replications and locations to use in various trials, and so forth. In the effort to speed up the breeding process time-wise, breeders face the challenge of making decisions on how to efficiently utilize the greenhouse, field, and winter nursery in advancing breeding materials. All these mentioned above are true reflections of the art part of plant breeding that make the breeders' job so fascinating. There is no doubt that plant breeding has played an important role in continued crop yield improvement over time. As the world population continues to increase and crop land decreases, breeders may face many new challenges ahead to keep increasing yield and meet new needs for specialty quality attributes by consumers.

## References

Acquaah G (2007) Principles of Plant Genetics and Breeding. Blackwell Publ, Malden, MA, USA.

Allard RW (1960) Principles of Plant Breeding. John Wiley, New York, USA.

Anai T, Yamada T, Kinoshita T, Rahman SM, Takagi Y (2005) Identification of corresponding genes for three low-[alpha]-linolenic acid mutants and elucidation of their contribution to fatty acid biosynthesis in soybean seed. Plant Sci 168: 1615–1623.

Athow KL, Probst AH (1952) The inheritance of resistance to frogeye leaf spot of soybeans. Phytopathology 42: 660–662.

Bailey MA, Mian MAR, Carter Jr TE, Ashley DA, Boerma HR (1997) Pod dehiscence of soybean: Identification of quantitative trait loci. J Hered 88: 152–154.

Bernard RL (1967) The inheritance of pod color in soybeans. J Hered 58: 165–168.

Bernard RL (1971) Two major genes for time of flowering and maturity in soybeans. Crop Sci 11: 242–244.

Bernard RL (1972) Two genes affecting stem termination in soybeans. Crop Sci 12: 235–239.

Bilyeu KD, Palavalli L, Sleper DA, Beuselinck PR (2003) Three microsomal omega-3 fatty-acid desaturase genes contribute to soybean linolenic acid levels. Crop Sci 43: 1833–1838.

Bilyeu K, Palavalli L, Sleper D, Beuselinck P (2005) Mutations in soybean microsomal omega-3 fatty acid desaturase genes reduce linolenic acid concentration in soybean seeds. Crop Sci 45: 1830–1836.

Bilyeu K, Palavalli L, Sleper DA, Beuselinck P (2006) Molecular genetic resources for development of 1% linolenic acid soybeans. Crop Sci 46: 1913–1918.

Boerma HR, Cooper RL (1975) Comparison of three selection procedures for yield in soybeans. Crop Sci 15: 225–229.

Boerma HR, Phillips DV (1983) Genetic implications of the susceptibility of Kent soybean to Cercospora sojina. Phytopathology 73: 1666–1668.

Bonato ER, Vello NA (1999) *E6*, a dominant gene conditioning early flowering and maturity in soybeans. Genet Mol Biol 22: 229–232.

Brim CA (1966) A modified pedigree method of selection in soybeans. Crop Sci 6: 220.

Brim CA (1973) Quantitative Genetics and Breeding. Am Soc Agron, Madison, WI, USA, pp 155–186.

Burton JW, Koinange EMK, Brim CA (1990) Recurrent selfed progeny selection for yield in soybean using genetic male sterility. Crop Sci 30: 1222–1226.

Burton JW, Carter TE, Wilson RF (1999) Registration of 'Prolina' soybean. Crop Sci 39: 294–295.

Buss GR, Ma G, Kristipati S, Chen P, Tolin SA (1999) A new allele at the *Rsv3* locus for resistance to soybean mosaic virus. In: HE Kauffman (ed) Proc World Soybean Res Conf, VI, Chicago, IL, USA, 4–7 Aug 1999. Superior Printing, Champaign, IL, USA, p 490.

Buttery BR, Buzzell RI (1973) Varietal differences in leaf flavonoids of soybeans. Crop Sci 13: 103–106.

Buttery BR, Buzzell RI (1975) Soybean flavonol glycosides: Identification and biochemical genetics. Can J Bot 53: 219–224.

Buzzell RI (1971) Inheritance of a soybean flowering response to fluorescent-daylength conditions. Can J Genet Cytol 13: 703–707.

Buzzell RI, Voldeng HD (1980) Inheritance of insensitivity to long daylength. Soybean Genet Newsl 7: 26–29.

Buzzell RI, Tu JC (1989) Inheritance of a soybean stem-tip necrosis reaction to soybean mosaic virus. J Hered 80: 400–401.

Byron DF, Orf JH (1991) Comparison of three selection procedures for development of early-maturing soybean lines. Crop Sci 31: 656–660.

Byth DE, Weber CR (1969) Two mutant genes causing dwarfness in soybeans. J Hered 60: 278–280.

Carter TE, Rufty TW (1993) A soybean plant introduction exhibiting drought and aluminum tolerance. In: G Kuo (ed) Adaptation of Vegetables and other Food Crops to Temperature and Water Stress. AVRDC Publ, Tainan, Taiwan, pp 335–346.

Chen P, Buss GR, Roane CW, Tolin SA (1991) Allelism among genes for resistance to soybean mosaic virus in strain differential soybean cultivars. Crop Sci 31: 305–309.

Chen P, Sneller CH, Mozzoni LA, Rupe JC (2007) Registration of 'Osage' Soybean. J Plant Reg 1(2): 89–92.

Chen P, Sneller CH, Purcell LC, Sinclair TR, King CA, Ishibashi T (2007) Registration of soybean germplasm lines R01-416F and R01-581F for improved yield and nitrogen fixation under drought stress. J Plant Reg 1(2): 166–167.

Cober ER, Voldeng HD (2001) A new soybean maturity and photoperiod-sensitivity locus linked to E1 and T. Crop Sci 41: 698–701.

Cooper RL (1990) Modified early generation testing procedure for yield selection in soybean. Crop Sci 30: 417–419.

Degago Y, Caviness CE (1987) Seed yield of soybean bulk populations grown for 10 to 18 years in two environments. Crop Sci 27: 207–210.

Domingo WE (1945) Inheritance of number of seeds per pod and leaflet shape in the soybean. J Agri Res 70: 251–268.

Erickson EA, Wilcox JR, Cavins JF (1988) Inheritance of palmitic acid percentages in two soybean mutants. J Hered 79: 465–468.

Fehr WR (1972) Inheritance of a mutation for dwarfness in soybeans. Crop Sci 12: 212–213.

Fehr WR (1980) Soybean. In: WR Fehr , HH Hadley (eds) Hybridization of Crop Plants. ASA, CSSA, Madison, WI, USA, pp 589–599.

Fehr WR (1991a) Principles of Cultivar Development, vol 1. Macmillan Publ, Iowa State Univ, Ames, Iowa, USA.

Fehr WR (1991b) Principles of Cultivar Development, vol 2. Macmillan Publ, Iowa State Univ, Ames, Iowa, USA.

Fehr WR, Hammond EG (1996) Soybean having low linolenic acid content and method of production. US Patent 5, 425, 534.

Fehr WR, Welke GA, Hammond EG, Duvick DN, Cianzio SR (1991a) Inheritance of reduced palmitic acid content in seed oil of soybeans. Crop Sci 31: 88–89.

Fehr WR, Welke GA, Hammond EG, Duvick DN, Cianzio SR (1991b) Inheritance of elevated palmitic acid content in seed oil of soybeans. Crop Sci 31: 1522–1524.

Fehr WR, Welke GA, Hammond EG, Duvick DN, Cianzio SR (1992) Inheritance of reduced linolenic acid content in soybean genotypes A16 and A17. Crop Sci 32: 903–906.

Funatsuki H, Ishimoto M, Tsuji H, Kawaguchi K, Hajika M, Fujino K (2006) Simple sequence repeat markers linked to a major QTL controlling pod shattering in soybean. Plant Breed 125(2): 195–197.

Gai J, Zhao T, Qiu J (1997) A review on the advances of soybean breeding since 1981 in China. In: Seed Industry and Agricultural Development. CAASS China Agri Press, Beijing, China, pp 168–174.

Goulden CH (1941) Problems in plant selection. Proc 7th Int Genet Congr, Edinburgh, pp 132–133.

Graef GL, Fehr WR, Hammond EG (1985) Inheritance of three stearic acid mutants of soybean. Crop Sci 25: 1076–1079.

Guimaraes EP, Fehr WR (1989) Alternative strategies of recurrent selection for seed yield of soybean. Euphytica 40: 111–120.

Gunduz I (2000) Genetic analysis of soybean mosaic virus resistance in soybean. PhD Dissert, Virginia Polytech Inst and State Univ, Blacksburg, VA, USA.

Hammond EG, Fehr WR (1983) Registration of A5 germplasm line of soybean. Crop Sci 23: 192.

Hartwig EE, Bromfield KR (1983) Relationships among three genes conferring specific resistance to rust in soybeans. Crop Sci 23: 237–239.

Hobbs TW, Schmitthenner AF, Kuter GA (1985) A new *Phomopsis* species from soybean. Mycologia 77(4): 535–544.

Hurburgh CR Jr, Brumm TJ, Guinn JM, Hartwig RA (1990) Protein and oil patterns in U.S. and world soybean markets. J Am Oil Chem Soc 67: 966–973.

Hymowitz T (1990) Soybean: The success story. In: J Janick, JE Simon (eds) Advances in New Crops. Timber Press, Portland, OR, USA, pp 159–163.

Hymowitz T (2004) Speciation and cytogenetics. In: HR Boerma, JE Specht (eds) Soybeans: Improvement, Production, and Uses. Agron Monogr, 3rd edn, No 16. ASA-CSSA-SSSA, Madison, WI, USA, pp 97–136.

Jackson EW, Fenn P, Chen P (2005) Inheritance of resistance to Phomopsis seed decay in soybean PI 80837 and MO/PSD-0259 (PI 56264). Crop Sci 45: 2400–2404.

Jackson EW, Feng C, Fenn P, Chen P (2006) Inheritance of resistance to purple seed stain caused by *Cercospora kikuchii* in PI 80837 soybean. Crop Sci 46: 1462-1466.

Johnson HW, Bernard RL (1962) Soybean genetics and breeding. Adv Agron 14: 149–221.

Jones DF, Singleton WR (1934) Crossed sweet corn. Conn Agri Exp Stn Bull 361: 487–536.

Keen NT, Buzzell RI (1991) New disease resistance genes in soybean against *Pseudomonas syringae* pv. *glycinea*: Evidence that one of them interacts with bacterial elicitor. Theor Appl Genet 81: 133–138.

Kiihl RAS, Hartwig EE (1979) Inheritance of reaction to soybean mosaic virus in soybean. Crop Sci 19: 372–375.

Kilen TC, Hartwig EE (1971) Inheritance of a light-quality sensitive character in soybeans. Crop Sci 11: 559–561.

Li XP, Tian AG, Luo GZ, Gong ZZ, Zhang JS, Chen SY (2005) Soybean DRE-binding transcription factors that are responsive to abiotic stresses. Theor Appl Genet 110: 1355–1362.

Liao H, Wong FL, Phang TH, Cheung MY, Li WY, Shao G, Yan X, Lam HM (2003) *GmPAP3*, a novel purple acid phosphatase-like gene in soybean induced by NaCl stress but not phosphorus deficiency. Gene 318: 103–11.

Luo GZ, Wang HW, Huang J, Tian AG, Wang YJ, Zhang JS, Chen SY (2005) A putative plasma membrane cation/protein antiporter from soybean confers salt tolerance in Arabidopsis. Plant Mol Biol 59: 809–820.

Ma G, Chen P, Buss GR, Tolin SA (1995) Genetic characteristics of two genes for resistance to soybean mosaic virus in PI 486355 soybean. Theor Appl Genet 91: 907–914.

McBlain BA, Bernard RL (1987) A new gene affecting time of maturity in soybeans. J Hered 78: 160–162.

Mclean RJ, Byth DE (1980) Inheritance of resistance to rust *Phakopsora pachyrhizi* in soybeans. Aust J Agri Res 31: 951–956.

Mukherjee D, Lambert JW, Cooper RL, Kennedy BW (1966) Inheritance of resistance to bacterial blight in soybeans. Crop Sci 6: 324–326.

Nagai I (1921) A genetico-physiological study on the formation of anthocyanin and brown pigments in plants. Tokyo Univ Coll Agri 8: 1–92.

Nagai I (1926) Inheritance in the soybean. Nogyo Oyobi Engei 1(14): 107–108.

Narvel JM, Fehr WR, Ininda J, Welke GA, Hammond EG, Duvick DN, Cianzio SR (2000) Inheritance of elevated palmitate in soybean seed oil. Crop Sci 40: 635–639.

Newman LH (1912) Plant Breeding in Scandinavia. Can Seed Growers Assoc, Ottawa, Canada.

Oltmans SE, Fehr WR, Welke GA, Cianzio SR (2004) Inheritance of low-phytate phosphorus in soybean. Crop Sci 44: 433–436.

Orf JH, Diers BW, Boerma HR (2004) Genetic improvement: Conventional and molecular strategies In: HR Boerma , JE Specht (eds) Soybeans: Improvement, Production, and Uses. Agron Monogr, 3rd edn, No 16. ASA-CSSA-SSSA, Madison, WI, USA, pp 417–450.

Owen FV (1927) Inheritance studies in soybeans. II. Glabrousness, color of pubescence, time of maturity, and linkage relations. Genetics 12: 519–529.

Palmer RG (1984) Genetic studies with T263. Soybean Genet Newsl 4: 40–42.

Pathan MS, Sleper DA (2008) Advances in soybean breeding. In: G Stacey (ed) Genetics and Genomics of Soybean. Springer, New York, USA, pp 113–133.

Patil A, Taware SP, Oak MD, Tamhankar SA, Rao VS (2007) Improvement of oil quality in soybean [*Glycine max* (L.) Merrill] by mutation breeding. J Am Oil Chem Soc 84: 1117–1124.

Piper CG, Morse WJ (1910) The soybean: History, varieties, and field studies. USDA Bureau of Plant Industry Bull 197, US Gov Print Office, Washington DC.

Piper TE, Fehr 1987 (1987) Yield improvement in a soybean population by utilizing alternative strategies of recurrent selection. Crop Sci 27: 172–178.

Porter KB, Weiss MG (1948) The effect of polyploidy on soybeans. J Am Soc Agron 40: 710–724.

Polson DE (1972) Day-neutrality in soybean. Crop Sci 12: 773–776.

Probst AH, Athow KL, Laviolette FA (1965) Inheritance of resistance to race 2 of *Cercospora sojina* in soybeans. Crop Sci 5: 332.

Purcell LC, Specht JE (2004) Physiological traits for ameliorating drought stress. In: HR Boerma, JE Specht (ed) Soybean: Improvement, Production, and Uses. Agron Monogr, 3rd edn, No 16. ASA-CSSA-SSSA, Madison, WI, USA.

Rahman SM, Takagi Y, Kinoshita T (1997) Genetic control of high stearic acid content in seed oil of two soybean mutants. Theor Appl Genet 95: 772–776.

Rahman SM, Kinoshita T, Anai T, Takagi Y (1999) Genetic relationship between loci for palmitate content in soybean mutants. J Hered 90: 423–428.

Rennie BD, Tanner JW (1989) Genetic analysis of low linolenic acid levels in the line PI 123440. Soybean Genet Newsl 16: 25–26.

Rennie BD, Zilka J, Crammer MM, Beversdorf WD (1988) Genetic analysis of low linolenic acid levels in the soybean line PI 361088B. Crop Sci 28: 655–657.

Ross AJ (1999) Inheritance of reduced linolenate soybean oil and its influence on agronomic and seed traits. MS Thesis, Iowa State Univ, Ames, Iowa, USA.

Ross AJ, Fehr WR, Welke GA, Hammond EG, Cianzio SR (2000) Agronomic and seed traits of 1%-linolenate soybean genotypes. Crop Sci 40: 383–386.

Sleper DA, Poehlman JM (2006) Breeding Field Crops. 5th edn. Blackwell, Ames, IA, USA.

Schnebly SR, Fehr WR, Welke GA, Hammond EG, Duvick DN (1994) Inheritance of reduced and elevated palmitate in mutant lines of soybean. Crop Sci 34: 829–833.

Sinclair JB (1982) Compendium of Soybean Diseases. 2nd edn. APS Press, St Paul, MN, USA.

Sinclair TR, Precell LC, King CA, Sneller CH, Chen P, and Vadez V. (2007) Drought tolerance and yield increase of soybean resulting from improved symbiotic N2 fixation. Field Crop Res 101: 68–71.

Soybean Genetics Committee (1995) Soybean genetics committee report. Soybean Genet Newsl 22: 11–14.

Stephens PA, Nickell CD, Kolb FL (1993) Genetic analysis of resistance to *Fusarium solani* in soybean. Crop Sci 33: 929–930.

Stoltzfus DL, Fehr WR, Welke GA, Hammond EG, Cianzio SR (2000a) A *fap5* allele for elevated palmitate in soybean. Crop Sci 40: 647–650.

Stoltzfus DL, Fehr WR, Welke GA, Hammond EG, Cianzio SR (2000b) A *fap7* allele for elevated palmitate in soybean. Crop Sci 40: 1538–1542.

Takahashi N (1934) Linkage relation between the genes for the forms of leaves and the number of seeds per pod of soybeans. Jpn J Genet 9: 208–225.

Takahashi Y, Fukuyama J (1919) Morphological and genetic studies on the soybean. Hokkaido Agri Exp Stn Rep 10.

Thompson JA, Bernard RL, Nelson RL (1997) A third allele at the soybean dt1 locus. Crop Sci 37: 757–762.

Upholf MD, Fehr WR, Cianzio SR (1997) Genetic gain for soybean seed yield by three recurrent selection methods. Crop Sci 37: 1155–1158.

Walker DR, Scaboo AM, Pantalone VR, Wilcox JR, Boerma HR (2006) Genetic mapping of loci associated with seed phytic acid content in CX1834-1-2 soybean. Crop Sci 46: 390–397.

Wang D, Graef GL, Procopiuk AM, Diers BW (2003) Identification of putative QTL that underlie yield in interspecific soybean backcross populations. Theor Appl Genet 108: 458–467.

Werner BK, Wilcox JR (1990) Recurrent selection for yield in *Glycine max* using genetic male sterility. Euphytica 50: 19–26.

Werner BK, Wilcox JR, Housley TL (1987) Inheritance of a ethyl methanesulfonate-induced dwarf in soybean and analysis of leaf cell size. Crop Sci 27: 665–668.

Wilcox JR (2001) Sixty years of improvement in publicly developed elite soybean lines. Crop Sci 49: 1711–1716.

Wilcox JR, Cavins JF (1985) Inheritance of low linolenic acid content of the seed oil of a mutant in *Glycine max*. Theor Appl Genet 71: 74–78.

Wilcox JR, Cavins JF (1987) Gene symbol assigned for linolenic acid mutant in the soybean. J Hered 78: 410.

Wilcox JR, Cavins JF (1990) Registration of C1726 and C1727 soybean germplasm with altered levels of palmitic acid. Crop Sci 30: 240.

Wilcox JR, Gnanasiri SP, Young KA, Raboy V (2000) Isolation of high seed inorganic P, low-phytate soybean mutants. Crop Sci 40: 1601–1611.

Williams LF (1950) Structure and genetic characteristics of the soybean. In: KS Markley (ed) Soybean and Soybean Products, vol 1. Interscience Publ, New York, USA, pp 111–114.

Wilson RF (2004) Seed composition. In: HR Boerma, JE Specht (eds) Soybeans: Improvement, Production, and Uses. Agron Monogr, 3rd edn, No 16. ASA-CSSA-SSSA, Madison, WI, USA, pp 621–678.

Woodworth CM (1921) Inheritance of cotyledon, seed-coat, hilum, and pubescence colors in soybeans. Genetics 6: 487–553.

Woodworth CM (1923) Inheritance of growth habit, pod color, and flower color in soybeans. J Am Soc Agron 15: 481–495.

Woodworth CM (1932) Genetics and breeding in the improvement of the soybean. Bull Agri Exp Stn 384: 297–404.

Woodworth CM (1933) Genetics of the soybean. J Am Soc Agron 25: 36–51.

Zabala G, Vodkin L (2003) Cloning of the pleiotropic *T* locus in soybean and two recessive alleles that differentially affect structure and expression of the encoded flavonoid 3' hydroxylase. Genetics 163: 295–309.

Zabala G, Vodkin LO (2007) A rearrangement resulting in small tandem repeats in the F3'5'H gene of white flower genotypes is associated with the soybean *W1* locus. Crop Sci 47 (suppl): S113–124.

# 3

# Identification of Genes Underlying Simple Traits in Soybean

*David Lightfoot*

## ABSTRACT

There are more than 450 loci in Soybase that underlie simple traits. There are more than 250 mutants in the type collection, in addition there are 43 pest resistance genes defined genetically, 68 altering growth, development symbiosis and fecundity, 78 underlying biochemical differences and 40 underlying pigmentation including chlorophyll. In contrast there are well over 1,000 genes identified as underlying trait controlled by 2–5 QTL in segregating populations. Methods to map, fine map and isolate the genes underlying the loci will be critical to future advances in soybean biotechnology. Here, the use of selective genotyping approaches, based on DNA pools, is reviewed. Critical advances in the analysis of simple traits are expected from high-throughput methods for genotyping. The genome sequence will allow more rapid marker to trait associations. Methods like targeted induced local lesion (TILLING) will allow the identification of genes underlying existing mutants. Advanced phenotype measurement methods will increase the spectrum of mutants identified. Examples of the methods for the identification of genes underlying simple traits are presented for both monogeneic and oligogeneic traits focusing on seed composition and disease resistance.

**Keywords:** selective genotyping; bulked segregant analysis; monogeneic traits; oligogeneic traits; TILLING

Plant Genomics and Biotechnology, Public Policy Institute 113, Department of Plant, Soil and General Agriculture, Southern Illinois University—Carbondale, Carbondale, IL 62901-4415, USA; e-mail: *ga4082@siu.edu*

## 3.1 Introduction

The development of a genome sequence for soybean (DOE JCSP, 2009) and high-throughput methods for marker scoring from worldwide advances in genetics like microarrays (Gupta et al. 2008; Hyten et al. 2008) and inexpensive Nextgen sequencing capacity (van Orsouw et al. 2007) provides tremendous power to those scientists engaged in the identification of genes underlying simple traits. In the next decade most simple gene traits will be associated with their underlying genes. However in paleo-tetrapolid soybean the phenotype of many mutants will be masked by genes in duplicated portions of the genome. Therefore, it is timely to review; the spectrum of allelic variations and mutants recorded over the last century; the phenotypes most amenable to identification by TILLING compared to those more tractable to identifications from fine mapping natural allelic variations; and principles and pitfalls of gene identification techniques for simple traits. Because the number of simply inherited traits is so large this chapter will focus on seed composition and disease resistance.

## 3.2 Summary of Single Gene Traits in Soybean

In soybean there are more than 450 loci that underlie simple traits (Palmer et al. 2004). Single genes can be separated into two broad classes those for which mutants are submitted to the Type Collection and those that were discovered as natural variation among cultivars or mutants within otherwise stable lines. There are more than 250 mutants in the type collection most in plant development, male sterility and color. Among the genes defined from natural variation the 43 pest resistance genes provide resistance to 14 microbial pests, two nematode species and five insect pests. Nodulation too has been a particularly tractable trait for mutagenesis and consequently there are 12 symbiosis related genes, though mostly non-nodulating loss of function lesions. Equally there are 25 mutants that cause chlorophyll deficiencies compared to just 15 for all other pigments. In comparison, despite massive investments, there are only six known genes underlying herbicide tolerance. However, by far the largest group of genes is that which includes 56 genes that alter growth and development. There are 250 loci defined in the Type collection and 56 other lesions recorded so that collection represents over 300 independent mutations in growth and development that are known. Therefore, the spectrum of simple traits shows clear biases with some traits highly mutable and others highly conserved.

Initial results from soybean TILLING projects (Cooper et al. 2008; Huang 2009) suggest a similar cross section of mutants can be observed. Initial targets for TILLING have included tractable targets in biochemical pathways (Table 3-1). However, the vast majority of lines show developmental changes

**Table 3-1** Some recent simple traits identified by mutation with known underlying genes.

| Traits | Gene/Allele | Population/s, or line/s | References |
|---|---|---|---|
| Reduced linolenate (3.6%) | A mutation in *GmFAD3A/ fan1* | C1640 from "Century" with EMS treatment | Wilcox et al. 1984; Chappell and Bilyeu 2006 |
| Reduced linolenate (4.1%) | A deletion in *GmFAD3A/ fan1*(A5) | A5 from "FA9525" with EMS treatment | Hammond et al. 1983; Bilyeu et al. 2003; Fehr 2007 |
| Reduced linolenate (5.6 %) | A single nucleotide mutation in *GmFAD3C/ fan2* | A23 from "FA47437" with EMS treatment | Primomo et al. 2002; Bilyeu et al. 2006; Fehr 2007 |
| Reduced linolenate (5.0%) | A mutation in *GmFAD3B/ fan3* | A26 from "A89-144003" with EMS treatment | Primomo et al. 2002; Bilyeu et al. 2006: Fehr 2007 |
| Low linolenate (1%) | Combination of *fan1, fan2* and *fan3* | A29 | Fehr 2007 |
| Reduced Palmitate (6.8%) | A single recessive allele *fap3* | A22 from "Asgrow A1937" with NMU treatment | Schnebly et al. 1994; Fehr 2007 |
| Reduced Palmitate (8.6%) | A mutant allele *fap1* | C1726 from "Century" with EMS treatment | Erickson et al. 1988; Fehr 2007 |
| Reduced Palmitate (5.7%) | Allele *sop1* (J3) | J3 from 'Bay' with X-ray irradiation | Takagi et al.1995 |
| Reduced Palmitate (3.5%) | Combination of *sop1 and fap1* | J3 x C1726 | Kinoshita et al. 1998 |
| Reduced Palmitate (6.4%) | A mutant allele *fap$_{nc}$* | N79-2077 | Cardinal et al. 2007 |
| Increased stearate (9%) | Mutations in *SACPD-C* (*fas*) | FAM94-41 and A6 | Zhang et al. 2008 |
| Reduced phytate P (~ 0.9 g kg$^{-1}$) | Combination of *lpa1-a* and *lpa2-a* | CX1834 | Wilcox et al. 2000; Walker et al. 2006; Scaboo et al. 2009; Maroof et al. 2009; Gillman et al. 2009 |
| Reduced phytate P (50% less) | A mutation in *myo-*inositol 1-phospate synthase gene (MIPS) | LR33 | Hitz et al. 2002 |
| Reduced phytate P (66.6% less) | A deletion in *MIPS1* | *Gm-lpa*-TW-1 | Yuan et al. 2007 |
| Reduced phytate P (46.3% less) | A mutation in *MIPS1* | *Gm-lpa*-ZC-2 | Yuan et al. 2007 |
| Hypernodulating (10 times more nodules) | The nodule autoregulation receptor kinase (*GmNARK*) | FN37 from fast neutron mutagenesis | Men et al. 2002 |

*Table 2-1 contd....*

*Table 3-1 contd....*

| Traits | Gene/Allele | Population/s, or line/s | References |
| --- | --- | --- | --- |
| Hypernodulating (20 times more nodules ) | *GmNARK* | SS2-2 from EMS mutagenesis | Lestari et al. 2006 |
| Atrazine resistance | *D* point mutation | gamma radiation | Atak et al. 2004 |

from color to growth pattern (Huang 2009). Each line carries 200–300 lesions in different genes so that purification of lesions by backcrossing will be a major task for the next decade. However, with a coordinated effort, the soybean community can look forward to a resource with an allelic series of mutants for every gene in the genome. Due to the paleo- tetraploid nature of the genome it is likely that most genes will have functional duplicates. Therefore, double, triple and quadruple mutations in gene families will be necessary to show many gene functions. Generating stacked mutations will be important, probably more so than in other plants (Bouchez and Bouchez 2001; Nawy et al. 2005).

## 3.3 Identifying Genes by Bulked Segregant Analysis

As noted in the Introduction, natural allelic variation within soybean gene pools is the second major source of simply inherited traits. In soybean there are about 1,000 loci that underlie simply inherited quantitative traits that are controlled by 2–5 genes (Lightfoot 2008). If a large portion of gene functions in soybean are to be identified from natural genetic variation, coupled with mapping before transformation or TILLING, efficient methods to map single gene traits are needed. Bulked segregant analysis (BSA) is a specialized form of selective genotyping (Darvasi and Soller 1992) that provides a rapid procedure for identifying markers in specific regions of the genome (Michelmore et al. 1991; Darvasi and Soller 1994). The only prerequisite for BSA is the existence of a population whose members contrast for a trait; either resulting from a single cross (standard QTL mapping) or from serial intercrosses (association mapping). In soybean, the method can be used for both qualitative traits and for detecting major genes underlying quantitative traits, even polygeneic traits like seed yield (Mansur et al. 1996; Yuan et al. 2002).

The BSA method involves comparing two pooled DNA samples from a segregating population (Meksem et al. 2001). Within each pool, or bulk, the small numbers of individuals selected are expected to have identical genotypes for a particular genomic region (target locus or region) but random genotypes at loci unlinked to the selected region. Pools can be selected from phenotype for locus discovery (Mansur et al. 1996) or by genotype for marker saturation of a particular region (Meksem et al. 2001). In the latter case, the pools can be fine tuned to saturate a region by the inclusion of recombination events flanking the target region. The two pools contrasting for a trait (e.g., resistant and susceptible to a particular disease) are analyzed to identify markers that distinguish them. Markers that are polymorphic between the pools may be genetically linked to the loci determining the trait used to construct the pools or may be error associations caused by a sampling error. Therefore, the size of pools and their accuracy of phenotypes or genotypes underlying their member genotype composition are critical factors in the success of the technique.

BSA has two immediate applications in developing genetic maps: (1) It provides a method to focus on regions of interest or areas sparsely populated with markers. (2) It is a method for rapidly locating genes that do not segregate in populations initially used to generate the northern US germplasm based composite genetic map (Song et al. 2004) or other high density maps (Dong 2003; Yamanaka et al. 2005; Lightfoot 2007). The BSA technique is advantageous in identifying markers associated with new traits without the need for full map construction. BSA is an extreme form of selective genotyping and so can be applied to traits under the control of one to several genes.

## 3.4 Identifying Genes by Selective Genotyping

Selective genotyping (Darvasi and Soller 1992) is a method for identifying small sets of genes underlying traits in which only individuals from the high and low phenotypic extremes are genotyped to get the most informative, quantitative trait values. Once marker trait linkages are detected these simple quantitative traits can be Mendelized to simple gene trait by developing near-isogenic lines (NILs) from recombinant inbred lines (RILs; Njiti et al. 1998; Triwitayakorn et al. 2005; Ruben et al. 2006; Afzal et al. 2008). The detection of quantitative trait loci (QTL) requires large sample sizes to attain reasonable power (Soller et al. 1976). It has been shown that the number of individuals genotyped to attain a given power can be decreased significantly, at the expense of a moderate increase in the number of individuals phenotyped (Darvasi and Soller 1992). The major limitation of this approach is that if the experiment is aimed at analyzing a number of traits, then by selecting the extremes of each trait one would select most of the population and thus no reduction in genotyping can be obtained. Selective genotyping is thus most appropriate for cases where only one trait is being analyzed. This conclusion is valid when selective genotyping is applied to QTL detection.

## 3.5 Examples of Methods for Bulked Segregant Analysis

Two groups (Mansur et al. 1996; Yuan et al. 2002) reported the use of BSA to identify major loci underlying the polygenic trait seed yield (Fig. 3-1; Table 3-2). What was surprising about this result was that loci with major effects were found underlying traits expected to be highly polygenic. The result was not specific to the three crosses used, since the loci were validated in other studies. The loci detected did not underlie the different disease resistances, often major yield determinates themselves. Instead it now seems likely that some QTLs of major effect in polygenic traits represent clusters of 3–4 linked genes all affecting the same trait (Afzal 2008; Lightfoot 2008). Gene clustering further empowers the BSA approach as it allows for locus identification followed by fine mapping to dissect and isolate the genes in the cluster.

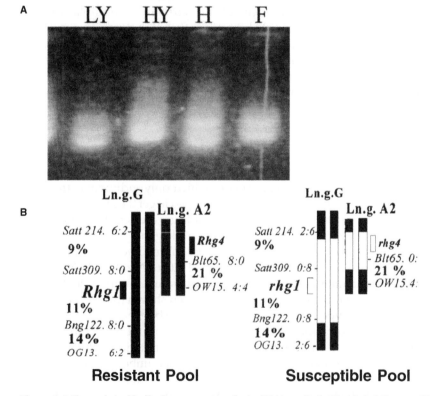

**Figure 3-1** Examples of bulked segregant analysis (BSA) analysis. Panel A: Microsatellite marker Satt326 amplified from genomic DNA separated on a 4% agarose gel and stained with ethidium bromide demonstrating BSA of a high seed yield pool (HY) vs. low seed yield pool (LY). Flyer (F) was the higher seed yield parent and Hartwig (H) was the lower seed yield parent. From Yuan et al. (2002) with permission. Panel B: Pool development to include recombinants in the flanking regions for linkage group A2 and G. From Meksem et al. (2001).

**Table 3-2** Detection of the QTLs underlying seed yield by Satt326 at two locations and the mean of four locations after BSA by ANOVA and interval maps in the complete population ($n$ = 92).

| Marker | LG | Location | $P$ | $R^2\%$ | LOD | Var % | Allelic mean±SEM (Mg/Ha) | |
|--------|----|----------|-----|---------|-----|-------|---------|---------|
| | | | | | | | Flyer | Hartwig |
| SATT326 | K | N98 | 0.0001 | 26.3 | 5.4 | 25.5 | 3.20+0.06 | 2.69±0.08 |
| | K | N99 | 0.0082 | 8.6 | 1.6 | 7.8 | 2.52 0.06 | 2.33±0.04 |
| | K | Mean | 0.0004 | 15.0 | 3.0 | 14.6 | 2.98 0.04 | 2.76±0.05 |

For example recently the BSA (Michelmore et al. 1991) technique was used to identify markers linked to SCN race 3 and race 14 resistant loci in the FxH RIL population. Here 20 (SCN race 3 and race 14) resistant and 20 (SCN race 3 and race 14) susceptible FxH RILs (lines) were genotyped. Population parents "Flyer" and "Hartwig" were also included for comparison. Five bulked DNA pools were generated from FxH population as follows (Table 3-3). RILs that were resistant to race 3 and race14 (pool 1), resistant to race 3 but susceptible to 14 (pool 2), susceptible to race 3 and 14 (pool 3) and two pools that were race 3 heterozygous but race 14 moderate resistant (as pool 2a and pool 2b). About 104 microsatellite markers were used for screening the pools (Kazi et al. 2007, 2009). To determine the markers liked to race 3 resistance loci, pools 2a or 2b were compared to pool 3 and for

**Table 3-3** Markers showing segregation among small pools constructed to identify all possible genes underlying resistance to SCN derived from Hartwig in the FxH population.

| Markers | LG | cM | Lane number and pool composition |
|---------|----|-----|----------------------------------|
| A. Satt151 | E | 44.9 | 1. SCN susceptible parent (Flyer) |
| B. Satt163 | G | 0.0 | 2. SCN resistant parent (Hartwig) |
| C. Satt399 | C1 | 76.2 | 3. Resistant to SCN race 3 and race 14 (pool 3) |
| D. Satt275 | G | 2.2 | 4. Race 3 and race 14 moderately resistant (pool 2a) |
| E. Sat_087 | K | 4.9 | 5. Resistant to race 3; race14 susceptible (pool 2b) |
| F. Sat_112 | H | 61.3 | 6. Resistance to race3, moderate to race 14 (pool 2c) |
| G. SIUCSat_122 | G | 8.6 | 7. Susceptible to race3 and race14 (pool 1) |

markers liked to race 14 resistance, pool 2 was compared to pool 1. Pool 1 comprised three race 3 and 14 resistant FxH RIL lines. Three race 3 heterozygous and race 14 moderately resistant FxH lines made pool 2a and 2c. The three FxH lines resistant to race 3 but susceptible to race 14 forms pool 2b. Pool 3 comprised four race 3 and 14 susceptible lines to make the susceptible pool (Fig. 3-2).

About 104 microsatellite markers were used to screen the pools. Parents were included for comparison. Twenty-one BARC-SSRs were found polymorphic among the FxH RILs. The DNA banding pattern of the resistant pool and resistant parent (Hartwig) was alike for 12 markers. Similarly the banding pattern of susceptible pool and susceptible parent (Flyer) were same for 12 markers as shown in Figure 3-3. In order to determine the race 3 resistant markers, pool 2a or 2c were compared to pool 3 and for race 14 resistant markers, pool 2 was compared to pool 1. The 21 polymorphic markers detected have a good potential to detect race 3 and race 14 resistant loci among FxH RIL (Fig. 3-4). Two out of these 21 markers were found significant in detecting QTL for SCN resistance using Mapmaker/QTL and QTL Cartographer. BSA identified molecular markers closely linked to the two major QTLs (Satt163 and Satt275) often associated with SCN resistance.

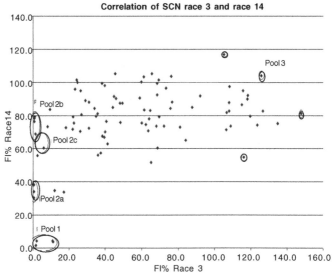

**Figure 3-2** Formation of five small pools to rapidly screen for potential markers associated with genomic regions underlying HgType 0 (race 3) and HgType 1.3.5 (race 14) resistance. Diamond sign indicates FxH RIL number. Pool 1 was resistant to both race 3 and race 14 (FxH RIL #33, 35, 93). Pool 2a was race 3 resistant and race 14 moderate resistant (FxH RIL #13, 25, 20). Pool 2b = race 3 resistant and race 14 susceptible (FxH RIL #39, 73, 18, 19). Pool 2c = race 3 resistant or moderately resistant to race 3 and race 14 susceptible (FxH RIL #13, 25, 20 and 60). Pool 3 was susceptible to both race 3 and race 14 (FxH RIL #1, 4, 27, 30).

Therefore, it was again demonstrated that while BSA has traditionally been used effectively for mapping genes, which account for nearly 100% variation in a trait (Michelmore et al. 1991), in soybean loci underlying less than 20% of variation have been detected.

Others have used BSA to map multigeneic SCN resistances. Ferdous et al. (2006) used BSA to identify *rhg1*-t a QTL on LG B1 interacting with *rhg1* to provide resistance to soybean cyst nematode race 3 in progeny derived from the soybean cultivar "Toyomusume". Cervigni et al. (2004) used BSA to confirm the location of *rhg1* in Hartwig with microsatellite markers and showed that it was a dominant gene for resistance to soybean cyst nematode race 3.

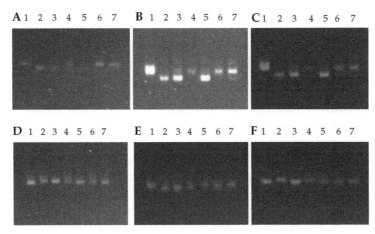

**Figure 3-3** Fluorogram showing similar banding pattern between the resistant parent and resistant pool and susceptible parent and susceptible pool after electrophoresis on 4% (w/v) agarose gel for 4 hours at 92 volts.

## 3.6 Examples of Identification of Loci Underlying Simple Traits

The BSA method has been used in the traditional manner to detect major loci underlying several resistance traits in soybean. Examples include, Paul et al. (2006) who used SSR markers linked a single locus potentially underlying much of soybean *Sclerotinia* stem rot resistance. Sandhu et al. (2005) found the correct location of the *Rps8* locus underlying resistance to *Phytophthora* root rot with SSRs after conventional mapping methods failed to show any association. Chowdhury et al. (2002) used RAPD markers to identify a locus underlying resistance to downy mildew disease in soybean when SSR markers were not polymorphic. Jeong et al. (2002) used AFLP markers to identify and fine map the soybean mosaic virus resistance locus, *Rsv3*. Hai-Chao et al. (2006) used AFLP and BSA to analyze the inheritance of resistance to soybean mosaic virus strain SC14 in soybean and show RSC14 and $R_{SC14Q}$ might be isogeneic loci. Jackson et al. (2008) used BSA and SSRs to map resistance to purple seed stain in PI 80837. Li et al. (2007) showed soybean aphid resistance genes in the soybean cultivars, Dowling and Jackson, map to linkage group M. As yet unpublished, Z.P. Shearin and colleagues (J.M. Narvel, H. Cheney, and H.R. Boerma) reported at the 2006 CSSA meeting that BSA was used for SSR mapping of genes in soybean conditioning resistance to stem canker.

BSA has been an effective tool to map the location of mutations in soybean. Men and Gresshoff (2004) used DAF with BSA to find a new marker closely linked to the soybean supernodulation *nts-1* locus. Karakaya et al. (2002) mapped the fasciation mutation to LG D1b with AFLP and SSR markers. Gijzen et al. (2003) showed that the soybean seed lustre phenotype and highly allergenic hydrophobic surface protein cosegregate and map to linkage group E.

Fine mapping and refined maps are a common use for BSA in soybean. Here bulk development is informed by preliminary marker data and intervals can be focused narrowly by bulk designs incorporating recombination events close to the target locus (Meksem et al. 2001). Kabelka et al. (2005) confirmed and localized two loci in *G. soja* that confer resistance to soybean cyst-nematode in soybean, including an *rhg1* like locus. The loci mapped to positions not previously found among *G. max* SCN resistance associated loci. Recently, SNP markers scored by the Goldengate technology have been combined with BSA to simultaneously identify and fine map the *Rpp* loci (Hyten et al. 2009). Showing another approach using selective genotyping SNPs were scored by Affymetrix chip hybridization patterns (Gupta et al. 2008) to locate both QTL underlying PRR resistance and the e-QTL they govern the transcript abundance from (Zhou et al. 2009).

In soybean a handful of simple genes underlying simple trait variations have been identified by molecular techniques (reviewed by Lightfoot 2008; Table 3-1). They include *Rpg1-b* that encodes a nucleotide binding leucine rich repeat (NB-LRR) protein for resistance to bacterial pustule (Ashfield et al. 2003). Another gene shown to be a NB-LRR was the gene for supernodulation GmNARK (Searle et al. 2003). The SCN resistance locus *rhg1* also contains a NB-LRR the kinase domain of which confers part of the resistance (Ruben et al. 2006; Afzal et al. 2008). Also cloned is the *T* locus that encodes flavonoid-3' monooxygenase (EC1.14.13.21) responsible for pubescence color (Toda et al. 2002; Zabala and Vodkin 2003). The recessive genes differ from the dominant by deletion of a single C nucleotide. There are mutant lines *I* for seed coat color was shown to be a member of the chalcone synthase family. By mutagenesis a number of genes were associated with enzymes controlling seed oil composition. Reduced linolenate fatty acid in seed oil was tracked with three mutations in *fan1*, *fan2* and *fan3*. Reduced palmitate was shown to be caused by two genes *sop1* and *fap1*.Reduced phytate *P* was shown to be caused by a mutation in myo-inositol 1-phospate synthase gene or a combination of mutations in homologues of the maize low phytic acid gene, *lpa1* (Hitz et al. 2002; Gillman et al. 2009).

Recent gene isolations include the *E4* and *E3/FT3* flowering time locus underlaid by phytochromes, GmphyA2 and GmphyA3, respectively (Liu et al. 2008; Watanabe et al. 2009); the *P34* gene was shown to underlay the major soybean allergen (Bilyeu et al. 2009); and raffinose synthase to underlie raffinose contents (Dierking and Bilyeu 2008). This rapid progress in identification of genes underlying simple gene traits is timely and welcome.

## 3.7 Conclusions

In conclusion, selective genotyping methods like BSA, and fine mapping techniques, layered on techniques like TILLING provide a powerful platform to identify fine map and subsequently isolate from the genome sequence candidate genes underlying simply inherited loci. Polymorphism can be found among gene pools or from mutations. By combining mapping with mutant analysis the soybean community can look forward to assigning a function to almost every gene perhaps within a few decades. Such rapid development of a functional genome map will rely on the low cost of 454 and SOLiD technologies exploiting the existing soybean genome sequence. A typical next generation gene function identification strategy will use the densest SNP chips available or selective genome sequencing to identify the candidate regions underlying trait inheritance (van Orsouw et al. 2007; Gupta et al. 2008; Hyten et al 2009). Region identification will be followed by analysis with a custom sequence capture chip for the high likelihood regions to capture DNA from pools. Finally captured DNA will be sequenced using the 454 system to find all underlying polymorphisms. Preliminary experiments suggest a single group can identify the functions of 3–5 genes a year within this paradigm. Under this paradigm with the expected reductions in cost and improvements in throughput the identification of a function for most of the approximately ten thousand unique gene families in soybean becomes a tractable goal.

## Acknowledgements

The continued support of SIUC, College of Agriculture and Office of the Vice Chancellor for Research to the author is appreciated. D. Hyten is thanked for access to unpublished work.

## References

Afzal AJ, Saini N, Srour A, Lightfoot DA (2008) The multigeneic *rhg1* locus: A model for the effects on root development, nematode resistance and recombination suppression. Nature Preceed hdl:10101/npre.2008.2726.1 (online).

Ashfield, T, Bocian, A, Held D, Henk AD, Marek LF, Danesh D, Lightfoot DA, Penuela S, Meksem K, Shoemaker RC (2003) Genetic and physical mapping of the soybean *rpg1-b* disease resistance gene reveals a complex locus containing several tightly linked families of NBS-LLR-genes. Mol Plant-Micr Interact 16: 817–826.

Atak C, Alikamano S, Ack L, Canbolat Y (2004) Induced of plastid mutations in soybean plant (*Glycine max* L. Merrill) with gamma radiation and determination with RAPD. Mutat Res 556: 35–44.

Bilyeu KD, Palavalli L, Sleper DA, Beuselinck PR (2003) Three microsomal omega-3 fatty-acid desaturase genes contribute to soybean linolenic acid levels. Crop Sci 43(5): 1833–1838.

Bilyeu K, Palavalli L, Sleper DA, Beuselinck P (2006) Molecular genetic resources for development of 1% linolenic acid soybeans. Crop Sci 46: 1913–1918.

Bilyeu K, Chengwei R, Nguyen H, Herman E, Sleper DA (2009) Association of a four basepair insertion in the P34 gene with the low allergen trait in soybean. Plant Genom: in press.

Bouche N Bouchez D (2001) Arabidopsis gene knockout: phenotypes wanted. Curr Opin Plant Biol 4: 111–117.

Cardinal AJ, Burton JW, Camacho-Roger AM, Yang JH, Wilson RF, Dewey RE (2007) Molecular analysis of soybean lines with low palmitic acid content in the seed oil. Crop Sci 47(1): 304–310.

Cervigni L, Gerardo D, Schuster I, Goncalves de Barros E, Alves Moreira M (2004) Two microsatellite markers flanking a dominant gene for resistance to soybean cyst nematode race 3 Euphytica 135: 99–105.

Chappel A, Bilyeu, KD (2006) A *GmFAD3A* mutation in the low linolenic acid soybean mutant C1640. *Plant Breed* 125: 535–536.

Chowdhury AK, Srinives P, Saksoong P, Tongpamnak P (2002) RAPD markers linked to resistance to downy mildew disease in soybean. Euphytica 128: 55–60.

Cooper JL, Till BJ, Laport RG, Darlow MC, Kleffner JM, Jamai A, El-Mellouki T, Liu S, Ritchie R, Nielsen N, Bilyeu KD, Meksem K, Comai L, Henikoff S (2008) TILLING to detect induced mutations in soybean. BMC Plant Biol 8: 9–19.

Darvasi A, Soller M (1992) Selective genotyping for determination of linkage between a marker locus and a quantitative trait locus. Theor Appl Genet 85: 353–359.

Darvasi A, Soller M (1994) Selective DNA pooling for determination of linkage between a molecular marker and a quantitative trait locus. Genetics 138: 1365–1373.

Dierking E, Bilyeu KD (2008) Association of a soybean raffinose synthase gene with low raffinose and stachyose seed phenotypes. Plant Genom 1(2):135–145.

Dong YS, Zhao LM, Liu B, Wang ZW, Jin ZQ, Sun H (2003) The genetic diversity of cultivated soybean grown in China. Theor Appl Genet 108: 931–936.

Erickson EA, Wilcox JR, Cavins JF (1988) Inheritance of altered palmitic acid percentages in two soybean mutants. J Hered 79: 465–468.

Fehr WR (2007) Breeding for modified fatty acid composition in soybean. Crop Sci 47: S72–S87.

Ferdous SA, Watanabe S, Suzuki-Orihara C, Tanaka Y, Kamiya M, Yamanaka N, Harada K (2006) QTL analysis of resistance to soybean cyst nematode race 3 in soybean cultivar Toyomusume. Breed Sci 56: 155–163.

Gijzen M, Weng C, Kuflu K, Woodrow L, Yu K, Poysa V (2003) Soybean seed lustre phenotype and surface protein cosegregate and map to linkage group E. Genome 46: 659–664.

Gillman JG, Bilyeu KD (2009) The low phytic acid phenotype in soybean line CX1834 is due to mutations in two homologues of the maize low phytic acid gene. Plant Genom: in press.

Gupta PK, Rustgi S, Mir RR (2008) Array-based high-throughput DNA markers for crop improvement. Heredity 101: 5–18.

Hai-Chao Li, Zhi H-J, Gai J-Y, Guo D-Q, Wang Y-W, Li K, Bai L, Yang H (2006) Inheritance and gene mapping of resistance to soybean mosaic virus strain SC14 in soybean. J Integr Plant Biol 48: 1466–1472.

Hammond EG, Fehr WR (1983) Registration of A5 germplasm line of soybean. Crop Sci 23: 192.

Hitz WD, Carlson TJ, Kerr PS, Sebastian SA (2002) Biochemical and molecular characterization of a mutation that confers a decreased raffinosaccharide and phytic acid phenotype on soybean seeds. Plant Physiol 128: 650–660.

Huang E (2009) Soybean TILLING. MSc Thesis, SIUC, Carbondale, Illinois, USA.

Hyten DL, Song Q, Choi IY, Yoon MS, Cregan PB (2008) High-throughput genotyping with the GoldenGate assay in the complex genome of soybean. Theor Appl Genet 116: 945–952.

Jackson EW, Feng C, Fenn P, Chen P (2008) Genetic mapping of resistance to purple seed stain in PI 80837 soybean. J Hered 99: 319–322.

Jeong SC, Kristipati S, Hayes AJ, Maughan PJ, Noffsinger SL, Gunduz I, Buss GR, Saghai-Maroof MA (2002) Genetic and sequence analysis of markers tightly linked to the soybean mosaic virus resistance gene, Rsv3. Crop Sci 42: 265–270.

Kabelka EA, Carlson SR, Diers BW (2005) Marker saturation and the localization of two soybean cyst nematode resistance loci from *Glycine soja* PI 468916. Crop Sci 45: 2473–2481.

Karakaya HC, Tang Y, Crega, PB, Knap HT (2002) Molecular mapping of the fasciation mutation in soybean, *Glycine max* (Leguminosae). Am J Bot 89: 559–565.

Kazi S, Njiti VN, Doubler TW, Yuan J, Iqbal MJ, Cianzio S, Lightfoot DA (2007) Registration of the Flyer by Hartwig recombinant inbred line mapping population. J Plant Reg 1: 175–178.

Kazi S, Shultz JL, Bashir R, Afzal AJ, Bond J, Arelli P, Lightfoot DA (2009) Identification of loci underlying seed yield and resistance to soybean cyst nematode race 2 in 'Hartwig'. Theor Appl Genet: in press.

Kinoshita T, Rahman SM, Anai T, Takagi Y (1998) Inter-locus relationship between genes controlling palmitic acid contents in soybean mutants. Breed Sci 48: 377–381.

Lestari P, Van K, Kim MY, Lee B-W, Lee S-H (2006) Newly featured infection events in a supernodulating soybean mutant SS2-2 by *Bradyrhizobium japonicum*. Can J Microbiol 52: 328–335.

Li Y, Hill CB, Carlson SR, Diers BW, Hartman GL (2007) Soybean aphid resistance genes in the soybean cultivars Dowling and Jackson map to linkage group M. Mol Breed 19: 25–34.

Lightfoot DA (2008) Soybean genomics: Developments through the use of cultivar Forrest. Int J Plant Genom 1–22: doi:10.1155/2008/793158.

Liu B, Kanazawa A, Matsumura H, Takahashi R, Harada K, Abe J (2008) Genetic redundancy in soybean photoresponses associated with duplication of the phytochrome A gene. Genetics 180(2): 995–1007.

Mansur LM, Orf JH, Chase K, Jarvik T, Cregan PB, Lark KG (1996) Genetic mapping of agronomic traits using recombinant inbred lines of soybean [Glycine max (L.) Merr.]. Crop Sci 36:1327–1336.

Maroof MS, Glover NM, Biyashev RM, Buss GR, Grabau EA (2009) Genetic basis of the low-phytate trait in the soybean line CX1834. Crop Sci 49: 69–76.

Meksem K, Pantazopoulos P, Njiti VN, Hyten DL, Arelli PR, Lightfoot DA (2001) 'Forrest' resistance to the soybean cyst nematode is bigenic: Saturation mapping of the *rhg1* and *Rhg4* loci. Theor Appl Genet 103: 710–717.

Men AE, Gresshoff PM (2004) DAF yields a cloned marker linked to the soybean (*Glycine max*) supernodulation *nts*-1 locus. J Plant Physiol 158: 999–1006.

Men AE, Laniya TS, Searle IR, Iturbe-Ormaetxe I, Gresshoff I, Jiang Q, et al. (2002) Fast neutron mutagenesis of soybean (*Glycine soja* L.) produces a supernodulating mutant containing a large deletion in linkage group H. Genom Lett 3: 147–155.

Michelmore RW, Paran I, Kesseli RV (1991) Identification of markers linked to disease-resistance genes by bulked segregant analysis: A rapid method to detect markers in specific genomic regions by using segregating populations. Proc Natl Acad Sci USA 88: 9828–9832.

Nawy T, Lee J-Y, Colinas J, Wang JY, Thongrad SC, Malamy JE, Birnbaum K, Benfey PN (2005) Transcriptional profile of the Arabidopsis root quiescent center. Plant Cell 17: 1908–1925.

Njiti VN, Doubler TW, Suttner RJ, Gray LE, Gibson PT, Lightfoot DA (1998) Resistance to soybean sudden death syndrome and root colonization by *Fusarium solani* f. sp. *glycines* in near-isogeneic lines. Crop Sci 38: 472–477.

Palmer RG, Pfeiffer TW, Buss GR, Kilen TC (2004) Qualitative genetics. In: HR Boerma, JE Specht (eds) Soybeans: Improvement, Production and Uses. 3rd edn. Agron Monogr 16, ASA, CSSA, SSSA, Madison, WI, USA, pp 137–234.

Paul C, Strutz D, Hartman GL (2006) Molecular markers linked to soybean Sclerotiana stem rot resistance using bulked segregant analysis. Am Phytopathol Soc Abstr 96: S91.

Primomo VS, Falk DE, Ablett GR, Tanner JW, Rajcan I (2002) Inheritance and interaction of low palmitic and low linolenic soybean. Crop Sci 42: 31–36.

Ruben E, Aziz J, Afzal J, Njiti VN, Triwitayakorn K, Iqbal MJ, Yaegashi S, Arelli P, Town C, Meksem K, Lightfoot DA (2006) Genomic analysis of the 'Peking' *rhg1* locus: candidate genes that underlie soybean resistance to the cyst nematode. Mol Genet Genom 276: 320–330.

Sandhu D, Schallock KG, Rivera-Velez N, Lundeen P, Cianzio S, Bhattacharyya MK (2005 Soybean Phytophthora resistance gene *Rps8* maps closely to the *Rps3* region. J Hered 96(5): 536–541.

Scaboo AM, Pantalone VR, Walker DR, Boerma HR, West DR, Walker FR, Sams CE (2009) Confirmation of molecular markers and agronomic traits associated with seed phytate content in two soybean RIL populations. Crop Sci 49: 426–432.

Schnebly SR, Fehr WR, Welke GA, Hammond EG, Duvick DN (1994) Inheritance of reduced and elevated palmitate in mutant lines of soybean, Crop Sci 34: 829–833.

Searle IR, Men AE, Laniya TS, Buzas DM, Iturbe-Ormaetxe I, Carroll BJ, Gresshoff PM (2003) Long-distance signaling in nodulation directed by a CLAVATA1-like receptor kinase. Science 299: 109–112.

Shearin ZP, Narvel JM, Cheney H, Boerma HR (2006) SSR mapping of genes in soybean conditioning resistance to stem canker (Abstr). Crop Sci Soc Am Meet, New Orleans, LA, USA, 59: 15.

Song QJ, Marek LF, Shoemaker RC, Lark KG, Concibido VC, Delannay X, Specht JE, Cregan PB (2004) A new integrated genetic linkage map of the soybean. Theor Appl Genet 109: 122–128.

Takagi Y, Rahman SM, Joo H, Kawakita T (1995) Reduced and elevated palmitic acid mutants in soybean developed by x-ray irradiation. Biosci Biotechnol Biochem 59: 1778–1779.

Toda K, Yang D, Yamanaka N, Watanabe S, Harada K, Takahashi R (2002) A single-base deletion in soybean flavonoid 3'-hydroxylase gene is associated with gray pubescence color. Plant Mol Biol 50(2): 187–196.

Triwitayakorn K, Njiti VN, Iqbal MJ, Yaegashi S, Town CD, Lightfoot DA (2005) Genomic analysis of a region encompassing QRfs1 and QRfs2: genes that underlie soybean resistance to sudden death syndrome. Genome 48: 125–138.

van Orsouw NJ, Hogers RC, Janssen A, Yalcin F, Snoeijers S, et al. (2007) Complexity reduction of polymorphic sequences (CRoPS): A novel approach for large-scale polymorphism discovery in complex genomes. PLoS ONE 2: e1172.

Watanabe S, Hideshima R, Xia Z, Tsubokura Y, Sato S, Nakamoto Y, Yamanaka N, Takahashi R, Ishimoto M, Anai T, Tabata S, Harada K (2009) Map-based cloning of the gene associated with soybean maturity locus E3. Genetics, PMID: 19474204 [epub ahead of print].

Walker DR,Scaboo AM, Pantalone VR, Wilcox JR, Boerma HR (2006) Genetic mapping of loci associated with seed phytic acid content in CX1834-1-2 soybean. Crop Sci 46: 390–397.

Wilcox JR, Cavins JF, Nielsen NC (1984) Genetic alteration of soybean oil composition by a chemical mutagen. J Am Oil Chem Soc 61: 97–100.

Wilcox JR, Premachandra GS, Young KA, Raboy V (2000) Isolation of high seed inorganic P, low-phytate soybean mutants. Crop Sci 40: 1601–1605.

Yamanaka N, Ninomiya S, Hoshi S, Tsubokura Y, Yano M, Nagamura Y, Sasaki T, Harada K (2001) An informative linkage map of soybean reveals QTL for flowering time, leaflet morphology and regions of segregation distortion. DNA Res 8: 61–72.

Yuan J, Njiti VN, Meksem K, Iqbal MJ, Triwitayakorn K, Kassem MA, Davis GT, Schmidt ME, Lightfoot DA (2002) Quantitative trait loci in two soybean recombinant inbred line populations segregating for yield and disease resistance. Crop Sci 42: 271–277.

Yuan F, Zhao H, Ren X, Zhu S, Fu X, Shu Q (2007) Generation and characterization of two novel low phytate mutations in soybean (*Glycine max* L. Merr.) Theor Appl Genet 115: 945–957.

Zabala G, Vodkin L (2003) Cloning of the pleiotropic T locus in soybean and two recessive alleles that differentially affect structure and expression of the encoded flavonoid 3' hydroxylase. Genetics 163(1): 295–309.

Zhang P, Burton JW, Upchurch RG, Whittle E, Shanklin J, Dewey RE (2008) Mutations in a δ-9-stearoyl-ACP-desaturase gene are associated with enhanced stearic acid levels in soybean seeds. Crop Sci 48(6): 2305–2313.

Zhou L, Mideros SX, Bao L, Hanlon R, Arredondo FD, Tripathy S, Krampis K, Jerauld A, Evans C, St Martin SK, Saghai Maroof MA, Hoeschele I, Dorrance AE, Tyler BM (2009) Infection and genotype remodel the entire soybean transcriptome. BMC Genom 10: 49–68.

# Molecular Genetic Linkage Maps of Soybean

*Sachiko Isobe** and *Satoshi Tabata*

## ABSTRACT

The development of molecular genetic linkage maps in soybean involves two essential components: DNA markers and mapping populations. Various types of DNA markers based on different principles have been developed using technologies available at various times, such as DNA hybridization, polymerase chain reaction (PCR) and DNA sequencing. DNA markers determined by these techniques have been utilized as a common resource in the soybean research community. Mapping populations, on the other hand, have been generated and used more independently based upon the interests and needs of individual researchers. In the initial stage, interspecific mapping populations between *Glycine max* and *Glycine soja* were used mainly to increase the number of polymorphic markers. Subsequently, intraspecific maps as well as integrated maps have been developed. The combination of known DNA markers and various mapping populations has molded a history of molecular linkage maps and related genetic analyses in soybean.

**Keywords:** DNA markers; RFLP; AFLP; RAPD; microsatellite; SSR; SNP

## 4.1 Overview

The development of molecular genetic linkage maps in soybean has been closely associated with DNA manipulation technologies available at any given time. The emergence of new technologies and tools, including DNA

Kazusa DNA Research Institute, 2-6-7 Kazusa-kamatari, Kisarazu, Chiba 292-0818, Japan.
*Corresponding author: *sisobe@kazusa.or.jp*

hybridization, polymerase chain reaction (PCR), high-throughput DNA sequencing and fluorescence detection systems, has facilitated the development of different types of DNA markers, as listed in Table 4-1. The first widely used DNA marker type was restriction fragment length polymorphisms (RFLPs), when hybridization technology and gene-derived probes (gene segments and/or cDNAs) were available. To cover the entire genome of soybean in a cost-effective manner, amplified fragment length polymorphism (AFLP) and random amplification of polymorphic DNA (RAPD) markers using degenerate or random primers were introduced. As DNA sequencing technology advanced in the mid-1990s, microsatellites or simple sequence repeats (SSRs) became the major marker type used because of their high degree of amplification and reproducibility. And, very recently, the so-called new generation of DNA sequencers can output an enormous amount of sequence data inexpensively, which is facilitating the generation of the most sophisticated type of DNA marker, single nucleotide polymorphisms (SNPs).

**Table 4-1** DNA markers used for genetic analysis in soybean.

| Abbreviations | Principle | Cost | Handling | Marker Quality |
|---|---|---|---|---|
| RFLP (Restriction fragment length polymorphism) | Hybridization | medium | complex | low |
| AFLP (Amplified fragment length polymorphism) | PCR | high | simple | medium |
| RAPD (Random amplification of polymorphic DNA) | PCR | high | simple | low |
| microsatellite or SSR (Simple sequence repeat) | PCR | medium | simple | high |
| SNP (Single nucleotide polymorphism) | PCR or Hybridization & extension | low | medium | high |

Another essential element in molecular mapping is mapping parents and populations. Various mapping populations in soybean have been developed independently based upon the interests and needs of individual researchers, i.e., the degree of polymorphism required and specific agronomic traits for analysis. While the information regarding DNA markers has been shared widely in the soybean research community and is used as a common system worldwide, the diversity of mapping populations has hindered the creation of a common basis for understanding the genetic systems in soybean. To overcome this, recent efforts have been made to develop new mapping populations with exchanged parents and to generate integrated genetic linkage maps.

## 4.2 Mapping Populations and Computer Software

$F_2$ populations or recombinant inbred lines (RILs) have been employed for the construction of linkage maps in soybean. The population sizes used for most of the soybean linkage maps range from 50 to 190 genotypes in the $F_2$ population, and from 94 to 330 genotypes in the RILs (Table 4-2). Although the first soybean linkage map with DNA-based markers was developed based on an $F_2$ population of an intraspecific cross (Apuya et al. 1988), interspecific crosses between *Glycine max* and *Glycine soja* constituted the major populations used in the initial stage of developing soybean linkage maps in order to increase the number of polymorphic markers. *G. soja* is thought to be the progenitor of the domesticated *G. max* species (Hymowitz and Newell 1981). *G. soja* is interfertile with *G. max* and therefore is considered to be a potential genetic resource for providing diversity for soybean genetics and breeding.

Linkage maps of soybean have been developed by several research groups in the United States as well as in East Asian countries, including Japan, China and Korea. In the history of map development, the molecular markers have been used commonly across all research groups, while the mapping populations have been developed independently among individual research groups in a particular country.

### 4.2.1 Mapping Populations Developed in the United States

More than 10 mapping populations have been established in the United States for the construction of linkage maps. Among these, three populations based on crosses between "Minsoy" × "Noir 1", "A81-356022" × "PI468.916", and "Clark" × "Harosoy" have been used as the main resources in the early stages of map construction.

### 4.2.1.1 Mapping Population "Minsoy" × "Noir 1"

The $F_1$ hybrid between "Minsoy" × "Noir 1" was generated by Reid Palmer of the Iowa State University, then 60 $F_2$ progenies were developed by Apuya et al. (1988). "Noir 1" was a variety developed in Hungary before being introduced into the United States (Nelson et al. 1987) that showed better agronomic traits than "Minsoy", which originated in China (Orf et al. 1999). The mapping population derived from the cross of "Minsoy" × "Noir 1" was used for the first RFLP linkage map in soybean (Apuya et al. 1988), and later it was used to develop an RFLP map (Lark et al. 1993). The progenies of this mapping population showed a high degree of variation that yielded offspring with better agronomic traits than their parents (Mansur et al.

**Table 4-2** Description of main soybean genetic linkage maps.

| Map style | Mapping population | | | No of linkage group | Total length (cM) | Number of mapped markers | | | | | | | | | Marker density | Reference |
|---|---|---|---|---|---|---|---|---|---|---|---|---|---|---|---|---|
| | Crossed variety | Structure | Number of genotypes | | | Classical | Isozyme | RFLP | RAPD | AFLP | Micro-satellite | SNP | Others | Total | | |
| Single | Minsoy x Noir 1 | $F_2$ | 50 | 4 | | | | 11 | | | | | | 11 | 0.0 | Apuya et al. (1988) |
| Single* | A81-356022 x PI 468.916 | $F_2$ | 60 | 26 | 1200 | | | 130 | | | | | | 130 | 9.2 | Keim et al. (1990) |
| Single* | A81-356022 x PI 468.916 | $F_2$ | 60 | 31 | 2147 | 4 | 5 | 243 | | | | | | 252 | 8.5 | Diers et al. (1992) |
| Single | Minsoy x Noir 1 | $F_2$ | 69 | 31 | 1550 | 5 | 2 | 132 | | | | | | 139 | 11.2 | Lark et al. (1993) |
| Single | E/I/Dupont Bonus x PI 81762 | $F_2$ | 68 | 21 | 2678 | | 600 | | | | | | | 600 | 4.5 | Rafalski and Tingey (1993) |
| Single* | A81-356022 x PI 468.916 | $F_2$ | 60 | 25 | 3771 | 3 | 4 | 365 | 11 | | | | | 383 | 9.8 | Shoemaker and Olson (1993) |
| Integrated 2 populations | | | | 25 | 2473 | 3 | 4 | 358 | 10 | | | | | 375 | 6.6 | Shoemaker and Specht (1995) |
| Single | Clark x Harosy | $F_2$ | 60 | 26 | 1056 | 13 | 7 | 110 | 8 | | | | | 138 | 7.7 | Shoemaker and Specht (1995) |
| Single | Clark x Harosy | $F_2$ | 60 | 29 | 1486 | 13 | 7 | 110 | 8 | | 40 | | | 178 | 8.3 | Akkaya et al. (1995) |
| Single | Minsoy x Noir 1 | RILs | 284 | 35 | 1981 | 1 | 6 | 224 | | | 45 | | | 276 | 7.2 | Mansur et al. (1996) |

| | | | | | | | | | | | | | |
|---|---|---|---|---|---|---|---|---|---|---|---|---|---|
| Single | Young x PI 416937 | F4 | 120 | 31 | 1600 | | 137 | | | | 137 | 11.7 | Lee et al. (1996a) |
| Integrated | 9 populations | | | 25 | 2539 | | 810 | | | | 810 | 3.1 | Shoemaker et al. (1996) |
| Single* | A81-356022 x PI 468.916 | F2 | 57 | | | | | | | | | | Diers et al. (1992) |
| Single* | C1640 x PI 479750 | F2 | 59 | | | | | | | | | | Brummer et al. (1995) |
| Single | Clark x Harosy | F2 | 60 | | | | | | | | | | Shoemaker and Specht (1995) |
| Single | Evans x PI 90763 | F2 | 115 | | | | | | | | | | |
| Single | Evans x PI 88788 | F2 | 102 | | | | | | | | | | |
| Single | Evans x PI 209.332 | RILs | 98 | | | | | | | | | | |
| Single | Evans x Peking | F2 | 110 | | | | | | | | | | |
| Single | Young x PI 416937 | F4 | 120 | | | | | | | | | | Lee et al. (1996a), Mian et al. (1996) |
| Single | PI 97100 x Coker 237 | F2 | 111 | | | | | | | | | | Lee et al. (1996b) |
| Single | PI437.654 x BSR-101 | RIL | 300 | 28 | 3441 | 10 | 165 | 25 | | 650 | 840 | 4.1 | Keim et al. (1997) |
| Integrated | 3 mapping populations | F2, RIL | 57-240 | – | – | 26 | 689 | 79 | 11 | 606 | 1421 | – | Cregan et al. (1999a) |
| Single | A81-356022 x PI0468.916 | F2 | 59 | 23 | 3003 | 3 | 501 | 10 | | 486 | 1004 | 3.0 | Cregan et al. (1999a) |
| Single | Minsoy x Noir 1 | RIL | 240 | 22 | 2787 | 10 | 209 | 0 | | 412 | 633 | 4.4 | Cregan et al. (1999a) |

*Table 4-2 contd....*

*Table 4-2 contd....*

| Map style | Mapping population | | | No of linkage group | Total length (cM) | Number of mapped markers | | | | | | | | | Marker density | Reference |
| | Crossed variety | Structure | Number of genotypes | | | Classical | Isozyme | RFLP | RAPD | AFLP | Micro-satellite | SNP | Others | Total | | |
| --- | --- | --- | --- | --- | --- | --- | --- | --- | --- | --- | --- | --- | --- | --- | --- | --- |
| Single | Clark x Harosy | F2 | 57 | 28 | 2534 | 14 | 7 | 95 | 57 | 11 | 339 | | | 523 | 4.8 | Cregan et al. (1999a) |
| Single | PI437.654 x BSR-101 | RIL | 330 | 35 | 3275 | | | 250 | 106 | | | | | 356 | 9.2 | Ferreira et al. (2000) |
| Single | Misuzudaizu x Moshidou Gong503 | F$_2$ | 190 | 33 | 1605 | 4 | | 247 | | | | | | 251 | 6.4 | Yamanaka et al. (2000) |
| Single | Noir 1 x BARC-2 | F$_2$ | 149 | 35 | 1400 | 4 | | 39 | 17 | 105 | 25 | | | 207 (190 marker) | 6.8 | Matthews et al. (2001) |
| Single | Misuzudaizu x Moshidou Gong503 | F$_2$ | 190 | 21 | 2909 | 5 | | 401 | 1 | | 96 | | | 503 | 5.8 | Yamanaka et al. (2001) |
| Single | Kefeng No.1 x Nannong 1138-2 | RIL | 184 | 21 | 3596 | 4 | | 229 | | | 219 | | | 452 | 8.0 | Zhang t al. (2004) |
| Integrated | 5 mapping populations | | | 20 | 2524 | 24 | 10 | 709 | 73 | 6 | 1015 | | 12 | 1849 | 1.4 | |
| Single* | A81-356022 x PI 468.916 | F2 | 59 | | | | | | | | | | | | | |
| Single | Clark x Harosy | F2 | 57 | | | | | | | | | | | | | |
| Single | Minsoy x Noir 1 | RIL | 240 | | | | | | | | | | | | | |
| Single | Minsoy x Archer | RIL | 233 | | | | | | | | | | | | | |
| Single | Archer x Noir 1 | RIL | 240 | | | | | | | | | | | | | |
| Integrated | 3 mapping populations | | | 20 | 2550 | 24 | 10 | 709 | 73 | 6 | 1014 | 1141 | | 2977 | 0.9 | Choi et al. (2007) |
| Single | Minsoy x Noir 1 | RIL | | | | | | | | | | | | | | |
| Single | Minsoy x Archer | RIL | | | | | | | | | | | | | | |
| Single | Evans x PI209332 | RIL | | | | | | | | | | | | | | |

| | | | | | | | | | | | | | | |
|---|---|---|---|---|---|---|---|---|---|---|---|---|---|---|
| Single | Misuzudaizu x Moshidou Gong503 | RIL | 94 | 20 | 2700 | 1 | 105 | | | 829 | | 935 | 2.9 | Hisano et al. (2007) |
| Single | Misuzudaizu x Moshidou Gong503 | $F_2$ | 190 | 20 | 3081 | 5 | 509 | 1 | 318 | 318 | 126 | 1277 | 2.4 | Xia et al. (2007) |
| Single* | Hwngkeum x IT182932 | RIL | 113 | | 2316 | | | | | 295 | 29 | 62 | 386 | 6.0 | Yang et al. (2008) |

* indicates interspecific cross between G. *max* and G. *soja*.

1993a). Subsequently, 284 $F_7$-derived RILs were generated from the $F_2$ plants for quantitative trait locus (QTL) mapping by Mansur et al. (1993b). These RILs have been used for various genetic analyses, including mapping of microsatellite markers on the linkage map (Mansur et al. 1996). A total of five integrated soybean linkage maps have been developed, four of which were based on integrated segregation data sets derived from "Minsoy" × "Noir 1" (Table 4.1). Recently, "Minsoy" and "Noir 1" were the source of two of five soybean varieties used in developing SNP markers, which were located on a consensus map constructed by Choi et al. (2007).

### 4.2.1.2 Mapping Population "A81-356022" × "PI468.916"

Sixty-two $F_2$ progenies were generated from a cross between "A81-356022", a soybean breeding line of the Iowa State University, and "PI468.913", a *G. soja* accession by W.R. Fehr of the Iowa State University. These parents were chosen because of their phenotypic differences (Carpenter and Fehr 1986; Graef et al. 1989), genetic diversity (Keim et al. 1989) and lack of chromosomal translocations (Palmer and Klein 1987). This mapping population was developed with the aim of obtaining more polymorphic markers by taking advantage of the wider diversity of the interspecific cross, and was used to develop an RFLP-based genetic linkage map (Keim et al. 1990; Diers et al. 1992; Shoemaker and Olson 1993). Subsequently, four integrated linkage maps were constructed using the segregation data from the "A81-356022" × "PI468.916" cross (Shoemaker and Specht 1995; Shoemaker et al. 1996; Cregan et al. 1999a; Song et al. 2004).

### 4.2.1.3 Mapping Population "Clark" × "Harosoy"

A "Clark" × "Harosoy" mapping population was developed by Shoemaker et al. (1995) for the purpose of integrating classical and isozyme markers into the public RFLP map. Prior to performing a parental cross, near-isogeneic lines (NILs) of the soybean cultivars "Clark" and "Harosoy", indicating reciprocal homozygous alleles (dominant/recessive) of classical and isozyme markers, were created. "Clark" and "Harosoy" are landmark varieties that have independently contributed a large percentage of genes to the northern cultivars (Gizlice et al. 1994). A total of seven pigmentation markers, six morphological markers and seven isozyme markers were segregated between "Clark" and "Harosoy". Six $F_1$ plants were generated between the cross, then 60 $F_2$ plants were developed. The segregation data from this mapping population were used to develop a microsatellite map (Akkaya et al. 1995) as well as three integrated maps (Shoemaker et al. 1996; Cregan et al. 1999a; Song et al. 2004).

## 4.2.2 Mapping Populations Developed in East Asia

An $F_2$ mapping population between "Misuzudaizu" and "Moshidou Gong 503" was developed in Japan (Yamanaka et al. 2000). "Misuzudaizu" is a variety bred in Japan, while "Moshidou Gong 503" was developed in the Northeast region in China from an interspecific cross between *G. max* and *G. soja*. "Misuzudaizu" and "Moshidou Gong 503" exhibit differences in various morphological traits such as flowering time, growth habits and seed storage protein components. A total of 190 $F_2$ plants were developed from a cross between "Misuzudaizu" and "Moshidou Gong 503", which were then used as a mapping population (Yamanaka et al. 2000, 2001; Xia et al. 2007). Based on the $F_2$ plants, 94 $F_8$-derived RILs were developed and used for map construction by Hisano et al. (2007).

A set of 184 $F_{2:7:10}$ RILs derived from a cross between the varieties "Kefeng No1" and "Nannong 1138" was developed in China for mapping of the newly developed microsatellite markers and QTL analysis (Zhang et al. 2004). "Kefeng No1" and "Nannong 1138" exhibit contrasting characteristics in agronomic traits, including four phenotype markers: W (flowering color) and *Rn1*, *Rn3*, and *Rsa* (resistance to soybean mosaic virus). A set of 113 $F_{12}$ RILs derived from $F_2$s of an interspecific cross between *G. max* "Hwangkeum" and *G. soja* 'IT 182932' were generated in Korea.

## 4.2.3 Computer Software for Map Construction

The computer program Linkage-1 (Suiter et al. 1983) was adopted for construction of the first soybean linkage map using RFLP markers (Table 4-1; Apuya et al. 1988). Since then, MapMaker (Lander et al. 1987; Lincoln et al. 1992a and 1992b) has been the program used predominantly for linkage analysis of single mapping populations. The JoinMap software (Stam 1993; VanOoijen and Voorrips 2001), capable of merging data from multiple mapping populations, was employed for the construction of the five integrated maps. Shoemaker and Specht (1995), Shoemaker et al. (1996), and Cregan et al. (1999a) developed population-specific linkage maps using MapMaker, then combined the linkage maps that were generated into an integrated linkage map using JoinMap. Meanwhile, Song et al. (2004) and Choi et al. (2007) constructed integrated linkage maps using only the JoinMap program.

## 4.3 Vicissitudes of Soybean Genetic Linkage Maps

Efforts toward the construction of DNA marker-based genetic linkage maps in soybean began in the 1980s. Progress in genetic analyses in soybean during this period was slow compared with other major crops, such as rice and maize, because of its amphidiploidy, lack of cytogenetic markers and lower genetic variation in the germplasms. Soybean genetic maps using classical markers were reported by Palmer and Kilen (1987) and Palmer and Kiang (1990), which had a total length of 420 cM with 57 markers and 530 cM with 49 markers, respectively. Since then, various genetic linkage maps have been reported in soybean, as summarized in Table 4.1. The history of map construction in soybean correlates with the progress in related technologies, such as hybridization, PCR, DNA sequencing, fluorescence detection and statistical algorithms.

### 4.3.1 Mapping with RFLP Markers

The first linkage map based on DNA markers was reported by Apuya et al. in 1988. Twenty-seven RFLP markers were developed using a lambda library of the soybean variety "Forrest" as a probe, and 11 of these markers were successfully located on 4 linkage groups by analyzing a mapping population of 50 $F_2$ plants of "Minsor" × "Noir 1'. The purpose of this study was to claim the usefulness of RFLP markers for developing soybean linkage maps; therefore, the number of mapped markers and the resulting linkage groups were far less than the classical linkage maps by Palmer and Kilen (1987).

A full-fledged trial of RFLP mapping was performed by Keim et al. (1990) for the first time. In order to obtain a large number of polymorphic markers, they adopted 60 $F_2$ plants of the interspecific mapping population "A81-356022" × "PI468.916". RFLP markers were developed using more than 500 randomly selected clones from a *Pst*I soybean genomic library (Keim and Shoemaker 1988) as probes. Consequently, 130 loci were mapped on 26 linkage groups, for a total of 1,200 cM in length.

For QTL analysis of seed protein and oil content, Diers et al. (1992) reconstructed an RFLP map with the mapping population used by Keim et al. (1990). 113 RFLP loci, as well as five loci for isozymes, three morphological loci, and a locus for one storage protein, were added to the previous RFLP map, resulting in a revised map with 252 marker loci. The number of linkage groups and the total length of the map were 31 and 2,147 cM, respectively, both of which extended the previous map of Keim et al. (1990). With a subsequent effort by Shoemaker and Olson (1993), a linkage map of 3,771 cM, which consisted of 25 linkage groups with a total of 383 marker loci, was developed. The number of mapped markers was almost three times higher than that of the previous map of Keim et al. (1990), while the average marker density showed little difference between the two maps. Meanwhile, Rafalski and Tingey (1993) developed an RFLP map with more than 600 RFLP loci mapped onto 21 linkage groups based on the population "E.I.

Dupont Bonus" × "G. soja PI 81762". The total length of this map was 2,678 cM, and the average locus density was one marker every 4.5 cM.

While interspecific mapping populations contributed enormously to the saturation of the soybean linkage map, intraspecific linkage maps have also been developed. After the first report of a soybean linkage map (Apuya et al. 1988), a map consisting of 132 RFLPs, two isozymes, four morphological markers and one biochemical marker using the same intraspecific mapping population, "Minsoy" × "Noir 1", was constructed (Lark et al. 1993). The map defined 1,550 cM of the soybean genome comprising 31 linkage groups. Although the map was less saturated with respect to marker loci compared with the interspecific map developed by Shoemaker and Olson (1993), this was compensated for as the populations used were directly connected with agronomic traits. Actually, QTL analysis on morphological and seed traits was performed in parallel with map construction (Mansur et al. 1993a, b). Lee et al. (1996a) developed an RFLP map based on an $F_4$-derived soybean population generated from the cross "Young" × "PI 416937". They adopted various cDNA and/or genomic clones from other legumes, including soybean, *Vigna radiata*, *Phaseolus vulgaris* L., *Archis hypogea* L. and *Medicago sativa*, as probes for RFLP analysis. As a result, a total of 137 RFLP loci were mapped on 31 linkage groups, representing 1,600 cM in length.

Yamanaka et al. (2000) constructed an RFLP map for QTL analysis of flowering time using an $F_2$ mapping population derived from a cross between "Misuzudaizu" and "Monshidou Gong 503". This linkage map consisted of 247 RFLP loci, including 92 markers based on soybean cDNA clones and four phenotypic loci, and was composed of 33 linkage groups. Later, the linkage map was reconstructed with additional RFLP and microsatellite markers in order to construct a functional linkage map with cDNA markers covering a large region of the soybean genome (Yamanaka et al. 2001). This linkage map accommodated 503 loci, including 189 RFLP loci derived from expressed sequence tag (EST) clones, and consisted of 20 major groups that are likely to correspond to the 20 soybean chromosomes, the total of which was 2,909 cM in length. In the same year, Matthews et al. (2001) reported a map that defined 39 EST-derived RFLP loci as well as 105 AFLPs, 25 microsatellites, 17 RAPDs and four morphological loci onto 35 linkage groups with a total length of 1,400 cM.

### 4.3.2 *Mapping with AFLP and RAPD Markers*

AFLP markers provide the ability for large-scale mapping at a lower cost than RFLP markers, and were considered efficient for genomes of moderate sizes, such as that of soybean in the 1990s. An RFLP linkage map with 355 loci was developed using a recombinant inbred population of 300 $F_{6:7}$ lines generated between parents "PI437.654" and "BSR-101" (Keim et al. 1994; Webb et al. 1995). "PI437.654" is an unadapted accession in the United States used for the introgression of soybean cyst nematode resistance genes, while "BSR-101" is an elite cultivar. Using this RFLP map as a 'scaffold' map, a total of 650 AFLP loci were assigned onto 28 linkage groups with 165 RFLP loci and 25 RAPD loci, the total of which represented a distance of 3,441 cM (Keim et al. 1997). Eighty-seven percent of AFLP bands were dominant marker alleles, and several AFLP loci were clustered on the map.

The AFLP marker system excels at defining a large set of markers on a single linkage map; however, it has the drawback of a low degree of transferability across multiple mapping populations. To overcome this, Meksem et al. (2001) demonstrated the conversion of AFLP bands to sequence-tagged-sites (STSs) in soybean. Subsequently, Xia et al. (2007) converted the AFLP markers to STS markers by cloning and sequencing the polymorphic AFLP bands from the "Misuzudaizu" × "Moshidou Gong 503" mapping population. In addition to 97 AFLP-derived STS loci, 19 BAC-end sequence-derived STS loci, 10 EST-derived STS loci, 318 AFLP loci, 318 microsatellite loci, 509 RFLP loci, and one RAPD locus were mapped onto 20 linkage groups, totaling a map distance of 3,080 cM.

Eleven RAPD markers were located on the soybean RFLP map for the first time in 1993 by Shoemaker and Olson. Since then, less effort has been paid to develop RAPD makers than other types of DNA markers, such as RFLPs, AFLPs and microsatellites, in soybean. There are concerns, however, whether such sophisticated markers can be realistically used in small-scale breeding programs and in developing countries with limited human, material and financial resources. Taking these concerns into account, Ferreira et al. (2000) located a total of 106 RAPD loci as well as 250 anchor RFLP loci onto 35 linkage groups representing 3,275 cM using the 330 $F_{6:7}$ RILs derived from a cross between "PI437.654" and "BSR-101", the same population that was used for construction of the AFLP-based map by Keim et al. (1997). Comparison of the loci detected by RFLP and RAPD markers indicates that both markers showed similar distribution along the entire genome and detected similar levels of polymorphism.

### 4.3.3 *Mapping with Microsatellite Markers*

Microsatellites, or SSRs, in the soybean genome were first assessed by Akkaya et al. (1992) and subsequently by Morgante and Olivieri (1993); both groups demonstrated that microsatellites exhibit a high level of length polymorphism. In 1994, Cregan et al. described a basic procedure for generating microsatellite markers that involved the selection of SSR-containing sequences from public DNA databases, and then cloning, sequencing and identifying microsatellite-containing genomic clones from the variety "Williams" (Cregan et al. 1994).

Akkaya et al. (1995) reported the mapping of 34 microsatellite loci on the soybean RFLP map based on the mapping population "Clark" × "Harosoy', previously reported by Shoemaker and Specht (1995). A total of 29 linkage groups for a total map length of 1,486 cM were generated with 178 marker loci, including 34 microsatellites derived from genomic DNA, 110 RFLPs, eight RAPDs, seven isozymes and 13 classical marker loci. The new map showed that genomic DNA-derived microsatellite markers were distributed rather evenly throughout the genome. In 1996, Mansur et al. described the generation of $F_7$ RILs derived from a cross of "Minsoy" × "Noir 1" and developed a map with 45 microsatellites, 224 RFLPs, six isozyme loci and one morphological locus for QTL analysis of agronomic traits (Mansur et al. 1996). The information on the primer pairs used to amplify microsatellites was obtained from the previous study by Akkaya et al. (1995), which permitted the generation of 15 common microsatellite markers for the two independent microsatellite maps.

A larger number of microsatellite markers was generated and mapped by Cregan et al. (1999b). They followed the same procedure as was used for construction of the two previous maps of Akkaya et al. (1995) and Mansur et al. (1996). However, introduction of the software OLIGO (National Biolabs, St. Paul, MN, USA) facilitated the design of a large set of primer pairs for amplification of the markers. Three mapping populations were used to obtain the segregation data sets, and a total of 606 microsatellite loci were mapped in one or more of three mapping populations. (The detailed features of the maps are described in Section 4.3.5). Later on, these microsatellite markers were remapped onto a single linkage map using the Chinese mapping population, Kefeng No1" × "Nannong 1138" , by Zhang et al. (2004). Together with four classical marker loci and 229 RFLP loci developed by Shoemaker et al. (1996), Zhang et al. (1997), Liu et al. (2000), and Wu et al. (2001), a total of 219 microsatellite marker loci were mapped onto 21 linkage groups, representing a total map length of of 3,596 cM.

A large set of microsatellite markers was developed by Hisano et al. (2007). Based on 63,676 nonredundant EST sequences retrieved from public DNA databases, a total of 6,920 primer pairs were designed to amplify microsatellites. Six hundred eighty of these markers, as well as 105 RFLP markers developed by Yamanaka et al. (2001), were subjected to linkage analysis using 94 RILs derived from a cross between "Misuzudaizu" and "Moshidou Gong 503". As a result, 693 microsatellite marker loci and 241 RFLP marker loci were mapped onto 20 linkage groups, which totaled 2,700.3 cM in length. Subsequently, Yang et al. (2008) generated 45 sequence-based (SB) markers, including microsatellite, SNPs and in-del marker systems, in order to construct a map in which markers were evenly distributed. As a result, a total of 386 loci were randomly mapped onto 20 linkage groups representing 2,316 cM.

## 4.3.4 SNP Mapping

In the initial phase of SNP development in soybean, SNPs originated mainly from single genes or DNA fragments with the aim of investigating gene structure and phylogenetic relationships: three SNPs on a 3,543-bp region of the $Gy_4$ glycinin locus (Scallon et al. 1987), two SNPs on a 789-bp sequence of cDNA encoding the $A_3B_4$ glycinin subunit (Zakharova 1989), and nine SNPs on a 400-bp segment of the RFLP probe A-199 (Zhu et al. 1995). In 2003, Zhu et al. assessed the frequency of SNPs in 143 DNA fragments based on the sequences of 25 soybean genotypes representative of the soybean germplasms in North America. It was reported that nucleotide diversity expressed as Watterson's è was 0.00053 and 0.00111 in coding and noncoding regions, respectively, which was lower than that estimated in the autogamous plant *Arabidopsis thaliana*.

In order to construct a soybean transcript map, Choi et al. (2007) developed SNP markers using six diverse genotypes, "Archer", "Minsoy", "Noir 1" , "Evans", "PI 209332" and "Peking", and mapped loci on three mapping populations, "Minsoy" × "Noir 1", "Minsoy" × "Archer" and "Evans" × " PI 209332". Based on a total of 2.44 Mb of aligned sequences obtained from 4,240 amplified STS DNA fragments, 5,551 SNPs were discovered with an average nucleotide diversity of $\theta = 0.000997$. *In silico* comparison of the observed genetic distances between adjacent genes versus the theoretical groups indicated that genes were clustered in the soybean genome. A total of 1,141 SNP loci along with 1,014 microsatellite, 709 RFLP, 73 RAPD, six AFLP, 24 classical and 10 isozyme marker loci were mapped on an integrated linkage map using the three mapping populations.

## 4.3.5 Integrated Maps

Genetic markers often show polymorphism in one population but not in another population, which hinders the efficient use of the developed markers. With the aid of the JoinMap program (Stam 1993; Van Ooijen and Voorrips 2001), segregation data from different mapping populations can be merged to construct a single genetic linkage map. This approach not only increases the number of markers, but also improves the accuracy and resolution of the map.

The first map integration in soybean was performed by Shoemaker and Specht (1995) in order to merge the linkage groups of the previously published maps with RFLP mapping. They constructed a population ("Clark" × "Harosoy", see Section 4.2.1.3) segregating for many markers known to be on the published map. Then, by combing segregation data sets of the "Clark" × "Harosoy" population and the interspecific ( "A81-356022" × "PI468.916" ) mapping population, a total of 375 marker loci, including 358 RFLP, 10 RAPD, four isozyme and three classical marker loci, were mapped onto 25 linkage groups, the total length of which was 2,473 cM. Subsequently, Shoemaker et al. (1996) integrated the segregation data of nine different mapping populations derived from two interspecific and seven intraspecific crosses, and defined the locations of 810 RFLP loci onto 25 linkage groups. Linkage groups contained up to 33 markers that were duplicated in other linkage groups, suggesting an ancient genome duplication event in soybean.

The multiplicity of RFLP loci has often complicated the process of comparing and merging linkage maps from different mapping populations. Cregan et al. (1999a) developed a set of microsatellite markers and mapped them using three existing mapping populations, "Minsoy" × "Noir 1 ", "A81-356022" × "PI468.916" and "Clark" × "Harosoy" (see Section 3.3). Map integration was carried out simply by comparing the positions of corresponding loci across the three mapping populations. As a result, a total of 633, 1,004, and 523 loci were mapped on the individual maps of "Minsoy" × "Noir 1", "A81-356022" × "PI468.916" and "Clark" × "Harosoy", respectively. The total lengths of the three individual maps ranged from 2,534 cM to 3,003 cM, while the number of linkage groups ranged from 22 to 28. Integration of the three maps produced a high-density map with a total of 1,421 marker loci including 606 microsatellite, 689 RFLP, 79 RAPD, 11 AFLP, 26 classical and 10 isozyme marker loci. This report was one of the milestones of linkage map construction in soybean, because each of the 20 consensus linkage groups was defined for the first time, and a basis for standardization of the identification of each linkage group was provided.

Song et al. (2004) reconstructed an integrated linkage map using five commonly used soybean populations, "Minsoy" × "Noir 1", "A81-356022" × "PI468.916", "Clark" × "Harosoy", "Minsoy" × "Archer", and "Archer" × "Noir 1" with newly developed 391 microsatellite markers and the published markers. The resulting map consisted of a total of 1,849 marker loci on 20 linkage groups, which totaled 1,849 cM in length. This report is another milestone in the history of soybean linkage mapping, because the map in this study corroborated to 20 linkage groups for the first time, which corresponds to the soybean haploid chromosome number. In 2007, Choi et al. reported an integrated map with SNP markers based on four mapping populations, details of which were described in Section 4.3.4. These two integrated linkage maps, constructed by Song et al. (2004) and Choi et al. (2007), have been regarded as the consensus maps for soybean.

## 4.4 Future Prospects

Despite the challenges of working with a relatively large and complex genome, the construction of genetic linkage maps in soybean has come a long way over the last 20 years. Currently, the number of linkage groups for soybean is

20, corresponding to its haploid chromosome number, and the total map length ranges from approximately from 2,500 cM to 3,000 cM. These numbers strongly suggest that the consensus linkage map of soybean has already been saturated. Recent advances in DNA sequencing technologies have enabled the production of larger sets of microsatellite and SNP markers (Song et al. 2004; Choi et al. 2007; Hisano et al. 2007; Xia et al. 2007), though these markers have not been defined to their consensus positions. Considering that the total number of DNA markers has reached 4,000 and is still increasing, the approach of splitting the chromosomes into "Bin boundaries", as was introduced in maize (Gardiner et al. 1993), might be effective.

Sequencing of the entire soybean genome is almost complete, which provides a substantial amount of genome sequence as well as physical maps of the entire genome (reviewed by Jackson et al. 2006). By taking advantage of these new data, integration of the genetic linkage map and the physical map has been attempted. Examples are the construction of linkage maps of two soybean cultivars, "Williams 82" and "Forrest" , both of which are targets of structural and functional genomics (reviewed by Jackson et al. 2006; Lightfoot et al. 2008), and the development of BAC-end sequence-based DNA markers (Shultz et al. 2007; Shoemaker et al. 2008). Information on the integrated physical and genetic maps can be retrieved from the Soybean Genome Database (SoyGD: *http://soybeangenome.siu.edu/*, Shultz et al. 2006). Further integration of the physical and genetic maps is expected to allow interactive development of soybean genetics and genomics, such as designing DNA markers to targeted regions and identifying genes through QTL detection; this would lead to a productive and systematic fusion of information on the gene structures and functions, and toward an understanding of the genetic systems in soybean.

## References

Akkaya MS, Bhagwat AA, Cregan PB (1992) Length polymorphisms of simple sequence repeat DNA in soybean. Genetics 132: 1131–1139.

Akkaya MS, Shoemaker RC, Specht JE, Bhagwat AA, Cregan PB (1995) Integration of simple sequence repeat DNA markers into a soybean linkage map. Crop Sci 35: 1439–1445.

Apuya NR, Frazier BL, Keim P, Roth EJ, Lark KG (1988) Restriction fragment length polymorphisms as genetic markers in soybean, *Glycine max* (L.) merrill. Theor Appl Genet 75: 889–901.

Brummer EC, Nickell AD, Wilcox JR, Shoemaker RC (1995) Mapping the Fan locus controlling linolenic acid content in soybean oil. J Hered 86: 245–247.

Carpenter JA, Fehr WR (1986) Genetic variability for desirable agronomic traits in populations containing *Glycine soja* germplasm. Crop Sci 26: 681–686.

Choi IY, Hyten DL, Matukumalli LK, Song Q, Chaky JM, Quigley CV, Chase K, Lark KG, Reiter RS, Yoon MS, Hwang EY, Yi SI, Young ND, Shoemaker RC, van Tassell CP, Specht JE, Cregan PB (2007) A soybean transcript map: gene distribution, haplotype and single-nucleotide polymorphism analysis.0Genetics 176: 685–696.

Cregan PB, Bhagwat AA, Akkaya MS, Ringwen J (1994) Microsatellite fingerprinting and mapping of soybean. Meth Mol Cell Biol 5: 49–61.

Cregan PB, Jarvik T, Bush AL, Shoemaker RC, Lark KG, Kahler AL, Kaya N, VanToai TT, Lohnes DG, Chung J, Specht JE (1999a) An integrated genetic linkage map of the soybean genome. Crop Sci 39: 1464–1490.

Cregan PB, Mudge J, Fickus ED, Marek LF, Danesh D, Denny R, Shoemaker RC, Matthews BF, Jarvik T, Young ND (1999b) Targeted isolation of simple sequence repeat markers through the use of bacterial artificial chromosomes. Theor Appl Genet 98: 919–928.

Diers BW, Keim P, Fehr WR, Shoemaker RC (1992) RFLP analysis of soybean seed protein and oil content. Theor Appl Genet 83: 608–612.

Ferreira AR, Foutz KR, Keim P (2000) Soybean genetic map of RAPD markers assigned to an existing scaffold RFLP map. J Hered 91: 392–396.

Gardiner JM, Coe EH, Melia-Hancock S, Hoisington DA, Chao S (1993) Development of a core RFLP map in maize using an immortalized $F_2$ population. Genetics 134: 917–930.

Gizlice Z, Carter Jr TE, Burton JW (1994) Genetic base for North American public soybean cultivars released between 1947 and 1988. Crop Sci 34: 1143–1151.

Graef GL, Fehr WR, Cianzio SR (1989) Relation of isozyme genotypes to quantitative characters in soybean. Crop Sci 29: 683–688.

Hisano H, Sato S, Isobe S, Sasamoto S, Wada T, Matsuno A, Fujishiro T, Yamada M, Nakayama S, Nakamura Y, Watanabe S, Harada K, Tabata S (2007) Characterization of the soybean genome using EST-derived microsatellite markers. DNA Res 14: 271–281.

Hymowitz T, Newell CA (1981) Taxonomy of the genus *Glycine*, domestication and uses of soybean. Econ Bot 35:272–288.

Jackson SA, Rokhsar D, Stacey G, Shoemaker RC, Schmutz J, Grimwood J (2006) Toward a reference sequence of the soybean genome: A multiagency effort. Crop Sci 46: S55–S61.

Keim P, Shoemaker RC (1988) Construction of a random recombinant DNA library that is primarily single copy sequence. Soybean Genet Newsl 15: 147–148.

Keim P, Shoemaker RC, Palmer RG (1989) RFLP diversity in soybean. Theor Appl Genet 77: 786–792.

Keim PB, Diers W, Olson TC, Shoemaker RC (1990) RFLP mapping in soybean: Association between marker loci and variation in quantitative traits. Genetics 126: 735–742.

Keim P, Beavis WD, Schupp JM, Baltazar BM, Mansur L, Freestone RE, Vahedian M, Webb DM (1994) RFLP analysis of soybean breeding populations: I. Genetic structure differences due to inbreeding methods. Crop Sci 34: 55–61.

Keim P, Schupp JM, Travis SE, Clayton K, Zhu T, Shi L, Ferreira A, Webb DM (1997) A high-density soybean genetic map based on AFLP markers. Crop Sci 37: 537–543.

Lander ES, Green P, Abrahamson J, Barlow A, Daly MJ, Lincoln SE, Newburg L (1987) MAPMAKER: An interactive computer package for constructing primary genetic linkage maps of experimental and natural populations. Genomics 1: 174–181.

Lark KG, Weisemann JM, Matthews BF, Palmer R, Chase K, Macalma T (1993) A genetic map of soybean (*Glycine max* L.) using an intraspecific cross of two cultivars: 'Minosy' and 'Noir 1'. Theor Appl Genet 86: 901–906.

Lee SH, Bailey MA, Mian MAR, Carter Jr TE, Ashley DA, Hussey RS, Parrott WA, Boerma HR (1996a) Molecular markers associated with soybean plant height, lodging, and maturity across locations. Crop Sci 36: 728–7350

Lee SH, Bailey MA, Mian MAR, Shipe ER, Ashley DA, Parrott WA, Hussey RS, Boerma HR (1996b) Identification of quantitative trait loci for plant height, lodging, and maturity in a soybean population segregating for growth habit. Theor Appl Genet 62: 516–523.

Lightfoot DA (2008) Soybean genomics: Developments through the use of cultivar "Forrest". Int J Plant Genom: 793158.

Lincoln S, Daley M, Lander E (1992a) Constructing genetic maps with MAPMAKER/EXP3.0. Whitehead Institute Technical Report 3rd edn.

Lincolon S, Daley M, Lander E (1992b) Mapping genes controlling quantative traits with MAPMAKER/QTL1.1. Whitehead Institute Technical Report 3rd edn.

Liu F, Zhuang BC, Zhang JS, Chen SY (2000) Construction and analysis of soybean genetic map. Acta Bot Sci 27: 1018–1026.

Mansur LM, Lark KG, Kross H, Oliveira A (1993a) Interval mapping of quantitative trait loci for reproductive, morphological, and seed traits of soybean (*Glycine max* L.). Theor Appl Genet 86: 907–913.

Mansur LM, Orf JH, Lark KG (1993b) Determining the linkage of quantitative trait loci to RFLP markers using extreme phenotypes of recombinant inbreds of soybean (*Glycine max* L. Merr.). Theor Appl Genet 86: 914–918.

Mansur LM, Orf JH, Chase K, Jarvik T, Cregan PB, Lark KG (1996) Genetic mapping of agronomic traits using recombinant inbred lines of soybean. Crop Sci 36: 1327–1336.

Matthews BF, Devine TE, Weisemann JM, Beard HS, Lewers KS, MacDonald MH, Park Y-B, Maiti R, Lin J-J, Kuo J, Pedroni MJ, Cregan PB, Saunders JA (2001) Incorporation of sequenced cDNA and genomic markers into the soybean genetic map. Crop Sci 41: 516–521.

Meksem K, Ruben E, Hyten D, Triwitayakorn K, Lightfoot DA (2001) Conversion of AFLP bands into high-throughput DNA markers. Mol Genet Genom 265: 207–214.

Mian Mar, Bailey MA, Ashley DA, Wells R, Carter Jr TE, Parrott WA, Boerma HR (1996) Molecular markers associated with water use efficiency and leaf ash in soybean. Crop Sci 36: 1252–1257.

Morgante M, Olivieri AM (1993) PCR—amplified microsatellites as markers in plant genetics. Plant J 3: 175–182.

Nelson RL, Amdor PJ, Orf JH, Lambert JW, Cavins JF, Kleiman R, Laviolette FA, Athow KL (1987) Evaluation of the USDA soybean germplasm collection: Maturity groups 000 to IV (PI 273.483 to PI 427.107). USDA Tech Bull 1718.

Orf JH, Chase K, Jarvik T, Mansur LM, Cregan PB, Adler FR, Lark KG (1999) Genetics of soybean agronomic traits: I. Comparison of three related recombinant inbred populations. Crop Sci 39: 1642–1651.

Palmer RG, Kilen TC (1987) Quantitative genetics and cytogenetics. In: JR Wilcox (ed) Soybeans: Improvement, Production, and Uses. Agron Monogr, 2nd edn. ASA-CCSA-SSSA, Madison, WI, USA, pp 135–209.

Palmer RG, Kiang YT (1990) Linkage map of soybean (*Glycine max* L. Merr.) In: SJ O'Brien (ed) Genetic Maps: Locus Maps of Complex Genomes. Cold Spring Harbor Lab Press, Cold Spring Harbor, NY, USA, pp 668–693.

Rafalski A, Tingey S (1993) RFLP map of soybean (*Glycine max*). In: SJ O'Brien (ed) Genetic Maps: Locus Maps of Complex Genomes. Cold Spring Harbor Lab Press, Cold Spring Harbor, NY, USA, pp 149–156.

Scallon BJ, Dickinson CD, Nielsen NC (1987) Characterization of a null-allele for the $Gy_4$ glycinin gene from soybean. Mol Gen Genet 208: 107–113

Shoemaker RC, Olson TC (1993) Molecular linkage map of soybean (*Glycine max* L. Merr.). In: SJ O'Brien (ed) Genetic Maps: Locus Maps of Complex Genomes. Cold Spring Harbor Lab Press, Cold Spring Harbor, NY, USA, pp 6131–6138.

Shoemaker RC, Specht JE (1995) Integration of the soybean molecular and classical genetic linkage groups. Crop Sci 35: 436–446.

Shoemaker RC, Polzin K, Labate J, Specht J, Brummer EC, Olson T, Young N, Concibido V, Wilcox J, Tamulonis JP, Kochert G, Boerma HR (1996) Genome duplication in soybean (*Glycine* subgenus *soja*).0Genetics 144: 329–338.

Shoemaker RC, Grant D, Olson T, Warren WC, Wing R, Yu Y, Kim H, Cregan P, Joseph B, Futrell-Griggs M, Nelson W, Davito J, Walker J, Wallis J, Kremitski C, Scheer D, Clifton SW, Graves T, Nguyen H, Wu X, Luo M, Dvorak J, Nelson R, Cannon S, Tomkins J, Schmutz J, Stacey G, Jackson S. (2008) Microsatellite discovery from BAC end sequences and genetic mapping to anchor the soybean physical and genetic maps. Genome 4: 294–302.

Shultz JL, Kurunam D, Shopinski K, Iqbal MJ, Kazi S, Zobrist K, Bashir R, Yaegashi S, Lavu N, Afzal AJ, Yesudas CR, Kassem MA, Wu C, Zhang HB, Town CD, Meksem K, Lightfoot DA (2006) The Soybean Genome Database (SoyGD): A browser for display of duplicated, polyploid, regions and sequence tagged sites on the integrated physical and genetic maps of *Glycine max*. Nucl Acids Res 34 (Database iss): D758–765.

Shultz JL, Kazi S, Bashir R, Afzal JA, Lightfoot DA (2007) The development of BAC-end sequence-based microsatellite markers and placement in the physical and genetic maps of soybean. Theor Appl Genet 114: 1081–1090.

Song QJ, Marek LF, Shoemaker RC, Lark KG, Concibido VC, Delannay X, Specht JE, Cregan PB (2004) A new integrated genetic linkage map of the soybean. Theor Appl Genet 109: 122–128.

Stam P (1993) Construction of integrated genetic linkage maps by means of a new computer package: JoinMap. Plant J 3: 739–744.

Suiter KA, Wendell JF, Case JS (1983) Linkage-1. J Hered 74: 203–204.

Van Ooijen JW, Voorrips RE (2001) JoinMap 3.0 software for the calculation of genetic linkage maps. Plant Res Int, Wageningen, The Netherlands.

Webb DM, Baltazar BM, Rao-Arelli AP, Schupp J, Clayton K, Keim P, Bravis, WD (1995) Genetic mapping of soybean cyst nematode race-3 resistance loci in the soybean PI 437.654. Theor Appl Genet 91: 574–581.

Wu XL, He CY, Wang YJ, Zhang ZY, Dongfang Y, Zhang JS, Chen SY, Gai JY (2001) Construction and analysis of a genetic linkage map of soybean. Acta Genet Sin 28: 1051–1061.

Xia Z, Tsubokura Y, Hoshi M, Hanawa M, Yano C, Okamura K, Ahmed TA, Anai T, Watanabe S, Hayashi M, Kawai T, Hossain KG, Masaki H, Asai K, Yamanaka N, Kubo N, Kadowaki K, Nagamura Y, Yano M, Sasaki T, Harada K ( 2007) An integrated high-density linkage map of soybean with RFLP, SSR, STS, and AFLP markers using A single $F_2$ population.0DNA Res 14: 257–269.

Yamanaka N, Nagamura Y, Tsubokura Y, Yamamoto K, Takahashi R, Kouchi H, Yano M, Sasaki T, Harada K (2000) Quantitative trait locus analysis of flowering time in soybean using a RFLP linkage map. Breed Sci 50: 109–115.

Yamanaka N, Ninomiya S, Hoshi M, Tsubokura Y, Yano M, Nagamura Y, Sasaki T, Harada K (2001) An informative linkage map of soybean reveals QTLs for flowering time, leaflet morphology and regions of segregation distortion. DNA Res 8: 61–72.

Yang K, Moon J-K, Jeong N, Back K, Kim HM, Jeong S-C (2008) Genome structure in soybean revealed by a genomewide genetic map constructed from a single population. Genomics 92: 52–59.

Zakharova ES, Epishin SM, Vinetski- YP (1989) An attempt to elucidate the origin of cultivated soybean via comparison of nucleotide sequences encoding glycinin B4 polypeptide of cultivated soybean, *Glycine max*, and its presumed wild progenitor, *Glycine soya*. Theor Appl Genet 78: 852–856.

Zhang DS, Dong W, Hhui DW, Chen SY, Zhang BC (1997) Construction of a soybean linkage map using an F2 hybrid population from a cultivated variety and a semi-wild soybean. Chin Sci Bull 42: 1326–1330.

Zhang WK, Wang YJ, Luo GZ, Zhang JS, He CY, Wu XL, Gai JY, Chen SY (2004) QTL mapping of ten agronomic traits on the soybean (*Glycine max* L. Merr.) genetic map and their association with EST markers. Theor Appl Genet 108: 1131–1139.

Zhu T, Shi L, Doyle JJ, Keim P (1995) A single nuclear locus phylogeny of soybean based on DNA sequence. Theor Appl Genet 90: 991–999.
Zhu YL, Song QJ, Hyten DL, Van Tassell CP, Matukumalli LK, Grimm DR, Hyatt SM, Fickus EW, Young ND, Cregan PB (2003) Single-nucleotide polymorphisms in soybean. Genetics 163: 1123–1134.

# 5

# Molecular Mapping of Quantitative Trait Loci

*Dechun Wang*[1*] and *David Grant*[2]

## ABSTRACT

Since the publication of the first soybean molecular linkage map in 1990, over 270 quantitative trait loci (QTL) mapping studies have been published and over 1,100 QTLs have been identified for over 80 traits. This chapter provides a review of QTLs identified in soybean with an emphasis on consensus QTLs identified in multiple mapping populations.

**Keywords:** Soybean; QTL; molecular mapping; linkage map

## 5.1 Introduction

A quantitative trait locus (QTL) is a genetic locus that affects a quantitative trait. The development of molecular markers and the subsequent constructions of molecular linkage maps enabled scientists to identify and to map a large number of QTLs in soybean. The first study of associative mapping of QTLs in soybean was based on the original soybean molecular linkage map (Keim et al. 1990a, b). From 1990 to 2007, over 270 papers (excluding reviews) on QTL mapping studies in soybean were published. Figure 5-1 shows the annual publications collected in the "Biological Abstracts" database (Thomson Scientific, Inc., 2008) on QTL mapping studies

[1]Department of Crop and Soil Sciences, Michigan State University, A384E Plant & Soil Sciences Building, East Lansing, MI 48824-1325, USA.
[2]USDA-ARS, Department of Agronomy, Iowa State University, Ames, IA 50011, USA.
*Corresponding author: *wangdech@msu.edu*

in soybean. The first big increase in the number of publications occurred in 1996, six years after the publication of the first molecular linkage map in soybean. The second large increase in the number of publications occurred in 2004, five years after the publication of an integrated soybean linkage map with a large number of user-friendly markers, mainly simple sequence repeat (SSR) markers (Cregan et al. 1999). The number of publications on QTL mapping studies in soybean ranged from 27 to 39 per year in the past four years. The number is expected to increase in the next five years due to the addition of more user-friendly and high-throughput markers to the integrated soybean map (Song et al. 2004; Choi et al. 2007).

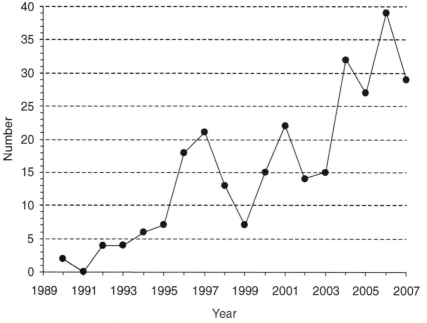

**Figure 5-1** Annual publications on QTL mapping studies in soybean.

## 5.2 Genetic Markers and Linkage Maps

A major objective of any QTL mapping study is to map QTLs underlying a trait of interest on a genetic linkage map, which is a linear map showing the relative positions of genetic markers. Therefore, genetic markers and linkage maps are essential for any QTL mapping study. Several types of genetic markers are available in soybean, including morphological, isozyme, restriction fragment length polymorphism (RFLP), random amplified polymorphic DNA (RAPD), amplified fragment length polymorphism (AFLP), simple sequence repeat (SSR), and single nucleotide polymorphism

(SNP) markers. The most abundant markers developed for soybean are RFLP markers (Apuya et al. 1988; Keim et al. 1989), SSR markers (Akkaya et al. 1995), AFLP markers (Keim et al. 1997), and SNP markers (Choi et al. 2007). The first soybean molecular linkage map with a significant coverage of the genome was published in 1990 by Keim et al. (1990a). This map contained 150 RFLP markers and three morphological markers. The map was further expanded to include 355 RFLP markers and 16 other types of markers by 1993 (Shoemaker and Olson 1993). By 1999, the map was expanded to have 501 RFLPs, 486 SSRs, and 27 markers of other types (Cregan et al. 1999). This map was integrated with two maps developed with two additional mapping populations (Cregan et al. 1999). The integrated map had 689 RFLPs, 606 SSRs, 79 RAPDs, and 47 markers of other types. Using the three mapping populations and two additional mapping populations, a new version of the integrated map was constructed in 2004 (Song et al. 2004). The new integrated map contained 1,015 SSRs, 709 RFLPs, 73 RAPDs, and 52 markers of other types, with a total map length of 2,523.6 cM (Song et al. 2004). The most recent version of the integrated map was published in 2007 (Choi et al. 2007). This map contains 2,989 markers, including 1,141 SNPs, 1,014 SSRs, 709 RFLPs, and 125 other types of markers. The map consists of 20 linkage groups with a total map length of 2,550.3 cM (Choi et al. 2007). The molecular markers, especially the SSR markers, from the integrated maps have been widely used in QTL mapping studies in soybean (Keim et al. 1990a; Diers et al. 1992; Wang et al. 2004a; Zhu et al. 2006; Neto et al. 2007).

In addition to the integrated linkage maps, several other molecular linkage maps were developed for soybean. A map with over 600 RFLP markers was developed by the DuPont Corporation (Rafalski and Tingey 1993). A map with 132 RFLP markers and eight other types of markers was developed by Lark et al. (1993). A map with 650 AFLPs, 165 RFLPs, and 25 RAPDs was developed by Keim et al. (1997). Liu et al. (2000) developed a map containing 100 RFLPs, 62 RAPDs, 42 AFLPs, 33 SSRs, and three other types of markers. Matthews et al. (2001) developed a map with 105 AFLPs, 39 RFLPs, 25 SSRs, 17 RAPDs, and four morphological markers. Yamanaka et al. (2001) developed a map with 401 RFLPs, 96 SSRs, and six other types of markers. Wu et al. (2001) constructed a map with 486 AFLPs, 196 RFLPs, 87 SSRs, 18 RAPDs, and five other types of markers. This map and the mapping population have been used in several QTL mapping studies in China (Wang et al. 2004b, c, d; Fu et al. 2007).

A detailed review of molecular markers, mapping populations, and computer software used in soybean linkage map constructions was provided by Drs. Sachiko Isobe and Satoshi Tabata in Chapter 4 of this book.

## 5.3 Target Traits in QTL Mapping Studies

Based on the data collected in the SoyBase database (Grant et al. 2008), there are 85 traits for which QTLs were identified (Table 5-1). Resistance to soybean cyst nematode (SCN) was the most studied trait followed by protein concentration, oil content, seed weight, plant height, and yield (Table 5-1). Thirty-one studies were carried out to identify QTLs for SCN resistance. The number of studies to identify QTLs for protein concentration, oil content, seed weight, plant height, and yield are 21, 20, 20, 19, and 17, respectively (Table 5-1). For 36 of the 85 traits listed in SoyBase, at least two QTL mapping studies were carried out to identify QTLs for each trait (Table 5-1).

## 5.4 QTLs Identified

According to SoyBase (Grant et al. 2008), over 1,100 QTLs in soybean have been identified (Table 5-1). For the most studied trait, SCN resistance, 99 QTLs were identified (Table 5-2). Some of the QTLs listed as separate QTLs in SoyBase appear to be the same QTL identified in the same population (e.g., SCN 29-1, SCN 29-4, and SCN 29-8 on linkage group G) while some other QTLs might be the same QTLs identified in different populations (see the "consensus QTL regions" section below). The amount of phenotypic variation accounted for by a single QTL varied from 1% to 97% (Table 5-2). The majority of the QTLs listed in SoyBase were not confirmed by separate studies. However, independent mapping studies with populations developed from different parents frequently identified QTLs for the same trait in a similar region on the integrated linkage map. The consistency of the QTL locations found in independent studies with different mapping populations indicates the existence of real QTLs in the concerned regions. For example, a QTL for SCN resistance was identified in the 0–37 cM region on linkage group G in 14 mapping populations. The major SCN resistance gene *rhg1* was found in this region (Concibido et al. 2004).

### 5.4.1 Consensus QTL Regions in Soybean

Using the integrated linkage map developed by Song et al. (2004) as a reference map, when QTLs identified in different mapping populations for the same trait are less than 10 cM from one another, the region containing these QTLs can be considered a consensus QTL region for the trait. These consensus QTL regions deserve further study to determine the true location of the QTL. Table 5-3 and Figure 5-2 summarize the consensus QTL regions for each trait based on the QTL data collected in SoyBase. It needs to be pointed out that when only a single marker that was associated with a QTL in a published study could be placed on the consensus map, an arbitrary 2 cM interval with 1 cM on either side of the marker was defined as the QTL region in SoyBase. Thus the true QTL position may be outside the arbitrary 2 cM region. For simplicity in summarizing the consensus QTL regions, the arbitrary 2 cM region was used in Table 5-3 and Figure 5-2, and in the text.

## 5.4.1.1 Consensus QTL Regions for Agronomic Traits

QTLs for yield were found in nine consensus genomic regions on seven linkage groups (LGs): C2, D2, I, J, K, L, and M (Table 5-3 and Fig. 5-2). The regions 39–47 cM on LG C2, 47–55 cM on LG D2, 11–22 cM on LG J, near 18–20 cM on LG M, and 35–40 cM on LG M were each found with two different mapping populations involving three or four mapping parents (the parents of the mapping populations) (Table 5-3). The 97–119 cM region on LG C2 was found with six mapping populations involving nine mapping parents (Table 5-3). The regions 31–37 cM on LG I, 36–50 cM on LG K, and 70–96 cM on LG L were each found with three or four different mapping populations involving five or six mapping parents (Table 5-3).

QTLs for maturity were found in eight consensus regions on six LGs: C1, C2, D1a, I, L, and M (Table 5-3 and Fig. 5-2). The regions 53–66 cM on LG C1, 37–49 cM on LG D1a, 31–36 cM on LG I, 54–68 cM on LG L, and near 18–20 cM on LG M were each found containing QTLs for maturity in two different mapping populations with three or four mapping parents (Table 5-3). The 111–125 cM region on LG C2 was found with five different mapping populations developed from eight mapping parents (Table 5-3). The region 88–96 cM on LG L and the region 32–40 cM on LG M were each found with three mapping populations developed from five or six mapping parents (Table 5-3).

QTLs for lodging were found in three consensus regions on two LGs: C2 and L (Table 5-3 and Fig. 5-2). All three regions, 107–116 cM on LG C2, 3–11 cM on LG L, and 68–101 cM on LG L were each found in three mapping populations developed from five or six mapping parents (Table 5-3).

QTLs for plant height were found in 10 consensus regions on eight LGs: C2, D1b, F, I, J, K, L, and M (Table 5-3 and Fig. 5-2). The 107–118 cM region on LG C2 was found in five mapping populations developed from nine mapping parents (Table 5-3). The regions 120–133 cM on LG D1b, 66–69 cM on LG F, 34–38 cM on LG I, 36–48 cM on LG K, 8–15 cM on LG L, and 34–44 cM on LG L were each found in two mapping populations developed from four mapping parents (Table 5-3). The regions 11–29 cM on LG J and 32–40 cM on LG M were each found in three mapping populations developed from six mapping parents (Table 5-3). The 68–114 cM region on LG L was found in six mapping populations developed from nine mapping parents (Table 5-3).

**Table 5-1** List of traits, the number of QTL mapping studies, and the number of QTLs identified for each trait in soybean based on the data collected in SoyBase.

| Trait name | No. of QTL mapping studies[a] | No. of QTLs[b] |
|---|---|---|
| Abnormal seedling | 1 | 3 |
| Acidic protein fraction | 1 | 4 |
| Alpha prime conglycinin protein fraction | 1 | 1 |
| Aluminum tolerance | 1 | 6 |
| Arabinose | 1 | 1 |
| Arabinose-galactose | 1 | 2 |
| Basic protein fraction | 1 | 2 |
| Beginning maturity | 2 | 5 |
| Beginning pod | 2 | 8 |
| Beta conglycinin protein fraction | 1 | 1 |
| Brown stem rot resistance | 6 | 17 |
| Canopy height | 2 | 3 |
| Canopy width | 1 | 3 |
| Carbon isotope discrimination | 1 | 5 |
| Cell wall polysaccharide | 1 | 1 |
| Chlorimuron ethyl sensitivity | 2 | 14 |
| Common cutworm resistance | 1 | 2 |
| Conglycinin protein fraction | 1 | 1 |
| Corn earworm resistance | 10 | 28 |
| Daidzein content | 1 | 2 |
| First flower | 10 | 32 |
| Flooding tolerance | 2 | 2 |
| Flowering time | 1 | 1 |
| Fructose content | 1 | 1 |
| Galactose content | 1 | 1 |
| Glycinin protein fraction | 1 | 1 |
| Glycitein content | 1 | 3 |
| Height/Lodging | 4 | 11 |
| Hypocotyl length | 1 | 3 |
| Iron efficiency | 12 | 36 |
| Javanese root-knot nematode resistance | 2 | 9 |
| Leaf area | 6 | 16 |
| Leaf ash | 1 | 11 |
| Leaf chlorosis | 1 | 2 |
| Leaf length | 5 | 15 |
| Leaf phosphorus content | 1 | 2 |
| Leaf width | 5 | 15 |
| Leaflet area | 1 | 6 |
| Leaflet shape | 5 | 8 |
| Linoleic acid content | 1 | 6 |
| Linolenic acid content | 1 | 7 |
| Lodging | 12 | 40 |
| Nitrogen accumulation at growth stage R5 | 1 | 17 |
| Oil content | 20 | 69 |
| Oil/protein ratio | 1 | 2 |
| Oleic acid concentration | 1 | 6 |
| Palmitic acid concentration | 2 | 5 |
| Peanut root-knot nematode resistance | 2 | |

| Trait name | No. of QTL mapping studies[a] | No. of QTLs[b] |
|---|---|---|
| Pectin concentration | 1 | 1 |
| Phomopsis seed decay | 2 | 2 |
| Photoperiod insensitivity | 2 | 3 |
| *Phytophthora sojae* partial resistance | 1 | 5 |
| Plant height | 19 | 82 |
| Pod dehiscence | 1 | 12 |
| Pod maturity date | 15 | 54 |
| Protein concentration | 21 | 79 |
| Reproductive period | 4 | 14 |
| Rhizoctonia rot and hypocotyl rot | 1 | 6 |
| Root necrosis | 1 | 4 |
| Salt tolerance | 1 | 1 |
| Sclerotinia stem rot | 6 | 91 |
| Seed abortion | 1 | 9 |
| Seed coat hardness | 1 | 7 |
| Seed filling period | 2 | 3 |
| Seed number | 2 | 3 |
| Seed set | 1 | 10 |
| Seed weight | 20 | 90 |
| Southern root-knot nematode resistance | 4 | 9 |
| Soybean cyst nematode resistance | 31 | 99 |
| Soybean looper resistance (229-M) | 1 | 1 |
| Specific leaf weight | 1 | 6 |
| Sprout yield | 1 | 4 |
| Stearic acid concentration | 1 | 1 |
| Stem diameter | 1 | 3 |
| Stem length | 1 | 1 |
| Sucrose concentration | 1 | 17 |
| Sudden death syndrome resistance | 8 | 33 |
| Tobacco budworm resistance (229-M) | 1 | 2 |
| Tobacco ringspot virus resistance | 1 | 1 |
| Trigonelline concentration (dry weight) | 1 | 2 |
| Trigonelline concentration (fresh weight) | 1 | 2 |
| Water use efficiency | 2 | 9 |
| Yield | 17 | 39 |
| Yield/Height | 4 | 16 |
| Yield/Seed weight | 3 | 7 |

[a] Number of QTL studies for each trait. Typically a QTL study represents a single segregating population. In some cases several different measures of a phenotypic trait were made in a single population and each measurement method was entered as a separate QTL study in SoyBase (Grant et al. 2008).

[b] Total number of QTLs for each trait in SoyBase (Grant et al. 2008). The count is not corrected for multiple reports of what appeared to be the same QTL identified in multiple QTL studies.

**Table 5-2** QTLs for soybean cyst nematode resistance listed in SoyBase by Grant et al. (2008).

| QTL | R² (%)[a] | LOD score | P-value | LG[b] | Start position[c] | End position[c] | Peak marker/ interval | Mapping parent 1 | Mapping parent 2 | Reference |
|---|---|---|---|---|---|---|---|---|---|---|
| SCN 2-3 | 1.0 | | 0.0008 | A1 | 7.8 | 9.8 | A487_1 | Hartwig | Williams 82 | Vierling et al. (1996) |
| SCN 18-1 | 7.4 | 2.78 | 0.0010 | A1 | 30.9 | 53.4 | A262_1, Satt300 | PI 438489B | Hamilton | Yue et al. (2001a) |
| SCN 1-1 | | | 0.0015 | A2 | 31.2 | 33.2 | A085_1 | M85-1430 | M83-15 | Concibido et al. (1994) |
| SCN 19-1 | 19.1 | 7.00 | 0.0010 | A2 | 45.6 | 53.1 | K400_2, T155_2 | PI 438489B | Hamilton | Yue et al. (2001a) |
| SCN 27-2 | 26.2 | 5.20 | | A2 | 47.6 | 49.6 | E(CCG)M(AAC)405 | Essex | Forrest | Meksem et al. (2001) |
| SCN 9-2 (1995) | 25.0 | | | A2 | 47.8 | 49.8 | I | Peking | Essex | Mahalingam et al. |
| SCN 3-1 | 9.0 | 5.80 | | A2 | 47.8 | 49.8 | I | PI 437654 | BSR101 | Webb et al. (1995) |
| SCN 29-5 | 17.7 | 14.50 | | A2 | 49.4 | 60.6 | Sat_400, Satt424 | Hamilton | PI 90763 | Guo et al. (2005) |
| SCN 8-5 | 23.2 | 5.10 | | A2 | 53.2 | 55.2 | BLT065_1 | Essex | Forrest | Chang et al. (1997) |
| SCN 13-2 | 40.0 | | 0.6400 | A2 | 53.2 | 55.2 | BLT065_1 | Flyer | Hartwig | Prabhu et al. (1999) |
| SCN 9-3 (1995) | 8.0 | | | A2 | 56.9 | 58.9 | S07a | Peking | Essex | Mahalingam et al. |
| SCN 30-3 | 29.0 | | 0.0005 | A2 | 59.6 | 61.6 | Satt424 | PI 437654 | Bell | Brucker et al. (2005) |
| SCN 8-4 | 15.1 | 2.80 | | A2 | 65.8 | 67.8 | OW15_400 | Essex | Forrest | Chang et al. (1997) |
| SCN 9-1 (1995) | 12.5 | | | A2 | 70.4 | 72.4 | A136_1 | Peking | Essex | Mahalingam et al. |
| SCN 26-1 | 9.5 | 3.71 | 0.0011 | B1 | 58.9 | 64.8 | A118_1, A006_1 | PI 89772 | Hamilton | Yue et al. (2001b) |
| SCN 2-1 | 91.0 | | 0.0001 | B1 | 63.8 | 65.8 | A006_1 | Hartwig | Williams 82 | Vierling et al. (1996) |
| SCN 23-1 | 16.6 | 6.83 | 0.0001 | B1 | 64.8 | 84.2 | A006_1, Satt583 | PI 89772 | Hamilton | Yue et al. (2001b) |
| SCN 24-1 | 6.8 | 2.78 | 0.0035 | B1 | 64.8 | 84.2 | A006_1, Satt583 | PI 89772 | Hamilton | Yue et al. (2001b) |
| SCN 17-1 | 12.7 | 4.20 | 0.0010 | B1 | 84.2 | 101.0 | Satt583, Sat_123 | PI 438489B | Hamilton | Yue et al. (2001a) |
| SCN 18-2 | 7.4 | 2.79 | 0.0010 | B1 | 84.2 | 101.0 | Satt583, Sat_123 | PI 438489B | Hamilton | Yue et al. (2001a) |
| SCN 20-1 | 11.0 | 2.70 | 0.0010 | B1 | 84.2 | 101.0 | Satt583, Sat_123 | PI 438489B | Hamilton | Yue et al. (2001a) |
| SCN 29-10 | 11.2 | 6.00 | | B1 | 102.6 | 124.0 | Satt359, Satt453 | Hamilton | PI 90763 | Guo et al. (2005) |
| SCN 2-2 | 1.0 | | 0.0001 | B1 | 125.0 | 127.0 | A567_1 | Hartwig | Williams 82 | Vierling et al. (1996) |
| SCN 17-2 | 11.7 | 2.75 | 0.0010 | B2 | 55.2 | 62.7 | A329_1, Satt168 | PI 438489B | Hamilton | Yue et al. (2001a) |

| | | | | | | | | | |
|---|---|---|---|---|---|---|---|---|---|
| SCN 19-2 | 8.1 | 2.56 | 0.0010 | B2 | 55.2 | 62.7 | A329_1, Satt168 | PI 438489B | Hamilton | Yue et al. (2001a) |
| SCN 10-1 | 21.0 | | | B2 | 97.5 | 99.5 | A593_1 | Peking | Essex | Qiu et al. (1999) |
| SCN 10-3 | 15.0 | | | B2 | 117.7 | 119.7 | T005_1 | Peking | Essex | Qiu et al. (1999) |
| SCN 11-3 | 9.0 | | | B2 | 117.7 | 119.7 | T005_1 | Peking | Essex | Qiu et al. (1999) |
| SCN 21-1 | 11.1 | 3.61 | 0.0010 | C1 | 18.6 | 21.0 | A059_1, A463_1 | PI 438489B | Hamilton | Yue et al. (2001a) |
| SCN 18-3 | 10.2 | 2.56 | 0.0010 | C1 | 21.0 | 24.1 | A463_1, Satt396 | PI 438489B | Hamilton | Yue et al. (2001a) |
| SCN 22-1 | 5.0 | 4.40 | | C2 | 1.0 | 1.0 | A121_1 | PI 46916 | A81356022 | Wang et al. (2001) |
| SCN 9-6 (1995) | 8.0 | | | C2 | 94.6 | 96.6 | A635_1 | Peking | Essex | Mahalingam et al. |
| SCN 17-3 | 7.1 | 6.80 | 0.0010 | C2 | 126.2 | 145.5 | Satt202, Satt371 | PI 438489B | Hamilton | Yue et al. (2001a) |
| SCN 20-2 | 8.3 | 3.05 | 0.0010 | C2 | 126.2 | 145.5 | Satt202, Satt371 | PI 438489B | Hamilton | Yue et al. (2001a) |
| SCN 19-3 | 10.7 | 5.47 | 0.0010 | D1a | 6.4 | 34.9 | A398_1, K478_1 | PI 438489B | Hamilton | Yue et al. (2001a) |
| SCN 20-3 | 9.4 | 4.17 | 0.0010 | D1a | 6.4 | 34.9 | A398_1, K478_1 | PI 438489B | Hamilton | Yue et al. (2001a) |
| SCN 21-2 | 7.4 | 4.14 | 0.0010 | D1a | 6.4 | 34.9 | A398_1, K478_1 | PI 438489B | Hamilton | Yue et al. (2001a) |
| SCN 26-2 | 7.8 | 3.30 | 0.0015 | D1a | 43.8 | 48.1 | Satt342, Satt368 | PI 89772 | Hamilton | Yue et al. (2001b) |
| SCN 23-2 | 9.7 | 4.59 | 0.0014 | D2 | 15.0 | 39.4 | B132_4, Satt372 | PI 89772 | Hamilton | Yue et al. (2001b) |
| SCN 16-1 | 41.0 | | 0.0010 | D2 | 86.3 | 88.3 | Satt082 | Hartwig | BR92-31983 | Schuster et al. (2001) |
| SCN 12-2 | 9.0 | | | E | 16.1 | 18.1 | A963_1 | Peking | Essex | Qiu et al. (1999) |
| SCN 22-3 | 23.0 | 3.50 | | E | 33.2 | 35.2 | Satt598 | PI 46916 | A81356022 | Wang et al. (2001) |
| SCN 29-9 | 12.5 | 7.20 | | E | 35.8 | 43.1 | Satt573, Satt204 | Hamilton | PI 90763 | Guo et al. (2005) |
| SCN 18-4 | 8.0 | 2.57 | 0.0010 | E | 37.3 | 45.1 | A656_1, Satt452 | PI 438489B | Hamilton | Yue et al. (2001a) |
| SCN 21-3 | 18.7 | 5.01 | 0.0010 | E | 37.3 | 45.1 | A656_1, Satt452 | PI 438489B | Hamilton | Yue et al. (2001a) |
| SCN 25-1 | 15.7 | 3.56 | 0.0053 | E | 51.0 | 70.2 | A135_3, Satt231 | PI 89772 | Hamilton | Yue et al. (2001b) |
| SCN 29-1 | 14.7 | 7.90 | | G | 0.0 | 12.5 | Satt163, Satt688 | Hamilton | PI 90763 | Guo et al. (2005) |
| SCN 29-4 | 28.1 | 22.10 | | G | 0.0 | 12.5 | Satt163, Satt688 | Hamilton | PI 90763 | Guo et al. (2005) |
| SCN 29-8 | 13.0 | 7.10 | | G | 0.0 | 12.5 | Satt163, Satt688 | Hamilton | PI 90763 | Guo et al. (2005) |
| SCN 4-1 | 26.2 | | 0.0001 | G | 0.8 | 2.8 | C006_1 | Evans | Peking | Concibido et al. (1997) |
| SCN 5-1 | 44.8 | | 0.0001 | G | 0.8 | 2.8 | C006_1 | Evans | PI 90763 | Concibido et al. (1997) |
| SCN 6-1 | 36.3 | | 0.0001 | G | 0.8 | 2.8 | C006_1 | Evans | PI 88788 | Concibido et al. (1997) |

*Table 5-2 contd....*

*Table 5-2 contd....*

| QTL | $R^2$ (%)[a] | LOD score | P-value | LG[b] | Start position[c] | End position[c] | Peak marker/ interval | Mapping parent 1 | Mapping parent 2 | Reference |
|---|---|---|---|---|---|---|---|---|---|---|
| SCN 13-1 | 6.4 | | 0.0760 | G | 0.8 | 2.8 | Satt038 | Flyer | Hartwig | Prabhu et al. (1999) |
| SCN 8-3 | 12.9 | 2.70 | | G | 1.0 | 3.0 | OI03_450 | Essex | Forrest | Chang et al. (1997) |
| SCN 4-4 | 28.1 | | | G | 1.8 | 8.6 | C006_1, Bng122_1 | Evans | Peking | Concibido et al. (1996) |
| SCN 5-3 | 52.7 | | | G | 1.8 | 8.6 | C006_1, Bng122_1 | Evans | PI 90763 | Concibido et al. (1996) |
| SCN 6-2 | 40.0 | | | G | 1.8 | 8.6 | C006_1, Bng122_1 | Evans | PI 88788 | Concibido et al. (1996) |
| SCN 7-1 | 51.4 | | | G | 1.8 | 8.6 | C006_1, Bng122_1 | Evans | PI 209332 | Concibido et al. (1996) |
| SCN 30-1 | 14.0 | | 0.0400 | G | 3.5 | 5.5 | Satt309 | PI 437654 | Bell | Brucker et al. (2005) |
| SCN 30-2 | 32.0 | | 0.0001 | G | 3.5 | 5.5 | Satt309 | PI 437654 | Bell | Brucker et al. (2005) |
| SCN 14-2 | 97.0 | | 0.0001 | G | 3.5 | 5.5 | Satt309 | Essex | Forrest | Meksem et al. (1999) |
| SCN 23-3 | 26.6 | 13.67 | 0.0001 | G | 4.5 | 5.8 | B053_1, Satt309 | PI 89772 | Hamilton | Yue et al. (2001b) |
| SCN 24-2 | 4.6 | 2.53 | 0.0095 | G | 4.5 | 5.8 | B053_1, Satt309 | PI 89772 | Hamilton | Yue et al. (2001b) |
| SCN 25-2 | 23.0 | 12.65 | 0.0001 | G | 4.5 | 5.8 | B053_1, Satt309 | PI 89772 | Hamilton | Yue et al. (2001b) |
| SCN 26-3 | 10.0 | 5.02 | 0.0001 | G | 4.5 | 5.8 | B053_1, Satt309 | PI 89772 | Hamilton | Yue et al. (2001b) |
| SCN 28-1 | 87.0 | 40.60 | | G | 4.5 | 8.6 | Satt309, Bng122 | Bell | Colfax | Glover et al. (2004) |
| SCN 28-3 | 64.0 | 17.70 | | G | 4.5 | 8.6 | Satt309, Bng122 | Bell | Colfax | Glover et al. (2004) |
| SCN 27-1 | 24.1 | 5.10 | | G | 4.8 | 6.8 | E(ATG)M(CGA)87 | Essex | Forrest | Meksem et al. (2001) |
| SCN 15-1 (2001) | | | | G | 5.8 | 35.5 | B053_1, A112_1 | PI 88287 | PI 89008 | Vaghchhipawala et al. |
| SCN 8-2 | 11.3 | 2.50 | | G | 7.6 | 9.6 | Bng122_1 | Essex | Forrest | Chang et al. (1997) |
| SCN 14-1 | 19.0 | | 0.0730 | G | 7.6 | 9.6 | Bng122_1 | Essex | Forrest | Meksem et al. (1999) |
| SCN 8-1 | 4.2 | 1.20 | | G | 10.6 | 12.6 | OG13_490 | Essex | Forrest | Chang et al. (1997) |
| SCN 3-2 | 22.0 | 15.40 | | G | 11.0 | 13.0 | PHP05354a, PHP05219a | PI 437654 | BSR101 | Webb et al. (1995) |
| SCN 1-3 | 36.0 | | 0.0001 | G | 23.1 | 25.1 | K069_1 | M85-1430 | M83-15 | Concibido et al. (1994) |
| SCN 17-4 | 15.8 | 9.08 | 0.0010 | G | 23.1 | 54.7 | A096_3, Satt130 | PI 438489B | Hamilton | Yue et al. (2001a) |
| SCN 18-5 | 12.8 | 7.52 | 0.0010 | G | 23.1 | 54.7 | A096_3, Satt130 | PI 438489B | Hamilton | Yue et al. (2001a) |
| SCN 19-4 | 13.6 | 4.46 | 0.0010 | G | 23.1 | 66.6 | Satt012, Satt130 | PI 438489B | Hamilton | Yue et al. (2001a) |

| QTL | R²[a] | LOD | P | LG[b] | Start | End | Marker(s) | Parent 1 | Parent 2 | Reference |
|---|---|---|---|---|---|---|---|---|---|---|
| SCN 2-4 | 1.0 | | 0.0018 | G | 34.5 | 36.5 | A112_1 | Hartwig | Williams 82 | Vierling et al. (1996) |
| SCN 20-4 | 5.8 | 2.03 | 0.0010 | G | 62.2 | 66.6 | Satt012, Satt199 | PI 438489B | Hamilton | Yue et al. (2001a) |
| SCN 22-2 | 27.0 | 4.80 | | G | 89.0 | 91.0 | A245_2 | PI 468916 | A81356022 | Wang et al. (2001) |
| SCN 29-3 | 6.7 | 3.00 | | G | 102.6 | 124.0 | Satt453, Satt359 | Hamilton | PI 90763 | Guo et al. (2005) |
| SCN 4-2 | 17.6 | | 0.0002 | G | 108.5 | 110.5 | A378_1 | Evans | Peking | Concibido et al. (1997) |
| SCN 10-5 | 12.0 | | | H | 120.3 | 122.3 | K014_1 | Peking | Essex | Qiu et al. (1999) |
| SCN 11-2 | 9.0 | | | H | 120.3 | 122.3 | K014_1 | Peking | Essex | Qiu et al. (1999) |
| SCN 10-4 | 13.0 | | | H | 123.1 | 125.1 | B072_1 | Peking | Essex | Qiu et al. (1999) |
| SCN 11-1 | 13.0 | | | H | 123.1 | 125.1 | B072_1 | Peking | Essex | Qiu et al. (1999) |
| SCN 12-1 | 11.0 | | | I | 37.1 | 39.1 | K011_1 | Peking | Essex | Qiu et al. (1999) |
| SCN 28-2 | 2.0 | 2.50 | | J | 65.0 | 67.8 | Satt244, Satt547 | Bell | Colfax | Glover et al. (2004) |
| SCN 28-4 | 7.0 | 3.40 | | J | 65.0 | 78.6 | Satt244, Satt431 | Bell | Colfax | Glover et al. (2004) |
| SCN 29-2 | 7.8 | 4.60 | | J | 67.8 | 75.1 | Satt547, Sat_224 | Hamilton | PI 90763 | Guo et al. (2005) |
| SCN 29-6 | 4.2 | 13.90 | | J | 67.8 | 75.1 | Satt547, Sat_224 | Hamilton | PI 90763 | Guo et al. (2005) |
| SCN 1-2 | 18.8 | | 0.0001 | J | 73.0 | 75.0 | B032_1 | M85-1430 | M83-15 | Concibido et al. (1994) |
| SCN 5-2 | | | 0.0001 | J | 73.0 | 75.0 | B032_1 | Evans | PI 90763 | Concibido et al. (1997) |
| SCN 29-7 | 4.0 | 3.00 | | L | 87.4 | 93.9 | Sat_286, Satt229 | Hamilton | PI 90763 | Guo et al. (2005) |
| SCN 3-3 | 7.0 | 4.80 | | M | 74.0 | 76.0 | PHP02275a, PHP02301a | PI 437654 | BSR101 | Webb et al. (1995) |
| SCN 4-3 | 14.3 | | 0.0001 | N | 33.9 | 35.9 | A280_1 | Evans | Peking | Concibido et al. (1997) |
| SCN 10-2 | 16.0 | | | | | | A018_3 | Peking | Essex | Qiu et al. (1999) |
| SCN 9-4 | 6.0 | | | | | | E01c | Peking | Essex | Mahalingam et al. (1995) |
| SCN 9-5 | 6.0 | | | | | | G15d | Peking | Essex | Mahalingam et al. (1995) |

[a]R² = Phenotypic variance explained by a QTL.

[b]LG = linkage group. The linkage group names are from the integrated map by Song et al. (2004).

[c]The start positions and end positions are from the integrated map by Song et al. (2004). When only a single marker that was associated with a QTL in a published study could be placed on the consensus map, an arbitrary 2 cM interval with 1 cM on either side of the marker was defined as the QTL region in SoyBase.

**Table 5-3** Soybean QTLs that were less than 10 cM apart but were identified in different populations for the same trait.

| Trait/QTL | LG[a] | Start position[b] | End position[b] | Mapping parent 1 | Mapping parent 2 | Reference |
|---|---|---|---|---|---|---|
| *Yield* | | | | | | |
| Sd yld 11-1 | C2 | 39 | 41 | Minsoy | Noir 1 | Specht et al. (2001) |
| Yld/SW 2-2 | C2 | 45 | 47 | Archer | Noir 1 | Orf et al. (1999b) |
| Sd yld 15-1 | C2 | 97 | 99 | BSR 101 | LG82-8379 | Kabelka et al. (2004) |
| Sd yld 5-1 | C2 | 107 | 109 | Archer | Noir 1 | Orf et al. (1999b) |
| Yld/Ht 2-1 | C2 | 107 | 109 | Archer | Minsoy | Orf et al. (1999b) |
| Sd yld 16-3 | C2 | 112 | 114 | IA2008 | PI 468916 | Wang et al. (2004a) |
| Sd yld 3-2 | C2 | 117 | 119 | PI 27890 | PI 290136 | Mansur et al. (1996) |
| Yld/Ht 4-2 | C2 | 117 | 119 | Minsoy | Noir 1 | Orf et al. (1999b) |
| Sd yld 5-2 | D2 | 47 | 49 | Archer | Noir 1 | Orf et al. (1999b) |
| Yld/Ht 2-4 | D2 | 53 | 55 | Archer | Minsoy | Orf et al. (1999b) |
| Sd yld 10-1 | I | 31 | 33 | Parker | PI 468916 | Yuan et al. (2002) |
| Sd yld 9-1 | I | 34 | 36 | A81356022 | PI 468916 | Yuan et al. (2002) |
| Sd yld 14-1 | I | 36 | 37 | A3733 | PI 437088A | Chung et al. (2003) |
| Yld/Ht 4-1 | J | 11 | 13 | Minsoy | Noir 1 | Orf et al. (1999b) |
| Yld/Ht 1-3 | J | 20 | 22 | PI 27890 | PI 290136 | Mansur et al. (1996) |
| Sd yld 16-1 | K | 36 | 38 | IA2008 | PI 468916 | Wang et al. (2004a) |
| Sd yld 12-1 | K | 46 | 47 | Essex | Forrest | Yuan et al. (2002) |
| Sd yld 13-1 | K | 47 | 50 | Flyer | Hartwig | Yuan et al. (2002) |
| Sd yld 8-1 | L | 70 | 72 | Archer | Minsoy | Orf et al. (1999a) |
| Sd yld 11-6 | L | 88 | 90 | Minsoy | Noir 1 | Specht et al. (2001) |
| Yld/Ht 1-1 | L | 91 | 93 | PI 27890 | PI 290136 | Mansur et al. (1996) |
| Yld/Ht 3-1 | L | 94 | 96 | Archer | Noir 1 | Orf et al. (1999b) |
| Sd yld 6-1 | M | 18 | 20 | Minsoy | Noir 1 | Orf et al. (1999b) |
| Yld/Ht 2-2 | M | 18 | 20 | Archer | Minsoy | Orf et al. (1999b) |
| Yld/SW 1-1 | M | 35 | 37 | Archer | Minsoy | Orf et al. (1999b) |
| Sd yld 3-1 | M | 38 | 40 | PI 27890 | PI 290136 | Mansur et al. (1996) |
| *Maturity* | | | | | | |
| Pod mat 1-1 | C1 | 53 | 55 | A81356022 | PI 468916 | Keim et al. (1990a) |
| Pod mat 8-5 | C1 | 64 | 66 | Archer | Minsoy | Orf et al. (1999b) |
| Pod mat 8-1 | C2 | 111 | 113 | Archer | Minsoy | Orf et al. (1999b) |
| Pod mat 13-4 | C2 | 111 | 113 | Minsoy | Noir 1 | Specht et al. (2001) |
| Pod mat 14-3 | C2 | 112 | 114 | IA2008 | PI 468916 | Wang et al. (2004a) |
| Pod mat 4-1 | C2 | 117 | 119 | PI 27890 | PI 290136 | Mansur et al. (1996) |
| Pod mat 1-5 | C2 | 123 | 125 | A81356022 | PI 468916 | Keim et al. (1990a) |
| Pod mat 1-2 | D1a | 37 | 39 | A81356022 | PI 468916 | Keim et al. (1990a) |
| Pod mat 13-2 | D1a | 47 | 49 | Minsoy | Noir 1 | Specht et al. (2001) |
| Pod mat 12-1 | I | 31 | 33 | Parker | PI 468916 | Yuan et al. (2002) |
| Pod mat 11-1 | I | 34 | 36 | A81356022 | PI 468916 | Yuan et al. (2002) |
| Pod mat 14-1 | L | 54 | 56 | IA2008 | PI 468916 | Wang et al. (2004a) |
| Pod mat 8-4 | L | 66 | 68 | Archer | Minsoy | Orf et al. (1999b) |
| Pod mat 13-6 | L | 88 | 90 | Minsoy | Noir 1 | Specht et al. (2001) |
| Pod mat 4-3 | L | 91 | 93 | PI 27890 | PI 290136 | Mansur et al. (1996) |
| Pod mat 9-2 | L | 94 | 96 | Archer | Noir 1 | Orf et al. (1999b) |
| Pod mat 8-2 | M | 18 | 20 | Archer | Minsoy | Orf et al. (1999b) |
| Pod mat 13-7 | M | 18 | 20 | Minsoy | Noir 1 | Specht et al. (2001) |
| Pod mat 14-4 | M | 32 | 34 | IA2008 | PI 468916 | Wang et al. (2004a) |

| Trait/QTL | LG[a] | Start position[b] | End position[b] | Mapping parent 1 | Mapping parent 2 | Reference |
|---|---|---|---|---|---|---|
| Pod mat 10-2 | M | 33 | 35 | Minsoy | Noir 1 | Orf et al. (1999b) |
| Pod mat 7-1 | M | 38 | 40 | PI 27890 | PI 290136 | Lark et al. (1994) |
| *Lodging* | | | | | | |
| Ldge 6-1 | C2 | 107 | 109 | Archer | Minsoy | Orf et al. (1999b) |
| Ldge 9-1 | C2 | 111 | 113 | Minsoy | Noir 1 | Specht et al. (2001) |
| Ldge 3-2 | C2 | 114 | 116 | PI 27890 | PI 290136 | Orf et al. (1999b) |
| Ldge 5-11 | L | 3 | 5 | PI 416937 | Young | Lee et al. (1996a) |
| Ldge 3-3 | L | 8 | 10 | PI 27890 | PI 290136 | Mansur et al. (1996) |
| Ldge 9-3 | L | 9 | 11 | Minsoy | Noir 1 | Specht et al. (2001) |
| Ldge 1-1 | L | 68 | 87 | PI 27890 | PI 290136 | Mansur et al. (1996) |
| Ldge 8-4 | L | 88 | 90 | Minsoy | Noir 1 | Orf et al. (1999b) |
| Ldge 4-2 | L | 88 | 90 | Coker237 | PI 97100 | Lee et al. (1996c) |
| Ldge 9-5 | L | 88 | 90 | Minsoy | Noir 1 | Specht et al. (2001) |
| Ldge 4-3 | L | 89 | 101 | Coker237 | PI 97100 | Lee et al. (1996c) |
| Ldge 3-1 | L | 91 | 93 | PI 27890 | PI 290136 | Mansur et al. (1996) |
| *Plant height* | | | | | | |
| Pl ht 8-1 | C2 | 107 | 109 | Archer | Minsoy | Orf et al. (1999b) |
| Pl ht 18-4 | C2 | 112 | 114 | IA2008 | PI 468916 | Wang et al. (2004a) |
| Pl ht 13-2 | C2 | 112 | 114 | Minsoy | Noir 1 | Specht et al. (2001) |
| Pl ht 11-1 | C2 | 116 | 118 | S100 | Tokyo | Mian et al. (1998) |
| Pl ht 6-3 | C2 | 116 | 118 | PI 27890 | PI 290136 | Lark et al. (1995) |
| Pl ht 6-12 | D1b | 120 | 122 | PI 27890 | PI 290136 | Lark et al. (1995) |
| Pl ht 5-5 | D1b | 131 | 133 | PI 416937 | Young | Lee et al. (1996a) |
| Pl ht 5-8 | F | 66 | 68 | PI 416937 | Young | Lee et al. (1996a) |
| Pl ht 11-3 | F | 67 | 69 | S100 | Tokyo | Mian et al. (1998) |
| Pl ht 12-1 | I | 34 | 36 | A81356022 | PI 468916 | Yuan et al. (2002) |
| Pl ht 16-1 | I | 36 | 38 | Essex | Williams | Chapman et al. (2003) |
| Pl ht 13-5 | J | 11 | 13 | Minsoy | Noir 1 | Specht et al. (2001) |
| Pl ht 6-6 | J | 20 | 22 | PI 27890 | PI 290136 | Lark et al. (1995) |
| Pl ht 5-9 | J | 27 | 29 | PI 416937 | Young | Lee et al. (1996a) |
| Pl ht 18-3 | K | 36 | 38 | IA2008 | PI 468916 | Wang et al. (2004a) |
| Pl ht 15-1 | K | 46 | 48 | Flyer | Hartwig | Yuan et al. (2002) |
| Pl ht 6-7 | L | 8 | 10 | PI 27890 | PI 290136 | Lark et al. (1995) |
| Pl ht 13-7 | L | 13 | 15 | Minsoy | Noir 1 | Specht et al. (2001) |
| Pl ht 5-12 | L | 34 | 36 | PI 416937 | Young | Lee et al. (1996a) |
| Pl ht 6-4 | L | 42 | 44 | PI 27890 | PI 290136 | Lark et al. (1995) |
| Pl ht 8-4 | L | 66 | 68 | Archer | Minsoy | Orf et al. (1999b) |
| Pl ht 1-1 | L | 68 | 87 | PI 27890 | PI 290136 | Mansur et al. (1993) |
| Pl ht 8-3 | L | 69 | 71 | Archer | Minsoy | Orf et al. (1999b) |
| Pl ht 6-1 | L | 86 | 88 | PI 27890 | PI 290136 | Lark et al. (1995) |
| Pl ht 4-2 | L | 88 | 90 | Coker237 | PI 97100 | Lee et al. (1996c) |
| Pl ht 13-8 | L | 88 | 90 | Minsoy | Noir 1 | Specht et al. (2001) |
| Pl ht 4-4 | L | 89 | 101 | Coker237 | PI 97100 | Lee et al. (1996c) |
| Pl ht 3-1 | L | 91 | 93 | PI 27890 | PI 290136 | Mansur et al. (1996) |
| Pl ht 5-10 | L | 100 | 102 | PI 416937 | Young | Lee et al. (1996a) |
| Pl ht 9-2 | L | 106 | 108 | Archer | Noir 1 | Orf et al. (1999b) |
| Pl ht 6-2 | L | 112 | 114 | PI 27890 | PI 290136 | Lark et al. (1995) |

*Table 5-3 contd....*

*Table 5-3 contd.*

| Trait/QTL | LG[a] | Start position[b] | End position[b] | Mapping parent 1 | Mapping parent 2 | Reference |
|---|---|---|---|---|---|---|
| Pl ht 18-6 | M | 32 | 34 | IA2008 | PI 468916 | Wang et al. (2004a) |
| Pl ht 13-9 | M | 33 | 35 | Minsoy | Noir 1 | Specht et al. (2001) |
| Pl ht 6-5 | M | 38 | 40 | PI 27890 | PI 290136 | Lark et al. (1995) |
| *Protein content* | | | | | | |
| Prot 2-1 | A1 | 93 | 95 | PI 27890 | PI 290136 | Mansur et al. (1996) |
| Prot 12-1 | A1 | 94 | 96 | Minsoy | Noir 1 | Specht et al. (2001) |
| Prot 21-1 | A2 | 145 | 147 | BSR 101 | LG82-8379 | Kabelka et al. (2004) |
| Prot 14-1 | A2 | 149 | 151 | M91-212006 | SZG9652 | Vollmann et al. (2002) |
| Prot 3-2 | B1 | 28 | 30 | A87296011 | C1763 | Brummer et al. (1997) |
| Prot 16-1 | B1 | 35 | 37 | Essex | Williams | Chapman et al. (2003) |
| Prot 4-11 | B2 | 28 | 30 | PI 416937 | Young | Lee et al. (1996b) |
| Prot 1-6 | B2 | 32 | 34 | A81356022 | PI 468916 | Diers et al. (1992) |
| Prot 4-10 | B2 | 43 | 46 | PI 416937 | Young | Lee et al. (1996b) |
| Prot 9-2 | C1 | 9 | 11 | Minsoy | Noir 1 | Orf et al. (1999b) |
| Prot 4-4 | C1 | 20 | 22 | PI 416937 | Young | Lee et al. (1996b) |
| Prot 12-2 | C1 | 32 | 34 | Minsoy | Noir 1 | Specht et al. (2001) |
| Prot 3-3 | C1 | 90 | 92 | A87296011 | C1763 | Brummer et al. (1997) |
| Prot 4-3 | C1 | 96 | 98 | PI 416937 | Young | Lee et al. (1996b) |
| Prot 21-2 | C1 | 123 | 125 | BSR 101 | LG82-8379 | Kabelka et al. (2004) |
| Prot 4-2 | C1 | 126 | 128 | PI 416937 | Young | Lee et al. (1996b) |
| Prot 17-1 | C2 | 117 | 119 | Essex | Williams | Hyten et al. (2004) |
| Prot 13-2 | C2 | 121 | 123 | Ma.Belle | Proto | Csanadi et al. (2001) |
| Prot 4-6 | E | 26 | 28 | PI 416937 | Young | Lee et al. (1996b) |
| Prot 3-6 | E | 30 | 32 | A87296011 | C1763 | Brummer et al. (1997) |
| Prot 18-1 | E | 30 | 32 | Coker 237 | PI 97100 | Fasoula et al. (2004) |
| Prot 1-8 | G | 89 | 91 | A81356022 | PI 468916 | Diers et al. (1992) |
| Prot 3-10 | G | 96 | 98 | A87296011 | C1763 | Brummer et al. (1997) |
| Prot 11-1 | I | 31 | 33 | Parker | PI 468916 | Yuan et al. (2002) |
| Prot 1-3 | I | 31 | 33 | A81356022 | PI 468916 | Diers et al. (1992) |
| Prot 3-12 | I | 31 | 33 | A87296011 | C1763 | Brummer et al. (1997) |
| Prot 10-1 | I | 34 | 36 | A81356022 | PI 468916 | Yuan et al. (2002) |
| Prot 15-1 | I | 36 | 37 | A3733 | PI 437088A | Chung et al. (2003) |
| Prot 1-2 | I | 38 | 40 | A81356022 | PI 468916 | Lark et al. (1994) |
| Prot 5-4 | K | 31 | 33 | Coker237 | PI 97100 | Lee et al. (1996b) |
| Prot 12-3 | K | 40 | 42 | Minsoy | Noir 1 | Specht et al. (2001) |
| Prot 12-4 | M | 33 | 35 | Minsoy | Noir 1 | Specht et al. (2001) |
| Prot 13-3 | M | 33 | 35 | Ma.Belle | Proto | Csanadi et al. (2001) |
| Prot 7-1 | M | 38 | 40 | Archer | Minsoy | Orf et al. (1999b) |
| *Oil content* | | | | | | |
| Oil 8-1 | A1 | 88 | 90 | Archer | Minsoy | Orf et al. (1999b) |
| Oil 4-3 | A1 | 91 | 93 | A87296011 | C1763 | Brummer et al. (1997) |
| Oil 3-2 | A1 | 93 | 95 | PI 27890 | PI 290136 | Mansur et al. (1996) |

| Trait/QTL | LG[a] | Start position[b] | End position[b] | Mapping parent 1 | Mapping parent 2 | Reference |
|---|---|---|---|---|---|---|
| Oil 13-1 | A1 | 94 | 96 | Minsoy | Noir 1 | Specht et al. (2001) |
| Oil 9-1 | C1 | 9 | 11 | Archer | Noir 1 | Orf et al. (1999b) |
| Oil 8-2 | C1 | 9 | 11 | Archer | Minsoy | Orf et al. (1999b) |
| Oil 5-1 | E | 23 | 25 | PI 416937 | Young | Lee et al. (1996b) |
| Oil 2-9 | E | 34 | 36 | A81356022 | PI 468916 | Diers et al. (1992) |
| Oil 17-2 | H | 86 | 88 | Coker 237 | PI 97100 | Fasoula et al. (2004) |
| Oil 19-2 | H | 89 | 91 | N87-984-16 | TN93-99 | Panthee et al. (2005) |
| Oil 14-3 | I | 22 | 24 | Ma.Belle | Proto | Csanadi et al. (2001) |
| Oil 12-1 | I | 31 | 33 | Parker | PI 468916 | Yuan et al. (2002) |
| Oil 11-1 | I | 31 | 33 | A81356022 | PI 468916 | Yuan et al. (2002) |
| Oil 13-4 | I | 34 | 36 | Minsoy | Noir 1 | Specht et al. (2001) |
| Oil 15-1 | I | 36 | 37 | A3733 | PI 437088A | Chung et al. (2003) |
| Oil 2-2 | I | 38 | 40 | A81356022 | PI 468916 | Diers et al. (1992) |
| Oil 4-11 | K | 98 | 100 | A87296011 | C1763 | Brummer et al. (1997) |
| Oil 14-2 | K | 104 | 106 | Ma.Belle | Proto | Csanadi et al. (2001) |
| Oil 18-2 | L | 34 | 36 | PI 416937 | Young | Fasoula et al. (2004) |
| Oil 2-7 | L | 36 | 38 | A81356022 | PI 468916 | Diers et al. (1992) |
| Oil 5-3 | L | 36 | 38 | PI 416937 | Young | Lee et al. (1996b) |
| Oil 3-1 | L | 91 | 93 | PI 27890 | PI 290136 | Mansur et al. (1996) |
| Oil 16-1 | L | 93 | 95 | Essex | Williams | Hyten et al. (2004) |
| Oil 9-3 | L | 94 | 96 | Archer | Noir 1 | Orf et al. (1999b) |
| Oil 16-2 | M | 35 | 37 | Essex | Williams | Hyten et al. (2004) |
| Oil/Prot 1-2 | M | 38 | 40 | PI 27890 | PI 290136 | Lark et al. (1994) |
| *Soybean cyst nematode resistance* | | | | | | |
| SCN 19-1 | A2 | 46 | 53 | Hamilton | PI 438489B | Yue et al. (2001a) |
| SCN 3-1 | A2 | 48 | 50 | BSR101 | PI 437654 | Webb et al. (1995) |
| SCN 9-2 | A2 | 48 | 50 | Essex | Peking | Mahalingam et al. (1995) |
| SCN 29-4 | A2 | 49 | 61 | Hamilton | PI 90763 | Guo et al. (2005) |
| SCN 8-5 | A2 | 53 | 55 | Essex | Forrest | Chang et al. (1997) |
| SCN 13-2 | A2 | 53 | 55 | Flyer | Hartwig | Prabhu et al. (1999) |
| SCN 30-3 | A2 | 60 | 62 | Bell | PI 437654 | Brucker et al. (2005) |
| SCN 2-1 | B1 | 64 | 66 | Hartwig | Williams 82 | Vierling et al. (1996) |
| SCN 24-1 | B1 | 65 | 84 | Hamilton | PI 89772 | Yue et al. (2001b) |
| SCN 20-1 | B1 | 84 | 101 | Hamilton | PI 438489B | Yue et al. (2001a) |
| SCN 29-7 | B1 | 103 | 124 | Hamilton | PI 90763 | Guo et al. (2005) |
| SCN 22-3 | E | 33 | 35 | A81356022 | PI 468916 | Wang et al. (2001) |
| SCN 29-6 | E | 36 | 43 | Hamilton | PI 90763 | Guo et al. (2005) |
| SCN 21-3 | E | 37 | 45 | Hamilton | PI 438489B | Yue et al. (2001a) |
| SCN 25-1 | E | 51 | 70 | Hamilton | PI 89772 | Yue et al. (2001b) |
| SCN 29-5 | G | 0 | 13 | Hamilton | PI 90763 | Guo et al. (2005) |
| SCN 4-1 | G | 1 | 3 | Evans | Peking | Concibido et al. (1997) |
| SCN 5-1 | G | 1 | 3 | Evans | PI 90763 | Concibido et al. (1997) |
| SCN 6-1 | G | 1 | 3 | Evans | PI 88788 | Concibido et al. (1997) |
| SCN 13-1 | G | 1 | 3 | Flyer | Hartwig | Prabhu et al. (1999) |
| SCN 8-3 | G | 1 | 3 | Essex | Forrest | Chang et al. (1997) |

*Table 5-3 contd....*

*Table 5-3 contd.*

| Trait/QTL | LG[a] | Start position[b] | End position[b] | Mapping parent 1 | Mapping parent 2 | Reference |
|---|---|---|---|---|---|---|
| SCN 7-1 | G | 2 | 9 | Evans | PI 209332 | Concibido et al. (1996) |
| SCN 30-2 | G | 4 | 6 | Bell | PI 437654 | Brucker et al. (2005) |
| SCN 23-3 | G | 5 | 6 | Hamilton | PI 89772 | Yue et al. (2001b) |
| SCN 15-1 | G | 6 | 36 | PI 88287 | PI 89008 | Vaghchhipawala et al. (2001) |
| SCN 3-2 | G | 11 | 13 | BSR101 | PI 437654 | Webb et al. (1995) |
| SCN 19-4 | G | 23 | 67 | Hamilton | PI 438489B | Yue et al. (2001a) |
| SCN 1-3 | G | 23 | 25 | M83-15 | M85-1430 | Concibido et al. (1994) |
| SCN 2-4 | G | 35 | 37 | Hartwig | Williams 82 | Vierling et al. (1996) |
| SCN 28-4 | J | 65 | 79 | Bell | Colfax | Glover et al. (2004) |
| SCN 29-2 | J | 68 | 75 | Hamilton | PI 90763 | Guo et al. (2005) |
| SCN 5-2 | J | 73 | 75 | Evans | PI 90763 | Concibido et al. (1997) |
| SCN 1-2 | J | 73 | 75 | M83-15 | M85-1430 | Concibido et al. (1994) |
| *Phytophthora resistance* | | | | | | |
| Phyto 1-1a | F | 16 | 18 | Conrad | Sloan | Burnham et al. (2003) |
| Phyto 1-1b | F | 16 | 18 | Conrad | Williams | Burnham et al. (2003) |
| Phyto 1-1c | F | 16 | 21 | Conrad | Harosoy | Burnham et al. (2003) |
| *Sudden death syndrome resistance* | | | | | | |
| SDS 8-2 | C2 | 120 | 122 | Douglas | Pyramid | Njiti et al. (2002) |
| SDS 2-6 | C2 | 131 | 133 | Essex | Forrest | Chang et al. (1996) |
| SDS 8-1 | G | 0 | 1 | Douglas | Pyramid | Njiti et al. (2002) |
| SDS 3-2 | G | 1 | 3 | Essex | Forrest | Chang et al. (1996) |
| *Brown stem rot resistance* | | | | | | |
| BSR 1-1 | J | 67 | 69 | BSR101 | PI 437654 | Lewers et al. (1999) |
| BSR 4-1 | J | 78 | 80 | Century | PI 437833 | Bachman et al. (2001) |
| BSR 3-1 | J | 78 | 80 | Century84 | L78-4094 | Bachman et al. (2001) |
| *Sclerotinia stem rot resistance* | | | | | | |
| Sclero 5-1 | A2 | 60 | 62 | S19-90 | Williams82 | Arahana et al. (2001) |
| Sclero 6-2 | A2 | 60 | 62 | Vinton81 | Williams82 | Arahana et al. (2001) |
| Sclero 2-2 | A2 | 60 | 62 | Corsoy79 | Williams82 | Arahana et al. (2001) |
| Sclero 4-1 | D1a | 109 | 110 | DSR173 | Williams82 | Arahana et al. (2001) |
| Sclero 5-3 | D1a | 109 | 110 | S19-90 | Williams82 | Arahana et al. (2001) |
| Sclero 3-5 | D1b | 118 | 120 | Dassel | Williams82 | Arahana et al. (2001) |
| Sclero 2-7 | D1b | 118 | 120 | Corsoy79 | Williams82 | Arahana et al. (2001) |
| Sclero 2-12 | F | 63 | 65 | Corsoy79 | Williams82 | Arahana et al. (2001) |
| Sclero 5-6 | F | 63 | 65 | S19-90 | Williams82 | Arahana et al. (2001) |
| Sclero 5-9 | G | 85 | 97 | S19-90 | Williams82 | Arahana et al. (2001) |
| Sclero 6-7 | G | 96 | 98 | Vinton81 | Williams82 | Arahana et al. (2001) |
| Sclero 2-20 | L | 54 | 56 | Corsoy79 | Williams82 | Arahana et al. (2001) |
| Sclero 3-14 | L | 54 | 56 | Dassel | Williams82 | Arahana et al. (2001) |
| Sclero 6-13 | O | 120 | 129 | Vinton81 | Williams82 | Arahana et al. (2001) |
| Sclero 3-19 | O | 120 | 129 | Dassel | Williams82 | Arahana et al. (2001) |
| Sclero 4-11 | O | 127 | 129 | DSR173 | Williams82 | Arahana et al. (2001) |

| Trait/QTL | LG[a] | Start position[b] | End position[b] | Mapping parent 1 | Mapping parent 2 | Reference |
|-----------|-------|-------------------|-----------------|------------------|------------------|-----------|
| *Corn earworm resistance* | | | | | | |
| CEW 3-1 | C2 | 90 | 102 | Cobb | PI 227687 | Rector et al. (1999) |
| CEW 8-3 | C2 | 111 | 113 | Archer | Minsoy | Terry et al. (2000) |
| CEW 8-1 | E | 2 | 4 | Archer | Minsoy | Terry et al. (2000) |
| CEW 7-1 | E | 8 | 10 | Minsoy | Noir 1 | Terry et al. (2000) |
| CEW 2-2 | H | 49 | 62 | Cobb | PI 171451 | Rector et al. (1999) |
| CEW 3-2 | H | 49 | 62 | Cobb | PI 227687 | Rector et al. (1999) |
| CEW 9-3 | H | 53 | 61 | Cobb | PI 229358 | Narvel et al. (2001) |
| CEW 6-2 | J | 15 | 17 | Cobb | PI 229358 | Rector et al. (2000) |
| CEW 7-4 | J | 20 | 22 | Minsoy | Noir 1 | Terry et al. (2000) |
| CEW 6-3 | M | 59 | 71 | Cobb | PI 229358 | Rector et al. (2000) |
| CEW 4-1 | M | 59 | 71 | Cobb | PI 171451 | Rector et al. (2000) |

[a]LG = linkage group. The linkage group names are from the integrated map by Song et al. (2004).

[b]The start positions and end positions are from the integrated map by Song et al. (2004). When only a single marker that was associated with a QTL in a published study could be placed on the consensus map, an arbitrary 2 cM interval with 1 cM on either side of the marker was defined as the QTL region in SoyBase.

### 5.4.1.2 Consensus QTL Regions for Seed Protein and Oil Content

QTLs for protein content were found in 13 consensus regions on 11 LGs: A1, A2, B1, B2, C1, C2, E, G, I, K, and M (Table 5-3 and Fig. 5-2). The regions 93–96 cM on LG A1, 145–151 cM on LG A2, 28–37 cM on LG B1, 28–46 cM on LG B2, 9–34 cM on LG C1, 90–98 cM on LG C1, 123–128 cM on LG C1, 117–123 cM on LG C2, 89–98 cM on LG G, and 31–42 cM on LG K were each found in two mapping populations developed from four mapping parents (Table 5-3). The 26–32 cM region on LG E and the 33–40 cM region on LG M were each found in three mapping populations developed from five or six mapping parents (Table 5-3). The 31–40 cM region on LG I was found in four mapping populations developed from seven mapping parents (Table 5-3).

QTLs for oil content were found in nine consensus regions on eight LGs: A1, C1, E, H, I, K, L, and M (Table 5-3 and Fig. 5-2). The 88–96 cM region on LG A1 was found in four mapping populations developed from seven mapping parents (Table 5-3). The regions near 9–11 cM on LG C1, 23–36 cM on LG E, 86–91 cM on LG H, 98–106 cM on LG K, 34–38 cM on LG L, and 35–40 cM on LG M were each found in two mapping populations developed from three or four mapping parents (Table 5-3). The 22–40 cM region on LG I was found in five mapping populations developed from nine mapping parents (Table 5-3). The 91–96 cM region on LG L was found in three mapping populations developed from six mapping parents (Table 5-3).

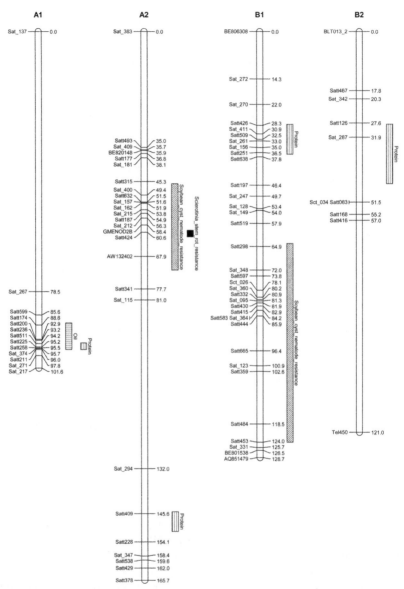

**Figure 5-2** Consensus genomic regions containing QTLs identified in multiple populations for the same trait. The linkage group names, marker names, and map distances are from the integrated map by Song et al. (2004). A bar to the right of a linkage group indicates a consensus QTL region. The trait affected by the QTL is shown to the right of the bar. All simple sequence repeat (SSR) markers in the consensus genomic region plus a 10 cM extension at the borders of the region are shown. All linkage groups are shown in their full lengths.

*Figure 5-2 contd....*

*Figure 5-2 contd....*

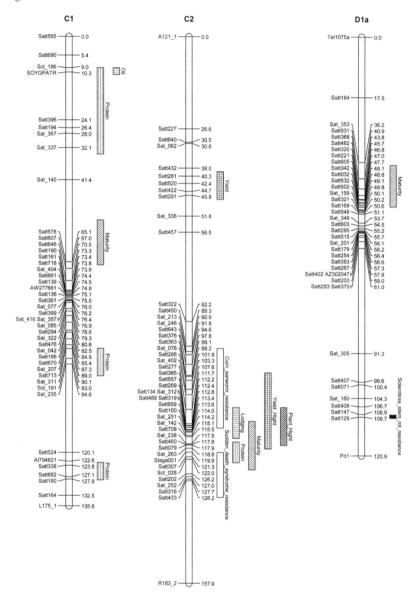

*Figure 5-2 contd....*

*Figure 5-2 contd....*

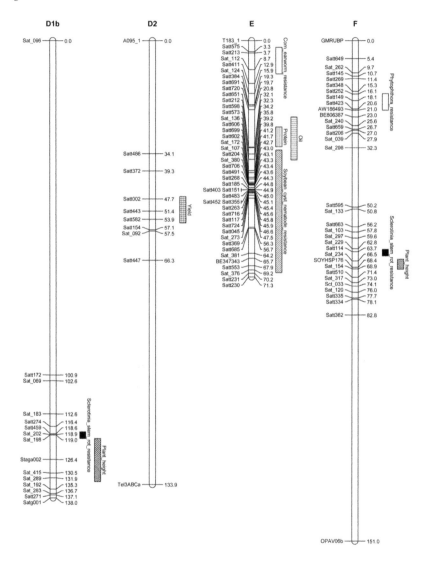

*Figure 5-2 contd....*

*Figure 5-2 contd....*

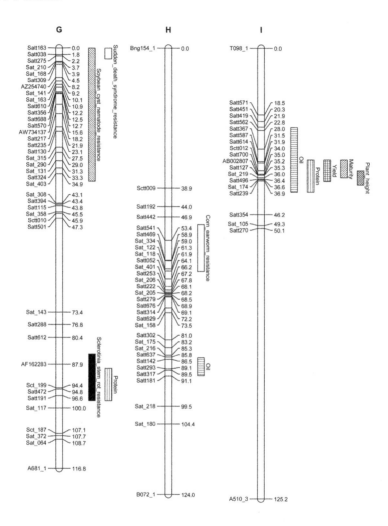

*Figure 5-2 contd....*

*Figure 5-2 contd....*

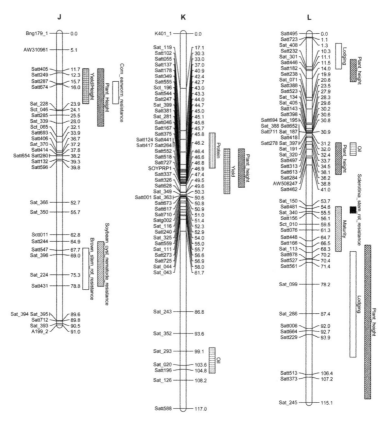

*Figure 5-2 contd....*

*Figure 5-2 contd....*

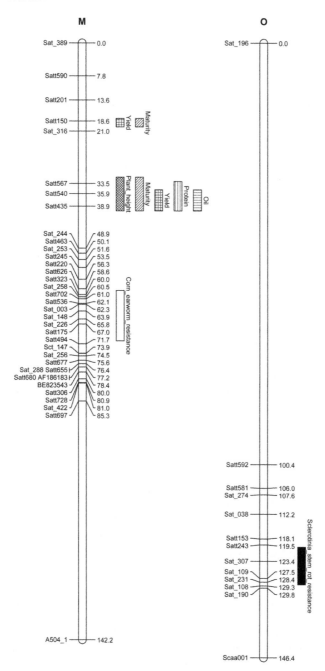

### 5.4.1.3 Consensus QTL Regions for Disease and Insect Resistance

QTLs for resistance to soybean cyst nematode (SCN) were found in five consensus regions on five LGs: A2, B1, E, G, and J (Table 5-3 and Fig. 5-2). The 46–72 cM region on LG A2 was found in seven mapping populations derived from 11 mapping parents (Table 5-3). This region contains the *Rhg4* SCN resistance gene. The regions 64–124 cM on LG B1 and 33–70 cM on LG E were each found in four mapping populations derived from six mapping parents (Table 5-3). The 0–37 cM region on LG G was found in 14 mapping populations derived from 20 mapping parents (Table 5-3). This region contains the *rhg1* SCN resistance gene. The 65–79 cM region on LG J was found in four mapping populations derived from seven mapping parents (Table 5-3).

One consensus region, 16–21 cM on LG F, was found to contain a QTL for partial resistance to *Phytophthora* root rot in three mapping populations developed from four mapping parents (Table 5-3 and Fig. 5-2). Two consensus regions, 120–133 cM on LG C2 and 0–3 cM on LG G, were found to contain QTLs for resistance to sudden death syndrome in two mapping populations developed from four mapping parents (Table 5-3 and Fig. 5-2). A consensus region, 67–80 cM on LG J, was found to contain a QTL for resistance to brown stem rot in three mapping populations developed from six mapping parents (Table 5-3 and Fig. 5-2).

QTLs for *Sclerotinia* stem rot resistance were found in seven consensus regions on seven LGs: A2, D1a, D1b, F, G, L, and O (Table 5-3 and Fig. 5-2). The near 60–62 cM region on LG A2 and the 120–129 cM region on LG O were each identified in three mapping populations developed from four mapping parents (Table 5-3). The regions near 109–110 cM on LG D1a, near 118–120 cM on LG D1b, near 63–65 cM on LG F, 85–98 cM on LG G, and near 54–56 cM on LG L were each identified in two mapping populations developed from three mapping parents (Table 5-3).

QTLs for corn earworm resistance were found in five consensus regions on five LGs: C2, E, H, J and M (Table 5-3 and Fig. 5-2). The regions 90–113 cM on LG C2, 2–10 cM on LG E, 15–22 cM on LG J, and 59–71 cM on LG M were each identified in two mapping populations developed from three or four mapping parents (Table 5-3). The 49–61 cM region on LG H was identified in three mapping populations developed from four mapping parents (Table 5-3).

### 5.4.2 Consensus QTL Regions Containing QTLs for Multiple Traits

In QTL mapping studies, it is common to find QTLs for different traits mapped to a common genomic region. In many cases, the traits affected by the co-localized QTLs are correlated, while in other cases, the trait correlation may not be obvious. While it remains to be resolved whether the co-localized QTLs are tightly linked QTLs or the same QTL with pleiotropic effects, it is useful to document the genomic regions containing QTLs for multiple traits. When the consensus QTL regions for each trait were compared, over 20 genomic regions were found containing QTLs for multiple traits (Table 5-3 and Fig. 5-2). Five genomic regions, 88–96 cM on LG A1, near 9–11 cM on LG C1, 26–36 cM on LG E, 31–40 cM on LG I, and 35–40 cM on LG M contain QTLs for both seed protein and seed oil contents (Fig. 5-2). Soybean protein content and oil content are highly negatively correlated with a correlation coefficient as high as –0.98 ($P < 0.001$) (Mansur et al. 1996). It is, therefore, expected to have some QTLs for the two traits mapped to common genomic regions. Four genomic regions, near 34–36 cM on LG I, 15–22 cM on LG J, 36–42 cM on LG K, and 35–40 cM on LG M, contain QTLs for both yield and plant height (Fig. 5-2). Yield was correlated (r = 0.59, $P < 0.001$) with plant height in the study by Mansur et al. (1996). Yield was also correlated (r = 0.48, $P < 0.001$) with maturity (Mansur et al. 1996). Three genomic regions, near 34-36 cM on LG I, near 18–20 cM on LG M, and 35–40 cM on LG M, contain QTLs for both yield and maturity (Fig. 5-2). Yield was negatively correlated (r = –0.58, $P < 0.001$) with protein content and positively correlated (r = 0.60, $P < 0.001$) with oil content (Mansur et al. 1996). Two genomic regions, near 34–36 cM on LG I and 35–40 cM on LG M, contain QTLs for yield, protein content, and oil content. Another region, 36–42 cM on LG K, contains QTLs for yield and protein content (Fig. 2). Lodging was highly correlated with plant height (r = 0.84, $P < 0.001$) (Mansur et al. 1996). Three genomic regions, 107–113 cM on LG C2, 8–11 cM on LG L, and 68–101 cM on LG L, contain QTLs for these two traits (Fig. 5-2).

QTLs for *Sclerotinia* stem rot resistance were co-localized with QTLs for plant height near 120 cM on LG D1b and near 65 cM on LG F, for maturity near 54–56 cM on LG L, for SCN resistance near 60–62 cM on LG A2, and for protein content in 89–98 cM on LG G (Fig. 5-2). *Sclerotinia* stem rot disease severity index was correlated with plant height (r = 0.54, $P < 0.001$) and maturity (r = 0.67, $P < 0.001$) in the study by Kim and Diers (2000). Shorter plant and earlier maturity were associated with greater resistance to *Sclerotinia* stem rot, which was considered an escape mechanism (Kim and Diers 2000). However, the QTLs co-localized with plant height on LG D1b and LG F and with maturity on LG L were for physiological resistance to *Sclerotinia* stem rot (Arahana et al. 2001). Therefore, the resistance associated with shorter plant and earlier maturity may also involve physiological resistance. A significant correlation (r = 0.40, $P < 0.05$) between *Sclerotinia* stem rot disease severity index and protein content was reported by Hoffman et al. (1998). There is no report of any correlation between *Sclerotinia* stem rot resistance and SCN resistance.

QTLs for SCN resistance were co-localized with QTLs for sudden death syndrome (SDS) near 4–6 cM on LG G, for brown stem rot resistance in 67–79 cM on LG J, for oil content in 33–36 cM on LG E, and for *Sclerotinia* stem rot resistance as described above. Coinheritance of SDS resistance with SCN resistance was reported by Chang et al. (1997) and the locus underlying the coinheritance was assigned to the region on LG G where the major SCN resistance was located (Chang et al. 1997). The region on LG J is known to contain multiple resistance genes to different pathogens (Bachman et al. 2001). Correlation of SCN resistance with oil content was not reported in the literature. Qiu et al. (1999) carried out a QTL mapping study in a population that was segregating for both SCN resistance and oil content. A marker on LG H was found to be associated with both SCN resistance and oil content.

QTLs for corn earworm resistance were co-localized with QTLs for plant height in 107–113 cM on LG C2 and in 15–22 cM on LG J. There is no known correlation between corn earworm resistance and plant height in soybean.

## 5.5 Future Opportunities and Limitations for QTL Discovery in Soybean

Over a thousand SNP markers were recently added to the integrated soybean linkage map (Choi et al. 2007) and research is ongoing to add several more thousands of SNP markers to the map (Perry Cregan, personal communication). High-throughput SNP genotyping systems have been developed and are commercially available. For example, the Illumina BeadStation 500 (Shen et al. 2005) can analyze 1,536 SNP loci in parallel in 192 DNA samples in three days (Perry Cregan, personal communication). The addition of thousands of SNP markers to the integrated linkage map and the availability of high-throughput SNP genotyping systems will significantly reduce the time needed to genotype mapping populations and accelerate QTL discovery in soybean.

The first draft sequence of the whole soybean genome was released in 2008 (JGI 2008). The availability of a whole-genome sequence will allow scientists to fine-map QTLs and eventually pinpoint the specific mutations that cause the phenotypic variations.

The development of new statistical approaches and computer software that allow joint analysis of multiple populations will also increase the power to identify real QTLs, especially QTLs with small effects. Bink et al. (2008) developed a pedigree-based approach that jointly analyzes the data from multiple populations that are related through their common ancestors in the pedigree. This approach is currently implemented in the computer software FlexQTL™. Jourjon et al. (2005) developed a computer software package, MCQTL, that can perform QTL mapping in multi-cross designs.

The major limitation to QTL discovery in soybean is the difficulty in obtaining accurate measurement of the traits with low heritability. For certain traits such as field resistance to *Sclerotinia* stem rot, reliable measurement is difficult to obtain even with efforts to provide the optimum conditions to induce the disease. Large experiments with multiple locations in multiple years are often required to obtain the phenotypic data.

# References

Akkaya MS, Shoemaker RC, Specht JE, Bhagwat AA, Cregan PB (1995) Integration of simple sequence DNA markers into a soybean linkage map. Crop Sci 35: 1439–1445.

Apuya NR, Frazier BL, Keim P, Roth EJ, Lark KG (1988) Restriction fragment length polymorphisms as genetic markers in soybean, *Glycine max* (L.) Merrill. Theor Appl Genet 75: 889–901.

Arahana VS, Graefa GL, Spechta JE, Steadmanb JR, Eskridgec KM (2001) Identification of QTLs for resistance to *Sclerotinia sclerotiorum* in soybean. Crop Sci 41:180–188.

Bachman MS, Tamulonis JP, Nickell CD, Bent AF (2001) Molecular markers linked to brown stem rot resistance genes, *Rbs1* and *Rbs2*, in soybean. Crop Sci 41: 527–535

Bink MCAM, Boer MP, ter Braak CJF, Jansen J, Voorrips RE, van de Weg WE (2008) Bayesian analysis of complex traits in pedigreed plant populations. Euphytica 161: 85–96.

Brucker E, Carlson S, Wright E, Niblack T, Diers B (2005) *Rhg1* alleles from soybean PI 437654 and PI 88788 respond differentially to isolates of *Heterodera glycines* in the greenhouse. Theor Appl Genet 111: 44–49.

Brummer EC, Graef GL, Orf J, Wilcox JR, Shoemaker RC (1997) Mapping QTL for seed protein and oil content in eight soybean populations. Crop Sci 37: 370–378.

Burnham KD, Dorrance AE, VanToai TT, St Martin SK (2003) Quantitative trait loci for partial resistance to *Phytophthora sojae* in soybean. Crop Sci 43: 1610–1671.

Chang SJC, Doubler TW, Kilo V, Suttner R, Klein J, Schmidt ME, Gibson PT, Lightfoot DA (1996) Two additional loci underlying durable field resistance to soybean sudden death syndrome (SDS). Crop Sci 36: 1684–1688.

Chang SJC, Doubler TW, Kilo VY, Abu Thredeih J, Prabhu R, Freire V, Suttner R, Klein J, Schmidt ME, Gibson PT, Lightfoot DA (1997) Association of loci underlying field resistance to soybean sudden death syndrome (SDS) and cyst nematode (SCN) race 3. Crop Sci 37: 965–971.

Chapman A, Pantalone VR, Ustun A, Allen FL, Landau-Ellis D, Trigiano RN, Gresshoff PM (2003) Quantitative trait loci for agronomic and seed quality traits in an $F_2$ and $F_{4:6}$ soybean population. Euphytica 129: 387–393.

Choi IY, Hyten DL, Matukumalli LK, Song QJ, Chaky JM, Quigley CV, Chase K, Lark KG, Reiter RS, Yoon MS, Hwang EY, Yi SI, Young ND, Shoemaker RC, Tassell CPv, Specht JE, Cregan PB (2007) A soybean transcript map: Gene distribution, haplotype and single-nucleotide polymorphism analysis. Genetics 176: 685–696.

Chung J, Babka HL, Graef GL, Staswick PE, Lee DJ, Cregan PB, Shoemaker RC, Specht JE (2003) The seed protein, oil, and yield QTL on soybean linkage group I. Crop Sci 43: 1053–1067.

Concibido VC, Denny RL, Boutin SR, Hautea R, Orf JH, Young ND (1994) DNA marker analysis of loci underlying resistance to soybean cyst nematode (*Heterodera glycines* Ichinohe). Crop Sci 34: 240–246.

Concibido VC, Young ND, Lange DA, Denny RL, Danesh D, Orf JH (1996) Targeted comparative genome analysis and qualitative mapping of a major partial-resistance gene to the soybean cyst nematode. Theor Appl Genet 93: 234–241.

Concibido VC, Lange DA, Denny RL, Orf JH, Young ND (1997) Genome mapping of soybean cyst nematode resistance genes in 'Peking', PI 90763, and PI 88788 using DNA markers. Crop Sci 37: 258–264.

Concibido VC, Diers BW, Arelli PR (2004) A decade of QTL mapping for cyst nematode resistance in soybean. Crop Sci 44: 1121–1131.

Cregan PB, Jarvik T, Bush AL, Shoemaker RC, Lark KG, Kahler AL, Kaya N, VanToai TT, Lohnes DG, Chung J, Specht JE (1999) An integrated genetic linkage map of the soybean genome. Crop Sci 39: 1464–1490.

Csanadi G, Vollmann J, Stift G, Lelley T (2001) Seed quality QTLs identified in a molecular map of early maturing soybean. Theor Appl Genet 103: 912–919.

Diers BW, Keim P, Fehr WR, Shoemaker RC (1992) RFLP analysis of soybean seed protein and oil content. Theor Appl Genet 83: 608–612.

Fasoula VA, Harris DK, Boerma HR (2004) Validation and designation of quantitative trait loci for seed protein, seed oil, and seed weight from two soybean populations. Crop Sci 44: 1218–1225.

Fu SX, Wang H, Wu JJ, Liu H, Gai JY, Yu DY (2007) Mapping insect resistance QTLs of soybean with RIL population. Hereditas (Beijing) 29: 1139–1143.

Glover KD, Wang D, Arelli PR, Carlson SR, Cianzio SR, Diers BW (2004) Near isogenic lines confirm a soybean cyst nematode resistance gene from PI 88788 on linkage group J. Crop Sci 44: 936–941.

Grant D, Imsande MI, Shoemaker RC (2008) SoyBase, The USDA-ARS Soybean Genome Database: *http://soybase.agron.iastate.edu.* (Accessed 2 August 2009).

Guo B, Sleper DA, Arelli PR, Shannon JG, Nguyen HT (2005) Identification of QTLs associated with resistance to soybean cyst nematode races 2, 3 and 5 in soybean PI 90763. Theor Appl Genet 111: 965–971.

Hoffman DD, Hartman GL, Mueller DS, Leitz RA, Nickell CD, Pedersen WL (1998) Yield and seed quality of soybean cultivars infected with *Sclerotinia sclerotiorum.* Plant Dis 82: 826–829.

Hyten DL, Pantalone VR, Sams CE, Saxton AM, Landau-Ellis D, Stefaniak TR, Schmidt ME (2004) Seed quality QTL in a prominent soybean population. Theor Appl Genet 109: 552–561.

JGI (2008) Phytozome: *Glycine max. http://www.phytozome.net/soybean* (Accessed 2 August 2009).

Jourjon MF, Jasson S, Marcel J, Ngom B, Mangin B (2005) MCQTL: Multi-allelic QTL mapping in multi-cross design. Bioinformatics (Oxford) 21: 128–130.

Kabelka EA, Diers BW, Fehr WR, LeRoy AR, Baianu IC, You T, Neece DJ, Nelson RL (2004) Putative alleles for increased yield from soybean plant introductions. Crop Sci 44: 784–791.

Keim P, Shoemaker RC, Palmer RG (1989) Restriction fragment length polymorphism diversity in soybean. Theor Appl Genet 77: 786–792.

Keim P, Diers BW, Olson TC, Shoemaker RC (1990a) RFLP mapping in soybean: Association between marker loci and variation in quantitative traits. Genetics 126: 735–742.

Keim P, Diers BW, Shoemaker RC (1990b) Genetic analysis of soybean hard seededness with molecular markers. Theor Appl Genet 79: 465–469.

Keim P, Schupp JM, Travis SE, Clayton K, Zhu T, Shi L, Ferreira A, Webb DM (1997) A high-density soybean genetic map based on AFLP markers. Crop Sci 37: 537–543.

Kim HS, Diers BW (2000) Inheritance of partial resistance to Sclerotinia stem rot in soybean. Crop Sci 40: 55–61.

Lark KG, Weisemann JM, Matthews BF, Palmer R, Chase K, Macalma T (1993) A genetic map of soybean (*Glycine max* L.) using an intraspecific cross of two cultivars: 'Minsoy' and 'Noir 1'. Theor Appl Genet 86: 901–906.

Lark KG, Orf J, Mansur LM (1994) Epistatic expression of quantitative trait loci (QTL) in soybean (*Glycine max* (L.) Merr.) determined by QTL association with RFLP alleles. Theor Appl Genet 88: 486–489.

Lark KG, Chase K, Adler F, Mansur LM, Orf JH (1995) Interactions between quantitative trait loci in soybean in which trait variation at one locus is conditional upon a specific allele at another. Proc Natl Acad Sci USA 92: 4656–4660.

Lee SH, Bailey MA, Mian MAR, Carter TE, Jr., Ashley DA, Hussey RS, Parrott WA, Boerma HR (1996a) Molecular markers associated with soybean plant height, lodging, and maturity across locations. Crop Sci 36: 728–735.

Lee SH, Bailey MA, Mian MAR, Carter TE, Jr., Shipe ER, Ashley DA, Parrott WA, Hussey RS, Boerma HR (1996b) RFLP loci associated with soybean seed protein and oil content across populations and locations. Theor Appl Genet 93: 649–657.

Lee SH, Bailey MA, Mian MAR, Shipe ER, Ashley DA, Parrott WA, Hussey RS, Boerma HR (1996c) Identification of quantitative trait loci for plant height, lodging, and maturity in a soybean population segregating for growth habit. Theor Appl Genet 92: 516–523.

Lewers KS, Crane EH, Bronson CR, Schupp JM, Keim P, Shoemaker RC (1999) Detection of linked QTL for soybean brown stem rot resistance in 'BSR 101' as expressed in a growth chamber environment. Mol Breed 5: 33–42.

Liu F, Zhuang BC, Zhang JS, Chen SY (2000) Construction and analysis of soybean genetic map. Acta Genet Sin 27: 1018–1026.

Mahalingam R, Skorupska HT (1995) DNA markers for resistance to *Heterodera glycines* I. Race 3 soybean cultivar Peking. Breed Sci 45: 435–443.

Mansur LM, Lark KG, Kross H, Oliveira A (1993) Interval mapping of quantitative trait loci for reproductive, morphological, and seed traits of soybean (*Glycine max* L.). Theor Appl Genet 86: 907–913.

Mansur LM, Orf JH, Chase K, Jarvik T, Cregan PB, Lark KG (1996) Genetic mapping of agronomic traits using recombinant inbred lines of soybean. Crop Sci 36: 1327–1336.

Matthews BF, Devine TE, Weisemann JM, Beard HS, Lewers KS, MacDonald MH, Park YB, Maiti R, Lin JJ, Kuo J, Pedroni MJ, Cregan PB, Saunders JA (2001) Incorporation of sequenced cDNA and genomic markers into the soybean genetic map. Crop Sci 41: 516–521.

Meksem K, Doubler TW, Chancharoenchai K, Njiti VN, Chang SJC, Arelli APR, Cregan PE, Gray LE, Gibson PT, Lightfoot DA (1999) Clustering among loci underlying soybean resistance to *Fusarium solani*, SDS and SCN in near-isogenic lines. Theor Appl Genet 99: 1131–1142.

Meksem K, Pantazopoulos P, Njiti VN, Hyten LD, Arelli PR, Lightfoot DA (2001) 'Forrest' resistance to the soybean cyst nematode is bigenic: saturation mapping of the *Rhg1* and *Rhg4* loci. Theor Appl Genet 103: 710–717.

Mian MAR, Ashley DA, Vencill WK, Boerma HR (1998) QTLs conditioning early growth in a soybean population segregating for growth habit. Theor Appl Genet 97: 1210–1216.

Narvel JM, Walker DR, Rector BG, All JN, Parrott WA, Boerma HR (2001) A retrospective DNA marker assessment of the development of insect resistant soybean. Crop Sci 41: 1931–1939.

Neto ALd-F, Hashmi R, Schmidt M, Carlson SR, Hartman GL, Li S, Nelson RL, Diers BW (2007) Mapping and confirmation of a new sudden death syndrome resistance QTL on linkage group D2 from the soybean genotypes PI 567374 and 'Ripley'. Mol Breed 20: 53–62.

Njiti VN, Meksem K, Iqbal MJ, Johnson JE, Kassem MA, Zobrist KF, Kilo VY, Lightfoot DA (2002) Common loci underlie field resistance to soybean sudden death syndrome in Forrest, Pyramid, Essex, and Douglas. Theor Appl Genet 104: 294–300.

Orf JH, Chase K, Adler FR, Mansur LM, Lark KG (1999a) Genetics of soybean agronomic traits: II. Interactions between yield quantitative trait loci in soybean. Crop Sci 39: 1652–1657.

Orf JH, Chase K, Jarvik T, Mansur LM, Cregan PB, Adler FR, Lark KG (1999b) Genetics of soybean agronomic traits: I. Comparison of three related recombinant inbred populations. Crop Sci 39: 1642–1651.

Panthee DR, Pantalone VR, West DR, Saxton AM, Sams CE (2005) Quantitative trait loci for seed protein and oil concentration, and seed size in soybean. Crop Sci 45: 2015–2022.

Prabhu RR, Njiti VN, Bell-Johnson B, Johnson JE, Schmidt ME, Klein JH, Lightfoot DA (1999) Selecting soybean cultivars for dual resistance to soybean cyst nematode and sudden death syndrome using two DNA markers. Crop Sci 39: 982–987.

Qiu BX, Arelli PR, Sleper DA (1999) RFLP markers associated with soybean cyst nematode resistance and seed composition in a 'Peking' X 'Essex' population. Theor Appl Genet 98: 356–364.

Rafalski A, Tingey S (1993) RFLP map of soybean (*Glycine max*). In: SJ O'Brien (ed) Genetic Maps: Locus Maps of Complex Genomes. Cold Spring Harbor Lab Press, New York, USA, pp 6149–6156.

Rector BG, All JN, Parrott WA, Boerma HR (1999) Quantitative trait loci for antixenosis resistance to corn earworm in soybean. Crop Sci 39: 531–538.

Rector BG, All JN, Parrott WA, Boerma HR (2000) Quantitative trait loci for antibiosis resistance to corn earworm in soybean. Crop Sci 40: 233–238.

Schuster I, Abdelnoor RV, Marin SRR, Carvalho VP, Kiihl RAS, Silva JFV, Sediyama CS, Barros EG, Moreira MA (2001) Identification of a new major QTL associated with resistance to soybean cyst nematode (*Heterodera glycines*). Theor Appl Genet 102: 91–96.

Shen R, Fan J-B, Campbell D, Chang W, Chen J, Doucet D, Yeakley J, Bibikova M, Garcia E-W, McBride C, Steemers F, Garcia F, Kermani B-G, Gunderson K, Oliphant A (2005) High-throughput SNP genotyping on universal bead arrays. Mutat Res 573: 70–82.

Shoemaker RC, Olson TC (1993) Molecular linkage map of soybean (*Glycine max* L. Merr.). In: SJ O'Brien (ed) Genetic Maps: Locus Maps of Complex Genomes. Cold Spring Harbor Lab Press, New York, USA, pp 6.131–6.138.

Song QJ, Marek LF, Shoemaker RC, Lark KG, Concibido VC, Delannay X, Specht JE, Cregan PB (2004) A new integrated genetic linkage map of the soybean. Theor Appl Genet 109: 122–128.

Specht JE, Chase K, Macrander M, Graef GL, Chung J, Markwell JP, Germann M, Orf JH, Lark KG (2001) Soybean response to water: a QTL analysis of drought tolerance. Crop Sci 41: 493–509.

Terry LI, Chase K, Jarvik T, Orf J, Mansur L, Lark KG (2000) Soybean quantitative trait loci for resistance to insects. Crop Sci 40: 375–382.

Thomson Scientific Inc (2008) Biological Abstracts®. *http://thomsonscientific.com/support/faq/wok3new/BiologicalAbstracts/* (Accessed 2 August 2009).

Vaghchhipawala Z, Bassuner R, Clayton K, Lewers K, Shoemaker R, Mackenzie S (2001) Modulations in gene expression and mapping of genes associated with cyst nematode infection of soybean. Mol Plant-Micr Interact 14: 42–54.

Vierling RA, Faghihi J, Ferris VR, Ferris JM (1996) Association of RFLP markers with loci conferring broad-based resistance to the soybean cyst nematode (*Heterodera glycines*). Theor Appl Genet 92: 83–86.

Vollmann J, Schausberger H, Bistrich H, Lelley T (2002) The presence or absence of the soybean Kunitz trypsin inhibitor as a quantitative trait locus for seed protein content. Plant Breed 121: 272–274.

Wang D, Arelli PR, Shoemaker RC, Diers BW (2001) Loci underlying resistance to Race 3 of soybean cyst nematode in *Glycine soja* plant introduction 468916. Theor Appl Genet 103: 561–566.

Wang D, Graef GL, Procopiuk AM, Diers BW (2004a) Identification of putative QTL that underlie yield in interspecific soybean backcross populations. Theor Appl Genet 108: 458–467.

Wang HL, Yu DY, Wang YJ, Chen SY, Gai JY (2004b) Mapping QTLs of soybean root weight with RIL population NJRIKY. Hereditas (Beijing) 26: 333–336.

Wang HL, Yu DY, Wang YJ, Chen SY, Gai JY (2004c) Mapping QTLs of soybean root weight with RIL population NJRIKY. Yichuan 26: 333–336.

Wang YJ, Dong Fang Y, Wang XQ, Yang YL, Yu DY, Gai JY, Wu XL, He CY, Zhang JS, Chen SY (2004d) Mapping of five genes resistant to SMV strains in soybean. Acta Genet Sin 31: 87–90.

Webb DM, Baltazar BM, Rao-Arelli AP, Schupp J, Clayton K, Keim P, Beavis WD (1995) Genetic mapping of soybean cyst nematode race-3 resistance loci in the soybean PI 437.654. Theor Appl Genet 91: 574–581.

Wu XL, He CY, Wang YJ, Zhang ZY, Dong Fang Y, Zhang JS, Chen SY, Gai JY (2001) Construction and analysis of a genetic linkage map of soybean. Acta Genet Sin 28: 1051–1061.

Yamanaka N, Ninomiya S, Hoshi M, Tsubokura Y, Yano M, Nagamura Y, Sasaki T, Harada K (2001) An informative linkage map of soybean reveals QTLs for flowering time, leaflet morphology and regions of segregation distortion. DNA Res 8: 61–67.

Yuan J, Njiti VN, Meksem K, Iqbal MJ, Triwitayakorn K, Kassem MA, Davis GT, Schmidt ME, Lightfoot DA (2002) Quantitative trait loci in two soybean recombinant inbred line populations segregating for yield and disease resistance. Crop Sci 42: 271–277.

Yue P, Arelli PR, Sleper DA (2001a) Molecular characterization of resistance to *Heterodera glycines* in soybean PI 438489B. Theor Appl Genet 102: 921–928.

Yue P, Sleper DA, Arelli PR (2001b) Mapping resistance to multiple races of *Heterodera glycines* in soybean PI 89772. Crop Sci 41: 1589–1595.

Zhu S, Walker DR, Boerma HR, All JN, Parrott WA (2006) Fine mapping of a major insect resistance QTL in soybean and its interaction with minor resistance QTLs. Crop Sci 46: 1094–1099.

# Molecular Breeding

*David R. Walker,[1]* Maria J. Monteros[2] and Jennifer L. Yates[3]*

## ABSTRACT

Marker-assisted selection (MAS) with DNA markers has played an increasingly important role in soybean breeding since the early 1990s. This is a result of improvements in technology, increasing saturation of the soybean genome with markers, the evergrowing number of trait loci that have been tagged with markers, the ability to accomplish certain objectives more efficiently or even exclusively using MAS, and the demonstrated power of MAS to improve and complement conventional breeding methods. The development and extensive utilization of PCR-based markers, particularly SSRs and SNPs, has been a major factor in making MAS feasible for soybean improvement. Although MAS cannot entirely replace phenotypic selection in cultivar development, it provides breeders with a tool to transfer useful genes from exotic germplasm with minimal linkage drag, select for resistance in the absence of a pest or pathogen, pyramid resistance genes into the same genetic backgrounds, identify heterozygotes carrying one allele of a beneficial recessive gene, and to assess the effect of specific combinations of alleles on a trait of interest. Recent and ongoing advances in soybean genomics and sequencing, genotyping technologies, and bioinformatics are making MAS increasingly efficient, accurate and affordable for public and private sector soybean breeders. Innovations at major seed companies have resulted in a large-scale adoption of MAS and in a shift from selection at one or

[1]USDA-ARS Soybean/Maize Germplasm, Pathology and Genetics Research Unit, Urbana, IL 61801, USA.
[2]The Samuel Roberts Noble Foundation, Ardmore, OK 73401, USA.
[3]Monsanto Company, Galena, MD 21635, USA.
*Corresponding author: *david.walker@ars.usda.gov*

two loci to genome-wide MAS. The expanding use of MAS has been accompanied, however, by an increase in intellectual property issues that restrict or complicate the use of DNA markers for certain breeding objectives.

**Keywords:** DNA markers; marker-assisted selection; soybean breeding; SSR; SNP

## 6.1 Introduction

In the late 1990s Nevin Young expressed a cautious optimism for the future of marker-assisted breeding (Young 1999). Although marker-assisted selection (MAS) for soybean cyst nematode (SCN; *Heterodera glycines*) resistance in soybean (*Glycine max* L. Merr.) was used as a case study on how genotype-based selection could be useful and cost-effective to a plant breeder, the reader was reminded that to get to that point, a great deal of time and money had been invested in refining the tools and techniques used. In addition, Nevin Young pointed out that crossovers occasionally occurred between the most important resistance locus and the nearest useful marker, and that no SCN-resistant public cultivars developed using MAS had yet been released.

Since the publication of Nevin Young's assessment, numerous advances have occurred in molecular marker and bioinformatics technologies, in the increased availability and density of markers, and through innovative genomics studies of soybean and other plants. Progress in mapping and identifying molecular markers associated with many agriculturally important traits provides the foundation for MAS in soybean. Markers can be used to improve efficiency and/or accuracy in plant breeding programs, and have distinct advantages over phenotypic selection for certain traits. Molecular markers have been used to determine genetic relatedness between accessions, to assist in the identification of novel sources of variation, to confirm the pedigree and identity of new varieties, to locate quantitative trait loci (QTLs) and genes of interest, and for marker-assisted breeding. Markers have also been used to investigate genes and gene interactions for a number of quantitative traits in several important crop species. More than 1,100 QTLs associated with a wide variety of soybean traits have been mapped using molecular markers (see Chapter 5 of this volume). The value and uses of various types of DNA markers have been shaped in large part by contemporary innovations in marker technologies that increased throughput and reduced costs per data point. Altogether these technological advances and research discoveries have enhanced the potential contribution that MAS can make to soybean genetic improvement. This chapter summarizes the role of MAS in soybean breeding at the beginning of the 21st century, and reviews new and developing technologies and strategies to leverage information gained from genomics research.

## 6.2 DNA Markers and a Brief History of MAS in Soybean Breeding

Since the feasibility of using molecular markers for MAS in soybean breeding is closely associated with the evolution of different types of DNA markers and the technologies available to use them, a brief review of marker types and their development for soybean appears warranted. Tanksley and Rick (1980) constructed the first isozyme-based genetic linkage map in tomato (*Lycopersicon esculentum*) and were among the first researchers to envision roles that molecular markers could potentially play in plant breeding and genetics research. Molecular mapping of soybean genes and QTLs increased dramatically when restriction fragment length polymorphism (RFLP) markers became available in sufficient numbers to construct genetic linkage maps (Apuya et al. 1988), and after the demonstration that DNA markers could be used to estimate the locations and importance of the QTLs underlying quantitative traits important to plant breeders (Tanksley et al. 1989; Keim et al. 1990). RFLP marker polymorphism involves DNA sequence differences between parents at sites recognized by restriction enzymes. Digestion of genomic DNA with an appropriate restriction enzyme results in differential cleavage to produce fragments which differ in length. These fragments are then separated using gel electrophoresis, and are detected through hybridization to a labeled DNA probe with a complementary sequence.

Prior to the late 1990s, RFLP-based mapping successfully identified the approximate locations and phenotypic contributions of a number of major genes and QTLs for a variety of traits, even though the mapping populations used often consisted of fewer than 150 individual plants or families (Orf et al. 2004). RFLP probes that annealed to two or more regions of the genome were also useful in identifying duplicated regions of the soybean genome, which can be viewed at the Soybean Breeders Toolbox website (*soybeanbreederstoolbox.org*). Although some MAS was accomplished using RFLPs, the application was hindered by the requirement for large amounts of relatively pure DNA, the labor- and time-intensive protocols, the need to use radioisotope-labeled probes, the annealing of probes to multiple sites in the soybean genome, and a low polymorphism level within *G. max* (Cregan et al. 1999).

The development of microsatellite, or simple sequence repeat (SSR), markers for soybean beginning in the mid-1990s resulted in a substantial improvement in the efficiency and feasibility of both molecular mapping and MAS (Morgante et al. 1994; Akkaya et al. 1995; Rongwen et al. 1995). Perry Cregan and colleagues working at the USDA-ARS Soybean Genomics and Improvement Laboratory in Beltsville, MD, pioneered efforts to develop polymerase chain reaction (PCR) primers designed to amplify segments of

DNA containing variable numbers of di- or tri-nucleotide tandem repeats, or microsatellites (Cregan et al. 1994, 1999; Akkaya et al. 1995). Sequence data for PCR primers complementary to regions flanking an SSR locus were subsequently made publicly available, and this fostered the rapid adoption of SSRs for gene and QTL mapping in the soybean breeding and genetics community.

An important feature of SSR and RFLP markers is that they are codominant, meaning that they allow heterozygotes to be distinguished from both homozygous classes. However, since SSR markers are detected through DNA amplification by PCR, they can be used with much smaller amounts of sample DNA than RFLPs. SSRs are particularly useful because multiple alleles can be detected at a single locus and because PCR primer sets are available for SSR loci distributed over a large portion of the soybean genome (Akkaya et al. 1992; Song et al. 2004). In addition, microsatellite repeats are more abundant in the genome than RFLPs, and they can be genotyped much more rapidly and efficiently. One of the most important advantages of SSRs over RFLPs is that it is easier to find multiple polymorphic markers on each molecular linkage group in *G. max* × *G. max* crosses (see also Chapter 4 of this volume). When a set of 13 SSR markers was used to characterize 96 cultivars, the average number of alleles at an SSR locus was 7.8, with a range of 5 to 17 (Song et al. 1999). The high polymorphism rates and widespread distribution of SSR markers in the genome made it possible to assemble consensus linkage maps using marker data from several independent populations (Cregan et al. 1999; Song et al. 2004). The Cregan et al. (1999) consensus map was quickly adopted by the public sector soybean molecular breeding community as a common reference. This map made it easier to select strategically spaced markers covering the soybean genome, and allowed researchers to discuss the locations of genes with a greater degree of precision. The updated map of Song et al. (2004), which included a total of 1,015 SSR markers, remains an important reference point for much of the current mapping and MAS work.

SSR markers are genotyped by separating PCR-generated amplicons differing in length using electrophoresis through polyacrylamide or agarose gels, or through capillary sequencers. The amplicons are typically detected with either ethidium bromide or a marker-specific dye that fluoresces at specific wavelengths. The high-throughput demands of commercial plant breeding programs made 96-capillary analyzers a good option for SSR genotyping because these machines were amenable to automation and nearly constant use. For public breeding programs with comparatively lower throughput, however, SSR alleles have been separated primarily on gels. Automated gel sequencers like the ABI 377 (Applied Biosystems, Foster City, CA) are now obsolete, but they were used extensively for mapping and MAS in some public soybean breeding and genetics research programs from about 1996 to 2005. Efforts to achieve multiplex PCR using more than one pair of SSR primers have been largely unsuccessful, so the more common practice is to conduct the PCRs for different markers independently and then pool the amplified products before they are loaded onto a gel or capillary. Throughput can be increased several fold by co-electrophoresis of two to four SSR markers labeled with different fluorescent tags and/or known to differ substantially in length. Public breeding programs have also used a less expensive alternative to semi-automated sequencing equipment by analyzing and visualizing the SSR fragments on traditional gel-based systems with ethidium bromide staining (Wang et al. 2003).

For certain traits, MAS was quickly recognized as a cost-effective alternative to phenotypic selection. This was especially true if phenotypic selection required assays that were expensive, time-consuming, or which could not be conducted until a plant was mature (e.g., selection for higher seed protein). Concibido et al. (2004) estimated that the cost per data point for MAS of soybean cyst nematode resistance was substantially less (US$0.25–$1.00) than the cost of phenotypic selection based on cyst counts (US$1.50–$5.00). Furthermore, MAS could be completed in 1–2 days, whereas the phenotypic assay would require 30 days. By the late 1990s, SSR markers were being widely used in the soybean breeding programs of large seed companies such as Pioneer and Monsanto, and by breeders at some public institutions like the University of Georgia and Michigan State University (Concibido et al. 2004; Orf et al. 2004).

Although SSRs have many advantages over RFLPs for MAS applications, they are still labor-intensive to resolve and score (Collard et al. 2005). Use of capillary sequencers reduced some of the labor requirements of gel-based sequencers, but the cost of purchasing and operating them was prohibitive for most public breeding programs. In addition, while the genome coverage provided by SSRs was a substantial improvement over that of RFLPs, large gaps in the genetic map remained. The map of Song et al. (2004) had 138 gaps exceeding 5 cM in length with no SSR markers, and 26 of these gaps were longer than 10 cM (Choi et al. 2007). This was obviously a limitation for fine-mapping and tagging genes or QTLs that are located within those intervals.

Many of the limitations of SSRs have been addressed through the development of single-nucleotide polymorphism (SNP) markers, another class of codominant DNA markers. Compared to SSRs, SNPs have the advantage of being amenable to high-throughput automated genotyping assays that allow samples to be genotyped more quickly and economically than with SSRs (Hurley et al. 2004). SNPs are more numerous and more widely distributed throughout the genome than SSRs, and their development in soybean has progressed at a rapid pace. Zhu et al. (2003) found approximately three SNPs per thousand base pairs of DNA sequenced.

Although the level of sequence diversity in cultivated soybean is low compared to that found so far in some other crop species, SNPs are nonetheless becoming important for MAS. In the 76.3 kilobase pairs (kbp) of DNA sequence analyzed by Zhu et al. (2003), the mean nucleotide diversity (θ) was 0.00097, with the frequency of SNPs in noncoding regions associated with genes being twice that found in coding sequences. For this reason, the authors suggested that SNP discovery should focus on the noncoding perigenic regions. More recent sequencing work by Choi et al. (2007) produced a similar estimate of average nucleotide diversity (θ = 0.000997). These researchers used 1,141 gene fragments containing 2,928 SNPs to generate the first transcript map of soybean. A number of these new markers mapped to one of the 138 gaps with lengths > 5 cM in the SSR map of Song et al. (2004), thus improving the potential to map both new and previously mapped alleles with greater accuracy in those regions. Many of the SNPs mapped by Choi et al. (2007) were discovered by re-sequencing sequence-tagged sites (STSs) developed from expressed sequence tag (EST) sequences.

A possible disadvantage of SNPs compared to SSRs is that they typically have only two alternative bases per locus, whereas SSR loci can have numerous alleles (i.e., variable numbers of tandem repeats; Yoon et al. 2007). It is therefore generally more difficult to tag a gene or a region containing a gene with a single SNP marker (unless the SNP itself affects the phenotype of interest). This limitation can be overcome by defining a *haplotype* composed of a specific set of DNA bases at several linked SNPs that span a target gene locus. With sufficient linkage disequilibrium (LD) between the markers and the desired allele, it is possible to select for a gene using the haplotype associated with the desired allele for a trait. In some cases, it is possible to identify haplotypes using a subset of SNPs called "tag SNPs". These capture a large portion of the total allelic variation in a haplotype block or region of high LD that is flanked by blocks showing historical recombination (Altshuler et al. 2005). Furthermore, the fact that most SNP loci have only two possible alleles actually facilitates automation of genotype analysis because it makes base-calling qualitative, whereas SSR genotyping is quantitative (i.e., the number of tandem repeats possible at an SSR locus can vary considerably; Koebner and Summers 2002). Studies have found that soybean has limited haplotype diversity in comparison with some other plant species that have been studied, a relatively high level of genome-wide linkage disequilibrium compared to plants like maize (*Zea mays* L.), and a genome that is a mosaic of only three or four unique haplotypes (Zhu et al. 2003; Rafalski and Morgante 2004). This is probably a reflection of a small number of domestication events from *G. soja* and/or the relatively small genetic base of North American soybean germplasm (Gizlice et al. 1993).

SNP genotyping assays typically utilize either PCR primer extension, a ligation reaction between two oligonucleotides, or hybridization of a probe that is sensitive to imperfect base pairing at the queried nucleotide position. Lee et al. (2004b) compared four detection methods using a Luminex 100 flow cytometer platform (Luminex, Austin, TX) for genotyping soybean SNPs. These evaluations took into consideration reliability, cost, and time required. Although the direct hybridization assay required the least time and expense, it was not as reliable with all four of the SNP loci tested as were single base extension of a PCR primer or allele-specific primer extension. The authors concluded, however, that the advantages of direct hybridization over the other assays might make it worthwhile to empirically adjust the reaction conditions to ensure accurate genotyping of large numbers of individual plants for MAS applications.

The TaqMan assay developed by Applied Biosystems (Foster City, CA) has proven highly reliable for genotyping SNP loci, and is being used by some large seed companies for very high-throughput applications. For public sector soybean breeders, however, the cost per data point of using this assay can be prohibitive for MAS applications. At present, a less expensive option may be to genotype SNPs using the HybProbe or Simple Probe assays developed for use with the Roche LightCycler 480 quantitative PCR thermal cycler (F. Hoffman-La Roche Ltd., Basel, Switzerland). This instrument is capable of detecting differences in the melting curves of amplicons containing one or more SNPs, and is compatible with a variety of assay formats, including TaqMan as well as Roche's own HybProbe and SimpleProbe assays. The Lightcycler480 was used to develop a SNP assay to define haplotypes and to distinguish the soybean rust resistant cultivar Hyuuga from the soybean ancestral genotypes and previously reported sources of rust resistance (Monteros et al. 2007a, b). It has also been used to distinguish soybean lines resistant to Southern root-knot nematode and frogeye leaf spot from those that are susceptible (Ha and Boerma 2008).

Two technologies for large-scale SNP genotyping currently receiving attention from soybean breeders and geneticists are the Illumina BeadStation 500®/GoldenGate assay system (Illumina Inc., San Diego, CA) and various real-time PCR thermal cyclers for which accurate and relatively affordable genotyping assays have been developed. The GoldenGate assay developed by Illumina allows simultaneous genotyping of up to 1,536 SNP loci, and can genotype a population of 192 individuals in just three days (Hyten et al. 2008). The cost of the equipment and software has thus far limited purchases to a few genotyping core facilities or laboratories where the expense can be defrayed to some extent through contract genotyping for other researchers. The current cost of the GoldenGate assay is also a limitation for its direct use in MAS, but its ability to rapidly identify SNPs tightly linked to important genes will be an important boost to the potential of MAS for multiple traits of interest. The efficacy of the GoldenGate assay for mapping single genes was demonstrated when bulked segregant analysis was used to map the

*Rpp3* locus conditioning rust resistance in soybean (Hyten et al. 2009). To optimize efficiency of this technology for QTL discovery and mapping, two distinct sets of SNP markers are being developed for use in the GoldenGate assay (Cregan et al. 2008). A "Universal 1,536 Soy Linkage Panel" consisting of SNPs distributed at 1.5–2.0 cM intervals across all linkage groups can be used to map QTLs in populations with high levels of linkage disequilibrium (Hyten et al. 2008). Another set consisting of more than 15,000 SNPs has been designed for gene discovery through association analysis using whole-genome scans and a custom high-density SNP array (Cregan et al. 2008).

The transition from SSRs to SNPs in public sector breeding programs has been hindered by the cost of equipment and reagents for SNP genotyping, the relatively limited number of SNP markers available until recently, and the lack of a genotyping method that is inexpensive, simple, quick, and highly reliable. Improvements in available technologies are being made on a regular basis, driven by the use of SNPs in the medical and human genetics communities for association analyses to find and map genes associated with mammalian diseases (Syvänen 2005; Choi et al. 2007). The development of simple and less expensive quantitative PCR-based genotyping assays has substantially reduced the cost per data point of SNP markers and is promoting increased use of SNPs in public breeding programs. For example, the cost of some quantitative PCR-based assays is low enough that they are now being used routinely in MAS for traits such as root-knot nematode resistance (H. Roger Boerma, pers. comm.). The cost of real-time PCR instruments that are substantially less expensive than the Illumina system can be shared by multiple programs and installed in genotyping core facilities where they are accessible to personnel from several laboratories. The relative simplicity of setting up a quantitative PCR assay and the availability of user-friendly software typically allow researchers to conduct their own genotyping, whereas use of the Illumina requires more extensive training.

SSRs still have certain advantages over SNPs for mapping genes and QTLs in soybean in some public sector laboratories. As mentioned earlier, this is due partly to their greater allelic diversity per marker and because several soybean breeding programs have already invested in SSR primer sets and genotyping equipment. Once SSRs within a few cM of a gene are identified, they can immediately be used for MAS. In university soybean marker laboratories, the trend until about 2008 was to use SSRs for mapping and a combination of SSRs and SNPs for MAS, while some large seed companies now rely primarily on SNP markers for both mapping and MAS. The release of the Williams 82 genome sequence ( see Chapter 10 of this volume), the availability of an ever increasing number of publicly available SNP markers (Cregan 2008), and decreases in the cost of re-sequencing DNA to identify new SNPs are increasing the use of SNPs in public sector breeding programs.

Other types of DNA markers have played relatively limited roles in soybean MAS. Random amplified polymorphic DNA (RAPD) markers and amplified fragment length polymorphisms (AFLPs) were used to some extent before SSRs became widely available, and have been used on a limited basis to map loci in regions with few polymorphic SSR markers, but have seldom been employed for MAS. Both of these classes of markers are usually dominant, so it is not possible to distinguish heterozygotes from one class of homozygotes. In addition, there have been repeatability problems with RAPDs, and the complex banding patterns on AFLP gels can be difficult to interpret. In some cases, amplicons generated by these markers have been sequenced to permit development of codominant sequence-tagged site (STS) markers more suitable for MAS and mapping studies. STS primers have also been developed by sequencing expressed sequence tags (ESTs), and these have the advantage of being from known coding regions.

## 6.3 Potential Uses for MAS in Soybean Breeding Programs

### 6.3.1 General Considerations

The uses and efficiency of MAS have grown in parallel with the increase in number of DNA markers and advances in the technologies that permit efficient and cost-effective genotyping. MAS can expedite development of new cultivars with specific traits, but phenotypic evaluation of inbred lines for those traits and for overall agronomic performance is still necessary to ensure that a potential cultivar will perform as expected, and that it will produce competitive yields in a range of environments (see Chapter 2 of this volume). Though numerous QTL mapping studies have been completed in soybean (for summaries, see Orf et al. 2004, and Chapter 5 of this volume), relatively few of these QTLs have been selected for using MAS on a large scale.

### 6.3.2 Specific Uses

Holland (2004) noted that the most frequent uses of MAS, in order from most common to least, have been for gene introgression, recovery of the recurrent parent genome, and in forward crosses for which high linkage disequilibrium exists between the gene(s) of interest and the markers flanking them. This assessment certainly reflects the situation in soybean breeding. Important uses of MAS in soybean are (i) selection of genetically diverse parents for making crosses; (ii) germplasm characterization and verification, including confirmation of hybrids derived from manual pollinations; (iii) marker-assisted introgression of useful genes; (iv) accurate selection of plants with genes conditioning low-heritability traits; (v) selection for resistance in the absence of a pathogen or pest; and (vi) gene pyramiding. Each of these topics is discussed below and is illustrated with examples of its use in soybean breeding programs. Also discussed is the growing use of pedigree analysis studies, in which a marker's effectiveness at tracking a desirable gene through an extended pedigree is evaluated.

### 6.3.2.1 Selection of Genetically Diverse Parents for Making Crosses

Marker data can be used to estimate relatedness between potential parents for breeding/mapping populations in order to maximize genetic diversity in the progeny, or to better sample variations present in exotic germplasm collections (Tanksley and McCouch 1997; Li et al. 2001; Cregan 2008). The identification of genetically diverse parents is of particular importance in soybean because modern North American soybean cultivars were developed from a narrow genetic base (Carter et al. 2004). Breeding has further reduced the genetic diversity among elite breeding lines and cultivars relative to that which existed among the founding ancestors (Gizlice et al. 1993). On the basis of pedigree analysis, Gizlice et al. (1996) determined that 80% of the genes found in public soybean cultivars released between 1947 and 1988 were derived from just 13 ancestral lines. Analysis of soybean cultivars using RFLPs generally detected only two alleles at most loci (Keim et al. 1990). In comparison, a study of a group of 20 inbred lines of maize (*Zea mays* L.) found an average of 4.5 different RFLP alleles at each locus (Melchinger et al. 1990). An analysis of losses in genetic diversity by Hyten et al. (2006) demonstrated that domestication of cultivated soybean (*G. max*) from *G. soja* was the bottleneck event in which most of the initial diversity was lost. Through the process of domestication, 81% of the rare alleles were lost, and significant changes in allele frequencies occurred in 60% of the genes. The implication of these findings is that novel or rare alleles are more likely to be found in agronomically inferior *G. soja* accessions rather than in adapted *G. max* germplasm (Cregan 2008). Introgression of *G. soja* genes into elite *G. max* genetic backgrounds naturally lengthens the amount of time needed to develop a high-yielding cultivar. Markers could play an important role in allowing breeders to tap this source of genetic diversity while minimizing transfer of undesirable alleles (Tanksley and McCouch 1997).

In an effort to identify sources of germplasm to broaden the soybean genetic base, Brown-Guedira et al. (2000) utilized RAPDs and SSR markers to evaluate the extent of genetic variation in plant introductions (PIs). Several groups of plant introductions distinct from the majority of North American soybean ancestors were identified and recommended to breeders who want to incorporate more genetic diversity into their soybean improvement programs. SSR and AFLP markers have also been used to evaluate Asian soybean accessions, and to show that Japanese and Chinese germplasm pools can be used as genetically distinct resources to enlarge the genetic base of the North American soybean population (Abe et al. 2003; Ude et al. 2003; Wang et al. 2006). Yamanaka et al. (2007) used SSR markers to evaluate the genetic relationships between Chinese, Japanese, and Brazilian soybean gene pools, and suggested that exchanges of these gene pools might be useful for increasing genetic variability in soybean breeding. SSR markers have also been used to compare genetic diversity among soybean PIs with resistance to *Phytophthora sojae* and to identify new alleles for resistance not currently present in American cultivars (Burnham et al. 2002). After the soybean aphid (*Aphis glycines* Matsumura) became a serious pest in the US, efforts to map resistance genes from different sources of resistance intensified. Chen et al. (2007) categorized the different sources of resistance by their SSR marker diversity in an attempt to better deploy resources towards mapping and introgressing unique aphid resistance genes.

### 6.3.2.2 Germplasm Characterization and Verification

DNA markers can be used to verify the identity of germplasm intended for use in crosses, and to resolve the true parentage of plants and cultivars in cases of ambiguity. Although morphological traits are often useful for distinguishing $F_1$ hybrids from inbred progeny (see Chapter 2 of this volume), molecular markers can be used to rule out the possibility of self-pollination if the two parents have similar morphological characteristics, or to confirm that seeds or immature plants are true hybrids. Although flower or hilum color may indicate cases of mistaken identity, molecular markers can provide a definitive test when phenotypic appearance is not distinct. Marker genotypes can also be used to confirm or establish the paternity of hybrids by providing molecular fingerprints that can help to identify candidate parents of an ambiguous plant or breeding line. For example, Narvel et al. (2001a) found that the alleles at four SSR markers linked to an insect resistance QTL on molecular linkage group (LG) M of 'Crockett' matched those of the insect resistant accession PI 229358 rather than the alleles of its supposed parent, PI 171451. In another case, marker data were crucial in convincing skeptics that the high-yielding cultivar N7001 was truly descended from an Asian germplasm accession (PI 416937) that is not well-adapted to North America (T. E. Carter, Jr., pers. comm.). Yoon et al. (2007) assembled a panel of 23 informative SNPs (BARCSoySNP23), which can be used effectively to identify cultivars.

### 6.3.2.3 Marker-Assisted Gene Introgression

Transfer of beneficial genes from one genotype to another can be facilitated and accelerated using linked markers. MAS for a specific trait relies on information previously gained from molecular mapping of QTLs and/or qualitative trait genes using DNA markers. If a backcross (BC) breeding approach is used to introgress one or a few genes from a donor parent into the genetic background of an agronomically superior recurrent parent, markers can be used to select BC-derived plants or seeds heterozygous for the gene(s) of interest, a strategy referred to as "foreground selection". Marker-assisted backcrossing (MABC) can also be used to accelerate recovery of the recurrent parent genome outside of the region(s) containing the gene(s) (Visscher et al. 1996), a strategy called "background selection".

Foreground selection, which uses linked markers to monitor the introgression and inheritance of an allele of interest, is the primary application of MAS in plant breeding (Frisch et al. 1999b). This technique is especially useful in backcrossing, where the objective is to introgress one or a few gene(s) from a donor parent (DP) into the genome of a recurrent parent (RP) (see also Chapter 2 of this volume). The DP is often an exotic and/or agronomically inferior germplasm accession that possesses a unique desirable trait, while the RP is typically a high-yielding, agronomically superior cultivar or line which lacks the trait sought from the DP. Selection can be based on the genotype of a single marker tightly linked to a locus or QTL associated with the trait, or based on the genotype at two or more marker loci flanking the trait locus.

Background selection, originally proposed by Tanksley and Rick (1980), is used to accelerate recovery of the RP genome during multiple backcrosses by selecting segregated individuals homozygous for the RP allele at markers on each molecular linkage group (LG) or chromosome (Frisch et al. 1999a). Background selection using markers flanking a gene introgressed from a donor parent can reduce the potential for linkage drag, the inadvertent co-introgression of deleterious alleles at loci linked to the introgressed gene. Linkage drag can be reduced by genotyping segregating individuals at markers flanking the introgressed gene in order to identify the segregant(s) with the smallest segment of DNA inherited from the DP. Dispersed markers on other LGs can also be used to identify plants with lower than average amounts of donor parent genome. Background selection would typically be used in conjunction with foreground selection using either markers or phenotypic selection. Minimizing the size of the introgressed segment is particularly important given that the soybean genome, like that of many other plants, is organized into gene-rich regions (gene space) and gene-poor regions composed largely of repetitive DNA sequences (Young et al. 2003; Choi et al. 2007). This arrangement increases the risk that a valuable allele from an unadapted germplasm source will be linked to one or more alleles at nearby loci that will detrimentally affect yield or other agronomically important characters. Frisch et al. (1999a) outlined various MAS background selection scenarios that considered different starting variables such as sample size and the position of the flanking markers relative to the gene of interest. Stuber et al. (1999) suggested limiting introgression of target DNA segments in maize to a maximum of two to four segments in order to reduce the effects of linkage drag.

Backcrossing can be very effective to improve a cultivar for disease resistance, for example, but the final product will be identical to the original cultivar for most other traits, including yield potential. Therefore, speed is critical in developing an improved line which can be released before the yield of the original cultivar is significantly surpassed by new releases from competitors. MAS is particularly useful in backcrossing programs to identify heterozygous plants possessing desirable alleles that are recessive or incompletely dominant (Ribaut and Hoisington 1998). Although breeders may use backcrossing at various points within the breeding process, the quickest way to introgress a gene into an elite background is to cross at each generation with the $F_1$ plants derived from the previous generation. In this situation, each backcross occurs between a heterozygous plant and the recurrent parent, and it is advantageous for the breeder to be able to focus crossing efforts on the 50% of the plants in each backcross generation that are heterozygous at the locus of interest. This is especially important in soybean because manual pollinations are tedious, frequently unsuccessful, and seldom produce more than two seeds per successful pollination. MABC allows identification of heterozygotes even if the introgressed gene is recessive, thus eliminating the need to conduct progeny testing in each generation to distinguish homozygotes from heterozygotes. In addition, MAS can be useful for selecting $BC_nF_1$ plants to backcross to the recurrent parent when a phenotype being selected for cannot be determined prior to flowering. Beckman and Soller (1986) estimated that the frequency of an introgressed allele after three generations of backcrossing with selection using a single linked marker would be 0.66, while the use of two markers flanking the introgressed allele (with a recombination frequency of 0.40 between markers) would increase this frequency to 0.85. In contrast, the frequency of the favorable allele introgressed from the donor parent would drop to about 0.06 after three backcrosses in the absence of MAS or phenotypic selection. In addition, MAS could also be useful following self-pollination of the final backcross-derived individuals to distinguish $BC_nF_2$ individuals homozygous for an introgressed dominant gene from the heterozygous individuals.

The development of a glyphosate-tolerant version of the University of Georgia cultivar Benning (released as 'H7242 RR') in less than five years is one example of how DNA markers for background selection can be used to accelerate recovery of the RP genome in a backcross program (Orf et al. 2004). The complete tolerance to glyphosate provided by a single copy of the transgene facilitated phentoypic selection of hemizygous plants, but SSR markers also allowed identification of the glyphosate-tolerant plants with the highest proportion of Benning genome in successive backcross generations. Three evenly spaced markers per LG (60 markers total) were used to fingerprint 30 glyphosate-tolerant $BC_2F_1$ plants, and a plant estimated to have a 91% Benning background (rather than the 87% average) was used for the third backcross to Benning. After this backcross, 54 glyphosate-tolerant lines estimated to be 99% similar to Benning were recovered. Following yield tests of 19 $BC_3F_{2:4}$ lines, seeds from three lines were composited to create a glyphosate-tolerant version of Benning (Orf et al. 2004). During the backcross program to develop H7242 RR, DNA was collected from a total of 202 plants, and the estimated cost for conducting the MAS (including labor, reagents and prorated equipment) was approximately US$2,500 (Orf et al. 2004).

MABC has been and continues to be an important strategy in soybean for transferring useful genes from *G. soja* and unadapted *G. max* germplasm (Concibido et al. 2003; Orf et al. 2004). Sebolt (2000) used RFLP and SSR markers to select plants with high-protein alleles from a *G. soja* accession. Other researchers have introgressed putative yield-enhancing alleles from unadapted PIs, followed by cycles of inbreeding or backcrossing to elite parents (Concibido et al. 2003; Kabelka et al. 2004; Guzman et al. 2007). The mapping of QTLs conditioning resistance to corn earworm (*Helicoverpa zea*) and velvetbean caterpillar (*Anticarsia gemmatalis*) in two different PIs (Rector et al. 1999, 2000) has allowed breeders to select for resistance while selecting against the remaining genome from the unadapted PI. Zhu et al. (2008) used markers linked to three QTLs from PI 229358 to develop insect-resistant near-isogenic lines of Benning . These lines have allowed better characterization of the individual and combined effects of the insect resistance QTLs while also providing germplasm with the PI 229358 resistance genes in backgrounds that are agronomically superior to that of the original PI parent. However utilization of genes from exotic sources has been hampered by the fact that crosses between an elite parent and an unadapted parent seldom produce progeny lines with much agronomic merit (Sleper and Shannon 2003).

Molecular markers can also be used to transfer transgenes from one genetic background to another, or to pyramid transgenes with native genes (Walker et al. 2002). Both the principle and technique are essentially the same as that used for the introgression of native genes and QTLs, except that a pair of PCR primers specific to the sequence of the transgene would be used. Detection of the transgene would be a simple plus-minus assay since there would be no corresponding homologous region in nontransgenic plants. Quantitative PCR could be used in lieu of a progeny test to determine whether a transgene-positive plant is homozygous for the transgene or hemizygous (i.e., carrying a copy of the transgene on only one chromosome from a homologous pair).

## 6.3.2.4 *Selection for Traits with Low Heritability*

Situations in which the phenotype is either difficult to assay or unreliable due to environmental effects present challenges in which MAS could improve selection efficiency (Melchinger 1990). A caveat for low heritability traits is that reliable phenotypic data are critical for identification of markers that can be used effectively for MAS. Adequate replication in multiple environments or assays is required to obtain accurate phenotypic data to map loci associated with quantitative traits and thereby identify associated markers. Overall, simulation studies comparing MAS with phenotypic selection for traits with relatively low heritability suggest that the additional genetic gain from MAS is likely to be highest in the early generations, and may rapidly decrease in subsequent generations (Stuber et al. 1999). Indeed, MAS for some traits may become less efficient than phenotype-based selection in the long term because the rate of fixation of unfavorable alleles at unselected QTLs with small effects is higher if stricter selection for QTLs with larger effects is applied through MAS in early generations (Hospital et al. 1997).

Modification of seed composition is a breeding objective for which MAS could provide greater breeding efficiency than phenotypic selection. Seed traits such as reduced levels of palmitic and linolenic acids or higher levels of stearic acid are increasingly important in the U.S. market (Wilson 2004; Zhang et al. 2008). Although some seed composition traits have relatively high heritabilities under controlled environmental conditions, they can be highly influenced by variation in field environments. Furthermore, selection is most effective in early generations, some of which are commonly grown in tropical winter nurseries with environmental conditions quite different from those in which the potential cultivars would ultimately be grown. Several studies have noted the environmental impact of temperature and differing precipitation patterns on seed protein content (Brummer et al. 1997; Yates 2006). Seed oil concentration and composition are also influenced by environmental conditions (Wilson 2004), and many of the enzymes that affect fatty acid composition in the seed are temperature-sensitive (Cheesbrough 1989). This could partially explain why Beuselinck et al. (2006) found discrepancies between genotypic versus phenotypic selection of lines containing genes that condition lower amounts of linolenic acid. Oleic acid content has also been shown to be environmentally sensitive, although the extent of environmental sensitivity was demonstrated to be dependent on genetic background in at least one study (Oliva et al. 2006). Monteros et al. (2008) used oleic acid content data taken from the same lines grown in two different environments to counteract the environmental sensitivity of the trait and thereby map QTLs affecting oleic acid content in the seed. Markers can thus improve the selection efficiency for such environmentally-sensitive traits by allowing selection for the genes that condition the desirable phenotype when the phenotype observed in a single environment could be misleading.

Efforts are continually made to identify and tag QTLs that increase seed yield, which is by far the most economically important trait in soybean. Although yield is a complex trait that requires extensive resources for phenotypic evaluation in diverse environments, several studies have mapped certain QTLs influencing yield to similar positions (Orf et al. 1999; Yuan et al. 2002; Smalley et al. 2004; Guzman et al. 2007). As QTL identities and positions start to coalesce, it should become feasible to enhance overall yield potential using MAS. Concibido et al. (2003) used markers to introgress a yield QTL on LG B2 from *G. soja* and to assess its effect in different genetic backgrounds. Although a yield advantage associated with this QTL was not observed in all of the genetic backgrounds tested, the work demonstrated that MAS could be used successfully to enhance yield in at least some backgrounds. However, the large effect that epistasis appears to have on yield and the relatively small contributions of individual yield QTLs, as well as large genotype × environment interactions, are likely to limit the impact that MAS can have on increasing soybean seed yield.

### 6.3.2.5 Selection for Resistance in the Absence of a Pathogen or Pest

Various situations can be envisioned in which phenotypic assays to screen breeding populations for resistance to a disease or pest would be undesirable or impossible. Screening for resistance to some diseases may be ineffective in early generations if there are not enough seeds from each line to conduct replicated testing. A breeder may also wish to avoid infecting plants with a virus, for example, if some of the susceptible plants possess other favorable genes for traits under selection. This would also be a greater concern with plants from early generations. Other disease or pest resistance assays may cause unacceptable delays in the breeding program, or phenotypic selection for resistance to some pathogens may be restricted to certain locations or seasons.

Development of North American cultivars with resistance to Asian soybean rust (caused by *Phakopsora pachyrhizi*) is an example of a breeding objective for which MAS could play an important role. The appearance of Asian soybean rust in the USA in November 2004 presented a challenge to soybean breeders wishing to develop resistant cultivars for the Midwest. Because *P. pachyrhizi* can only survive the winter in living host tissue, large-scale phenotypic screening must be conducted in the Southeast, close to areas where the pathogen overwinters. The natural summer photoperiods at those latitudes cause soybeans from early MGs (i.e., MGs 000 to V) to mature before useful phenotypic data for disease ratings can be obtained in most years. Tagging rust resistance genes with markers would allow breeders to select for resistance in the absence of the pathogen, to distinguish between plants homozygous or heterozygous for any particular resistance gene, and to track inheritance of resistance genes.

### 6.3.2.6 Pyramiding Genes

Markers can be used to combine genes conditioning the same trait ("pyramiding") or genes that condition different traits ("stacking"). For convenience, we will refer to both procedures as pyramiding, though Dekkers and Hospital (2002) used the term "genotype building" to refer to the process of combining favorable alleles from two parental lines, and reserved the term "pyramiding" for combining favorable alleles originating from more than two parental lines.

One of the most useful applications of MAS is in pyramiding beneficial alleles that improve the same trait, particularly in cases where one of the genes would mask the presence of the other genes (Melchinger 1990; Huang et al. 1997). An example would be where the objective is to combine genes that condition moderate levels of resistance with a major resistance ($R$) gene or a transgene that confers race-specific resistance, or to pyramid two or more $R$ genes. Resistance conditioned by a single $R$ gene is likely to be overcome by novel biotypes of a pathogen if the gene is widely deployed as the sole resistance gene in new cultivars. The durability, level, and/or range of resistance could theoretically be increased by supplementing the major gene with other resistance genes (Nelson 1978; Melchinger 1990; Saghai Maroof et al. 2008). The high level of resistance conferred by the major $R$ gene would phenotypically mask the presence of additional resistance alleles at other loci, making it difficult to pyramid the genes using phenotypic assays. Markers linked to the individual genes could be used to select plants or families possessing multiple resistance genes and to combine those genes in the same genetic background, as Huang et al. (1997) demonstrated with bacterial blight resistance genes in rice.

There are numerous other examples of potential pyramiding applications of MAS in soybean. A breeder could use MAS or MABC to combine the soybean rust resistance genes *Rpp1* (Hyten et al. 2007b) with either *Rpp?*(Hyuuga) (Monteros et al. 2007b) or *Rpp3* (Hyten et al. 2009). Historically, the resistance of cultivars with single *Rpp* genes has not remained effective for more than a few years, but a combination of two or more *Rpp* genes should be more difficult for the rust pathogen, *P. pachyrhizi*, to overcome. SNP assays for MAS of traits such as southern root-knot nematode resistance allow breeders to easily select for resistance alleles at both major and minor QTLs, thereby pyramiding multiple genes to achieve a higher level of resistance (Ha et al. 2007). *Rsv1, Rsv3, and Rsv4* together condition resistance to all strains of soybean mosaic virus (SMV), and could be pyramided to obtain comprehensive SMV resistance (Saghai Maroof et al. 2008). With the identification of different soybean aphid biotypes (Kim et al. 2008), efforts to pyramid different aphid resistance genes such as those from PI 567541B (Zhang et al. 2009), PI 243540 (Mian et al. 2008), and "Dowling" or "Jackson" (Hill et al. 2006a, b) will be critical to ensure broad-spectrum protection from this pest. Walker et al. (2004, 2006) used markers to combine an insect resistance allele from the Japanese soybean accession PI 229358 with a *cry1Ac* transgene that encodes a *Bacillus thuringiensis* (Bt) protein toxic to lepidopteran pests of soybean. Phenotypic selection of single plants with a combination of the native gene and transgene would have been more time-consuming and less accurate than selection using markers. MAS could be equally useful for combining genes conditioning resistance to different pests or pathogens if conducting separate phenotypic assays with each would be impractical or impossible.

Markers can also be useful for pyramiding QTL alleles to modify a trait that does not have a high heritability. An example of this would be the levels of seed fatty acids such as oleic acid, which can be influenced by maternal effects as well as changes in the environment (Brim et al. 1968; Pantalone et al. 2004). MAS could be used to pyramid mapped genes that condition higher oleic acid levels in the seed. Although genotypic selection for increased oleic acid requires the use of multiple markers to select for the highest level of oleic acid, the alternative of using phenotypic selection for oleic acid may be more challenging, given the complex nature of the trait (Monteros et al. 2008).

### 6.3.2.7 Pedigree Studies and Retrospective Assessments of Phenotypic Selection Using Markers

While these applications of markers are not technically MAS, they are examples of how DNA marker data can guide breeders in choosing between MAS and phenotypic selection to achieve certain breeding objectives. The effect of allele substitution at previously mapped QTLs can be confirmed by demonstrating co-inheritance of a specific allele with a certain phenotype. This has been a successful approach for tracking QTL inheritance in the southern North American elite germplasm, which traces its ancestry to slightly fewer lines than the northern North American germplasm (Delannay et al. 1983; Gizlice et al. 1993). Pedigree studies have mainly focused on resistance to disease and pests, including frogeye leaf spot (*Cercospora sojina*; Missaoui et al. 2007), peanut root-knot nematode (*Meloidogyne arenaria*; Yates et al. 2006), southern root-knot nematode (*M. incognita*; Ha et al. 2004), bacterial pustule (*Xanthomonas campestris* pv. glycines; Narvel et al. 2001a), and the Mexican bean beetle [*Epilachna varivestis* (Mulsant); Narvel et al.

2001b]. Alleles conditioning tolerance to abiotic stresses such as salt tolerance have also been tracked through pedigree analysis studies (Lee et al. 2004a). The strength of association between the desirable phenotype and the specific allele from the original resistant or tolerant parent through its descendants is a good indication of that marker's potential efficacy for MAS across multiple populations and generations. Marker pedigree studies also provide information on how much of the total genetic variance a QTL must explain to result in a high likelihood of it being retained during phenotypic selection.

DNA markers closely linked to mapped QTLs have also been used to conduct retrospective analyses of conventional breeding efforts based on phenotypic selection for a trait. Narvel et al. (2001b) assessed the progress that six independent public breeding programs had made over three decades in developing insect-resistant lines and cultivars. In this study, SSR markers linked to four QTLs associated with lepidopteran resistance were used to survey the inheritance of genomic segments from an insect-resistant plant introduction. This study confirmed the importance of a QTL near Satt220 and Satt536 on LG M which had been retained in at least 13 out of 15 lines and cultivars selected visually for resistance to coleopteran and/or lepidopteran pests. Graphical genotypes constructed from genotypic data for markers on LG M showed that most of the lines and cultivars still contained >10 cM of donor parent DNA flanking the estimated location of this resistance QTL. This potential for linkage drag could explain in part why development of high-yielding insect resistant cultivars using phenotypic selection has been largely unsuccessful (Lambert and Tyler 1999). Some lines or cultivars were found to also contain the PI-derived resistance alleles at QTL on either LG G or LG H, but none appeared to possess the full complement of insect resistance QTLs mapped in PI 229358. Thus, despite three decades of phenotype-based breeding for insect resistance in several breeding programs, none of the released germplasm had both the full resistance of the PI and the yielding ability of some susceptible cultivars that were available at the time of release (Narvel et al. 2001b).

## 6.4 Considerations and Limitations

Although genotype-based selection using molecular markers can be more efficient than phenotypic selection for some traits, MAS has limitations (Lande and Thompson 1990; Eathington et al. 2007). Ongoing efforts to map additional genes and QTLs associated with many important traits, and to fine-map other loci using recently developed markers will gradually reduce some of these concerns, soybean breeders should be aware of these limitations as well as the advantages of MAS (Hospital et al. 1992).

Although computer simulation studies have indicated that MAS could be more effective than phenotypic selection, these simulations were based on assumptions about trait heritability, the proportion of additive genetic variance that could be explained by the markers, the distance of the marker(s) from the gene(s), and the type and intensity of selection (Staub and Serquen 1996; Hospital et al. 1997; Knapp 1998). Three major limitations to practical applications of MAS include the following: (i) the full complement of genes determining a trait is rarely known, particularly for traits with a low heritability; (ii) most markers are near, but not within a QTL, so crossovers between the marker(s) and a QTL may occur; and (iii) the success of MAS is dependent upon the initial quality of the phenotypic data and the QTL mapping studies that first established the marker-trait association (Dekkers and Hospital 2002). Related to the third factor, the utility of MAS depends on the level of linkage disequilibrium in a population, the size of the populations required to detect traits with low heritability, and sampling errors in the estimation of relative weights in the selection indices (Stuber et al. 1999). Less scientific factors may also come into play; two trains of thought that Morris et al. (2003) highlighted in considering the pros and cons of MAS and phenotypic selection were the time savings possible with the former and the lower costs associated with the latter, but these generalizations do not apply to all traits. The cost and time variables are constantly changing as well; as genotyping becomes more high-throughput and routine, the availability of quality phenotypic data that is not confounded by genotype × environment variation becomes the bottleneck (Xu and Crouch 2008). Making decisions about whether to use MAS and what to use it for can be complicated, and while simulation studies provide useful suggestions, they may not accurately reflect the tools and knowledge actually available to a breeder for a particular trait.

Even if QTL studies are properly conducted, there is a discrepancy between the number of QTL studies and the number of actual cases in which MAS is regularly employed for trait introgression and genome recovery. Although over 10,000 marker-trait associations have been reported in the literature (Bernardo 2008), very few are routinely used for MAS in breeding programs. Xu and Crouch (2008) demonstrated this gap by graphing the number of articles with the keywords "quantitative trait locus" versus the number with "marker-assisted selection", noting the former outpaced the latter by a factor of three. Tanksley and Nelson (1996) suggested that the two main reasons for this were that (i) QTL discovery and variety development have been independent efforts, and that (ii) breeding-related QTL studies have focused on the manipulation of quantitative traits in elite germplasm. More recently, articles have highlighted and cautioned about the difference between QTL studies that attempt to account for all of the genetic variation underlying a trait versus those that focus on the few QTLs that could be effectively used for MAS in a breeding program (Bernardo 2008). Stuber et al. (1999) proposed that another important factor has been the influence of genetic background on QTL expression (i.e., epistasis), which means that the effect of a QTL may not be the same in populations derived from different crosses. This third factor appears to have been the primary reason why Reyna and Sneller (2001) observed no significant benefits from three yield QTLs of the northern cultivar Archer in near isogenic lines derived from crosses to two southern U.S. cultivars.

Holland (2004) stated that examples where MAS has been or is soon expected to be an important part of mainstream forward crossing breeding programs share two important features, both of which have been discussed in some detail in previous sections. First, the markers are tightly linked to a small number of loci with relatively large effects on traits that are difficult or costly to phenotype. Second, specific marker alleles are associated with desired alleles at target loci consistently across multiple breeding populations. An example of MAS in soybean breeding which meets those criteria is selection for resistance to the soybean cyst nematode (SCN). In a review of the progress of SCN research in soybean, Concibido et al. (2004) described the factors that contributed to the success of MAS for this trait. Resistance alleles at the *rhg1* and *Rhg4* loci have been consistently detected in multiple SCN-resistant sources, and markers are available , with few exceptions, that can distinguish between resistant and susceptible lines from any number of populations. Although SCN resistance has a relatively high heritability, it is tedious, time-consuming, and expensive to screen for in the greenhouse. In addition, multiple alleles are required to condition the highest level of SCN resistance, and this complicates the development of highly resistant cultivars through phenotypic selection alone. However, there are currently some intellectual property obstacles to using MAS in selecting for this particular trait, as described below.

### 6.4.1 Technical Considerations

As mentioned above, MAS efficiency is ultimately contingent upon a tight linkage of the marker to the gene(s) conditioning the trait under selection, and identification of tightly linked markers requires QTL mapping studies that utilize accurate phenotypic and genotypic data. An ideal situation would be to have one or more SNP markers actually located within the gene of interest. These studies require mapping populations large enough to reduce the risk of detecting false positives and to allow detection of epistatic interactions that affect complex traits. Numerous studies have highlighted the importance of sample size in QTL mapping studies, and have indicated that a full understanding of the numbers and individual effects of the QTLs influencing many quantitative traits cannot be described using populations with fewer than 1,000 individuals (Asíns 2002; Holland 2007). MAS has sometimes fallen short of its promise because the QTLs that condition a desirable phenotype and the interactions among them have not been adequately described (Holland 2004). However, this is not a problem with MAS per se, but rather with the ability to correctly identify tightly linked polymorphic markers.

Precision mapping and high-throughput MAS have been greatly aided by recent and ongoing technological advances that have improved the efficiency of processes such as DNA extraction and genotype scoring. These innovations improve the accuracy and precision of QTL mapping and MAS, and yield results with a quicker turnaround. For example, tissue grinders have been designed that uniformly pulverize plant tissue from several hundred samples at once, allowing a nearly five-fold increase in the number of samples that can be processed in a day (Dreher et al. 2003). The use of robots and/or multichannel pipettors can accelerate pipetting procedures and improve uniformity among samples. Some of the technological advances have been driven by the extensive interest and success in using SNPs to map and tag disease-related genes in mammalian genomes (Cardon and Abecasis 2003). As mentioned in an earlier section, this interest in SNPs is due not only to their abundance in the genome, but also to the potential for automation of SNP-based genotyping. While SNP markers hold great promise for MAS in soybean breeding programs, the relatively limited number of markers publicly available and the prohibitive cost of equipment initially delayed their widespread adoption by the public sector. This is now changing with increasing access to real-time thermal cyclers with SNP genotyping capabilities, and with technologies such as the GoldenGate assay, which is able to evaluate 1,536 SNPs on 192 different lines in only a few days (Hyten et al. 2008). Although the GoldenGate assay is not economical for MAS, its capacity to quickly identify SNP markers tightly linked to an important locus will enhance the efficiency of MAS in soybean.

The cost and availability of microarrays for genotyping applications are also approaching levels that are reasonable for many programs. The ability to genotype thousands of SNPs on an array for any given individual in an increasingly shorter timeframe means that better marker-trait associations can be identified for use in MAS (Syvänen 2005). With an approximately five-fold reduction in the size of each spot, microarray chips can now accommodate about 6.4 million 25-mer probes, which is about 25 times greater than what was previously possible (Zhu and Salmeron 2007). The complete genome of different lines can be effectively contrasted and compared using such a gene chip in approximately two days.

The ability to screen a higher number of individuals with more markers may mean that soybean breeders are better equipped to apply MAS for several genes or for several traits simultaneously. Soybean growers expect varieties to have multiple disease resistance traits as well as high and stable yield performance, and breeders are relying more heavily on markers in early-generation selection for these traits. MAS is also advantageous with high-value traits that are complex, such as high oleic acid. The markers linked to the six QTLs mapped by Monteros et al. (2008) are predictive of the high oleic acid phenotype, and have been confirmed in different environments and in different genetic backgrounds. Walker et al. (2006) reported that the lowest levels of phytic acid among lines derived from CX1834-1-2, a low-phytate mutant donor line, occurred only in lines that were homozygous for recessive alleles at two loci associated with seed phytate content. Markers linked to each of the loci could be used to distinguish double heterozygotes from single heterozygotes and homozygotes in each generation of backcross lines. In addition to circumventing the environmental effects on fatty acid content and the dominance effects on phytic acid content, MAS permits selection before flowering. Without MAS, seed trait phenotypes could be evaluated only after the next round of backcrossing had been completed "blindly". As with the low phytate trait, aphid resistance in PI 567598B is conditioned by two recessive genes (Mensah et al. 2008), and the use of MAS facilitates introgression of the resistance alleles into adapted germplasm.

Much of the cost and time required for MAS is in the DNA isolation procedure, but once the DNA has been obtained, it can be genotyped at multiple marker loci linked to QTLs associated with the same or different traits. Relative to conducting multiple phenotypic assays on a plant or family, simultaneous selection for several genes using MAS may be the more cost-effective option. Furthermore, the cost of DNA isolation may also be reduced by using less stringent purification methods than those used to obtain DNA for mapping projects. MAS typically requires only enough DNA to screen samples with five or fewer PCR-based markers, and long-term storage and stability are seldom a concern. Most of the commonly used DNA extraction protocols, such as the CTAB-based method of Keim et al. (1988), are designed to produce relatively pure DNA that can be stored at –80°C for a year or more, and some creative techniques have been developed to increase throughput from the CTAB-based method (Flagel et al. 2005). Simpler procedures have frequently proven adequate for rapid screening of segregating populations, however, since the DNA is often discarded within a few weeks after it is isolated and used for genotyping to select the best lines. The methods that Kang et al. (1998) and Kamiya and Kiguchi (2003) developed to rapidly extract DNA from a single seed permit genotyping even before the seeds are planted. These methods were readily adapted at the University of Georgia, where MAS often begins with DNA extraction from quarter-seed chips removed from the end of a seed opposite the embryo. The remainder of the seed can be planted either before or after it has been genotyped. "Chipping" seeds by hand is a tedious process, but designing a simple machine to do it has proven difficult because of the need to avoid damaging the embryo, and because of the variation in sizes among soybean seeds. Nevertheless, both Monsanto and Pioneer have succeeded in automating this process, reserving the ¾-seed portions for planting while analyzing DNA extracted from the ¼-seed chips for multiple traits. Indeed, Andrew Nickell of Monsanto cited the development of a single-seed chipper as one of the innovations that has had a major impact on increased use of MAS in that company's soybean breeding program (*www.stewartseeds.com/ pdf/products/RR2Yield Article Final Andy Nickell 4-1-08.pdf*; verified 5 Feb 2009). Using MAS in this manner not only conserves land, since undesirable genotypes are culled prior to planting, but also eliminates the time-consuming steps of tagging and collecting leaf tissue from plants in the field (Xu and Crouch 2008).

In the examples cited above, the markers used for MAS were tightly linked to genes that had a large influence on a trait. MAS is naturally less effective if it is limited to QTLs that only condition a small portion of the trait variation within a population (Staub and Serquen 1996; van Berloo and Stam 2001). Even with high-heritability traits in which marker-trait relationships are easily detected, it is still difficult to fully resolve the QTL position to an interval of less than 10 cM (Asíns 2002). While two or more markers that flank the QTL in this confidence interval can be used in MAS, there is still a chance that recombination could separate the favorable allele from the selected marker alleles, and 10 cM corresponds to a large segment of introgressed DNA. Phenotypic confirmation of traits in selected lines is therefore important after the appropriate marker alleles have been transferred into homozygous plants or lines.

### 6.4.2 Financial Considerations

While some breeding programs, particularly in private industry, have embraced marker-assisted selection, others are still considering the relative benefits of MAS versus phenotypic selection. Morris et al. (2003) compared the costs of both strategies for introgressing a single dominant gene into an inbred maize line. This model involves the assumption that the presence of the introgressed gene could be detected phenotypically in each generation. The authors reported that MAS saved approximately three cycles of backcrossing, but cost approximately 175 times more than phenotypic selection. Furthermore, this assessment did not include the cost of equipment or of mapping the QTLs to find useful markers for MAS. However, they estimated that it would be possible to accrue approximately US$133,623 in benefits from releasing a product two years earlier, so all or most of the MAS-related expenses could be recovered. This would be particularly important for private industry, since the company that is the first to release a cultivar or hybrid with a novel trait often establishes market dominance. In addition, if a breeder has a small window of time in which to make important selections, then he/she may be willing to spend more in order to get critical data a few weeks sooner. It should also be noted that introgression of a partially dominant or recessive gene using phenotypic selection could substantially increase the cost and time requirements in a conventional breeding approach.

Commercial breeding programs at major seed companies have been quicker than most of their public counterparts to supplement phenotypic selection with MAS for several reasons, but cost is probably foremost among them (Eathington et al. 2007). While new techniques and machines have greatly reduced the cost per data point for genotyping, the expense of purchasing and maintaining state-of-the-art technology is prohibitive for most public breeding programs. A breeder operating on a limited budget must continue to work in the field and greenhouse regardless of whether MAS is used to supplement selection efforts, so a new plot combine may seem to be a wiser investment than additional laboratory equipment. In contrast, large, multinational seed companies have the necessary capital to establish and operate central, high-throughput genotyping centers that can extract and fingerprint DNA samples submitted by multiple breeding programs. Much of the necessary equipment can be used to genotype samples from a variety of crop plants, and many operations can be performed using robots, further improving the throughput and economy of scale that can be achieved at such facilities. Operation and maintenance of sophisticated genotyping equipment can be assigned to expert technicians, and in-house engineers and computer programmers can develop specialized machinery and bioinformatics programs if similar requirements are not commercially available. The ability of major seed companies to purchase large quantities of reagents in bulk also reduces costs per data point.

The start-up costs for using MAS in public breeding programs can be substantially reduced, however, if a breeder has access to local or regional genotyping facilities. Many universities have established genomics core facilities in which DNA genotyping and sequencing equipment is maintained and shared by multiple research groups who would be unable to purchase the instruments independently. Regional genotyping centers are likely to play an increasingly important role in public soybean breeding, as they already do in public wheat breeding programs (Anderson et al. 2007). Such genotyping centers can operate high-throughput equipment to reduce the cost per data point to breeders, and can recoup part of the costs associated with purchasing and maintaining this equipment by charging a fee for custom genotyping of samples submitted by breeders. The establishment of regional genotyping centers at the University of Georgia and at the University of Missouri—Columbia was a keystone objective in the 2007 SoyCAP (Soybean Coordinated Agricultural Policy) proposal. Some small private-sector soybean breeding programs have similarly developed arrangements with universities or other companies to gain access to modern genotyping equipment (Sleper and Shannon 2003). New technology drives down the cost of genotyping, making it more affordable for small breeding programs. For example, melting curve analysis, a technique for SNP detection, incurs approximately 60% of the cost of the traditional TaqMan assay (Chantarangsu et al. 2007) and 50% of the cost of SSR markers (Ha and Boerma 2008), but has the speed and automation of real-time PCR applications.

### 6.4.3 Breeding Strategies

MAS is an important tool for breeders to develop better breeding populations from which to select high-yielding varieties, but its proper place in a breeding program involves a consideration of the interaction of MAS with different breeding strategies. Various studies have outlined the optimal breeding strategy to use in combination with MAS, particularly for pyramiding multiple genes and for selection for complex traits. Ribaut and Betrán (1999) proposed using a single large-scale application of MAS in an early generation to fix favorable alleles from the two parents at specific loci. The high selection pressure imposed would require the breeder to screen large populations in order to maintain adequate allelic variability at unselected loci in subsequent generations. This is because selecting for multiple traits simultaneously through MAS can decrease genetic variability for other traits if the selection intensity is high. Each time a marker is used to select for a favorable trait, a region of the genome that is in linkage disequilibrium with the marker is transferred with it, in essence "fixing" the region for that particular haplotype. If any of those regions contain an allele that influences yield, the breeder will not have the opportunity to select for or against it, which is a disadvantage if the haplotype is associated with reduced yield. Since yield is a complex trait and is conditioned by multiple interacting genes, an increase in the number of fixed regions means that fewer unique interactions are possible, many of which could have resulted in higher yield potential. This is particularly critical in the $F_2$ generation, when the genomes of the individual plants are highly heterozygous, and recombination is more effective at producing different combinations of genes. This is also a stage at which the use of MAS can be very effective, since many quantitative traits cannot be accurately phenotyped in the $F_2$ generation. However, if MAS is applied too intensively or for too many traits at this stage, too few $F_2$ plants will be selected, with the result that the number of unique recombination events represented in later generations would be drastically reduced. This is a disadvantage because it limits the probability of finding those rare recombinants in which transgressive segregation for traits in which markers are not yet highly predictive, such as yield, have occurred. This problem continues into the next generation if MAS is applied for additional traits (Koeber and Summers 2003). While the scheme proposed by Ribault and Betrán (1999) could be quite effective for the improvement of maize and other plants that are relatively easy to cross, the large starting population sizes required would be more difficult to generate in soybean.

Ishii and Yonezawa (2007) calculated the optimum MAS procedures for pyramiding genes from multiple donor lines under various conditions and using different breeding strategies. In a case where there is no redundancy in the markers that would be used to genotype the different donor lines, the authors established three guidelines. First, when genes from four or more donor lines are to be pyramided into the same genetic background, backcrosses to recover the recurrent parent background should be conducted separately for each donor line. Second, the plants obtained via the backcross should be crossed in a schedule with a symmetrical structure and marker disposition, if possible. In a case where genes are to be pyramided together by crossing inbred donors (i.e., with no backcrossing involved), the schedule used for crossing the donors should have a tandem structure in which the donors with the fewest markers are crossed first. The authors provide additional guidance on how to maximize efficiency, and on the likelihood of recovering plants with the desired genotype in cases where some markers are redundant between donor parents.

Liu et al. (2003) proposed a method for conducting MAS for QTLs with epistatic effects based on simulation studies. They contended that a considerable loss in genetic response to MAS occurs when epistasis underlying selection is neglected, and that when epistasis is present, MAS is often more effective than selection based solely on additive or additive and dominance effects. MAS considering breeding values calculated from known QTL effects and the genotypes of individuals, and which the authors define as including additive × additive epistasis, tended to result in good responses and low standard errors in these simulation studies.

Traditional QTL mapping studies have limited power to accurately estimate epistatic interactions between loci. The techniques of MARS (marker-assisted recurrent selection) and genome-wide or genomic selection are gaining momentum as strategies to enrich populations for alleles that condition more complex and low heritability traits such as yield (Bernardo 2008). The apparent power of these techniques, estimated through simulation studies, may be the result of the increased ability of these strategies to capture favorable epistatic interactions among alleles. A key advantage of MARS and genome-wide selection is the ability to use historic phenotypic data accumulated in breeding programs over several years versus generating phenotypic data from dedicated QTL mapping populations. This ability has been highly developed in genome selection strategies by the use of "training populations", which are populations of key germplasm accessions used to estimate marker effects for key traits (Heffner et al. 2009). Such data will eventually be used in early generation populations to select for alleles that have been associated in multiple locations and years with desirable traits such as higher yield. In a comparison of the two methods, genome-wide selection appeared to have a slight edge on MARS in at least one study (Bernardo and Yu 2007).

### 6.4.4 Intellectual Property Issues

Beyond the genetic and technical obstacles that can limit the use of MAS in soybean breeding, intellectual property rights restrict the use of MAS for certain traits in the public sector. Both Pioneer Hi-Bred International, Inc., and Monsanto Technology LLC hold patents that restrict the use of MAS for the cyst nematode resistance genes *rhg1* and *Rhg4* (Webb 1996, 2003; Hauge et al. 2006). These restrictions, especially on selecting for the major *rhg1* resistance allele on LG G, undoubtedly reduce the extent to which MAS is used in public-sector breeding programs. However, in February 2007 Monsanto announced that it would provide academic researchers and public institutions free access to its marker technology for *rhg1*. Pioneer has held intellectual property rights for its soybean markers since 1996, including restrictions on using MAS to select for brown stem rot resistance. Approved and pending patents complicate the decisions that breeders in the public sector must make in determining whether MAS is a good option for attaining certain breeding objectives.

## 6.5 Current Use of MAS in Soybean Breeding

The degree to which MAS is being used in soybean breeding programs varies considerably from one program to another. Markers are being used extensively for soybean breeding in the private sector (Concibido et al. 2003; Eathington et al. 2007), and by public institutions in South America as well as in the United States and Canada (Moraes et al. 2006). A survey conducted by the authors revealed that about half of the North American public soybean breeding programs used MAS in 2007. Table 6-1 lists traits for which several public North American soybean breeding programs are employing MAS. Markers are being used to enhance seed quality, resistance to pests and pathogens, tolerance to abiotic stresses, yield, and maturity date. All of these traits fit one or more of the previously mentioned criteria that favor MAS as an alternative to phenotypic selection, including difficult or expensive phenotypic assays and the inability to select for the trait on

**Table 6-1** Traits for which marker-assisted selection were being used in public breeding programs in 2007.

| Trait | Univ. of Georgia | Univ. of Illinois | Univ. of Missouri-Columbia | Univ. of Arkansas | Univ. of Tennessee | Agric. & Agri-Foods Canada |
|---|---|---|---|---|---|---|
| **I. Seed quality** | | | | | | |
| Protein | × | × | | × | × | × |
| Oleic acid | × | | × | × | × | |
| Linolenic acid | × | | × | | × | |
| Phytic acid | × | | | | × | |
| Kunitz trypsin inhibitor | × | | | | | |
| Sugars | | | | × | | |
| Calcium | | | | × | | |
| Hardness | | | | × | | |
| **II. Pest and disease resistance** | | | | | | |
| Soybean cyst nematode | | × | × | | | × |
| Southern root-knot nematode | × | | | | | |
| Reniform nematode | × | | | | | |
| Frogeye leaf spot | | | | | | |
| Phomopsis | | | | × | | |
| Pythium | | | | × | | |
| Sudden death syndrome | | × | | | | |
| Soybean rust | × | × | | | | |
| Purple stain (*Cercospora*) | | | | × | | |
| Soybean mosaic virus | | | | × | | × |
| Lepidopteran insects | × | | | | | |
| Soybean aphid | | × | | | | × |
| **III. Abiotic stress tolerance** | | | | | | |
| Drought | × | | | | | |
| Flooding | | | | × | | |
| **IV. Agronomic traits** | | | | | | |
| Yield | × | | | | | |
| Maturity | | | | | | × |
| **V. Genetic diversity** | | | | × | | |
| **VI. Hybrid and backcross** | | | | | | |
| confirmation | × | | | | × | × |

immature plants. Several public sector breeding programs also use MAS for the confirmation of $F_1$ and backcross hybrids (Table 6-1). Manual pollination of soybean flowers to make crosses is tedious and yields few seeds, so if a breeder cannot confirm that a plant is a hybrid on the basis of morphological traits, confirmation using polymorphic markers can save substantial time and resources.

Some public soybean breeding programs that use MAS do their genotyping locally, often using equipment at a core genotyping facility operated by a university. Since much of the equipment and bioinformatics software required to genotype soybean can also be used for a variety of research on other organisms, this is a convenient arrangement for breeders at a university. Some core genotyping facilities are staffed by technicians who genotype submitted samples on a fee per sample/data point basis. At other facilities, persons working in the breeding program conduct the genotyping themselves, and the program is charged a fee to use the equipment. An advantage of the latter arrangement is that students and postdoctoral associates get more hands-on experience with machines and protocols. Such participatory genotyping facilities have worked well at places like the University of Georgia and the University of Illinois. Contract-based genotyping at regional centers may be the most cost-effective option for breeders who wish to use MAS on a limited basis with a minimal investment in equipment and reagents. This option is likely to become increasingly attractive as greater numbers of genes and QTLs are tagged with SNPs because the high levels of automation and throughput that can be achieved at larger genotyping facilities will reduce both the cost per data point and turnaround time.

Large seed companies like Monsanto, Syngenta, and Pioneer Hi-Bred began to use MAS in commercial plant breeding programs for several crop species in the late 1990s, and the use of markers has increased rapidly since 2000. MAS has been used in the private sector to select for resistance to several pests and diseases, including soybean cyst nematode, brown stem rot, and Phytophthora root rot (Sleper and Shannon 2003). The structure and evolution of marker-assisted recurrent selection programs at Monsanto Co. was summarized by Eathington et al. (2007), and illustrates how such programs differ from public breeding programs. A major difference is that contemporary commercial breeding programs typically involve integrated cooperation among teams with clearly delineated responsibilities, ranging from marker development to bioinformatics and software development, whereas public breeding programs typically require a few individuals to perform a broad range of tasks. At Monsanto, a breeding technology organization is responsible for duties such as evaluating new technologies for breeding, generating molecular marker fingerprints, and statistical support. Technologies and data then flow from this organization to a line development breeding group that analyzes the molecular data prior to making selections for the next breeding cycle (Eathington et al. 2007). Use of MAS at Monsanto has expanded dramatically since genotyping at the Ankeny, IA facility switched to SNP markers in 2000. This is largely due to the ease with which the entire genotyping process, from DNA extraction through allele calling, can be automated. The number of molecular marker data points collected grew 40-fold between 2000 and 2006, and was marked by a six-fold decrease in the cost per data point (Eathington et al. 2007). Development of information technology systems to manage molecular and phenotypic data and development of integrated molecular marker decision-making systems have also been critical to streamlining the breeding programs at the large seed companies (Xu and Crouch 2008). Evaluation of marker-assisted recurrent selection relative to conventional selection over a one-year period has proven the potential for MAS to significantly enhance gains in seed yield for soybean (Eathington et al. 2007).

Since 2000, a number of innovations at Pioneer and Monsanto have coalesced to enable a dramatic increase in the use of MAS in their soybean breeding programs. Pioneer released new MAS-derived Pioneer Y Series™ cultivars on a limited scale in 2008, with plans for full commercial launch of over 30 cultivars from MGs 0 through VII in 2009. These cultivars were developed using a system that Pioneer calls Accelerated Yield Technology (AYT™) (*www.pioneer.com/AYT/ayt_adv.pdf*; verified 12 Feb 2009). MAS has also played a key role in Monsanto's development of its Roundup Ready 2 Yield™ soybean cultivars, in part by ensuring that the glyphosate tolerance transgene is located in a region of the genome associated with a favorable yield haplotype (*www. stewartseeds.com/pdf/products/ RR2Yield_Article_Final_Andy_Nickell_4-1-08.pdf*). Both of these companies have implemented a radical change in the MAS paradigm, moving from selection at one or a few loci to selection for marker haplotypes from throughout the genome that are historically associated with increased yield or other important traits. Heffner et al. (2009) termed this approach "Genomic Selection". In April 2008, Monsanto reported that advances in high-throughput genotyping of every seed was allowing them to evaluate almost six times as many lines as they had been able to manage during the development of the first generation of Roundup Ready soybean cultivars (*www.stewartseeds.com/pdf/products/RR2Yield_Article_Final_Andy_Nickell_4-1-08.pdf*). While increased densities of SNP markers, and improved DNA isolation and genotyping methods have clearly been major factors in making genomic selection possible and cost-effective, the current level of success could not have been achieved without equally important innovations from information technology teams. Advances in computational capacity and programs allow the breeding programs to manage and mine enormous amounts of data to identify combinations of marker alleles associated with improved agronomic or resistance traits. Innovations like the use of bar coding to label samples and retrieve them later have also been important, as sample tracking may currently be a bigger rate-limiting factor than genotyping (Xu and Crouch 2008). The ability to screen large numbers of seeds gives breeders a better method to search for the proverbial needles in the haystack, but as Xu and Crouch (2008) pointed out, the development of decision support tools that breeders can use to translate genotype data into selections in a short window of time has also been critical to successful large-scale adoption of MAS in private-sector breeding programs.

## 6.6. Prospects for the Future: Better Technologies and Better Techniques

Continuing advances in genomics, mapping techniques, genotyping technologies, and bioinformatics will improve the selection efficiency that is possible with MAS and make it competitive with phenotypic selection for an increasing number of valuable soybean traits. By mid-2006, more than 2,007 published markers had been listed at the SoyBase website, and an additional 1,060 unpublished SNPs had been mapped (Jackson et al. 2006). The third version of the soybean consensus linkage map published by Choi et al. (2007) included 1,141 new SNP markers, and as of January 2008, assays had been designed for a total of 3,456 SNPs (Hyten et al. 2008). Increased saturation of the soybean linkage map with these new markers will make it possible to accurately tag more beneficial alleles than heretofore possible.

The progression of QTL analyses beyond genetic mapping in a single biparental population will also increase the potential value of MAS by facilitating fine-mapping of QTLs and identification of rare alleles with favorable effects on a trait. It is now possible to mine data from previous QTL mapping studies conducted in multiple populations to consolidate multiple QTLs into clusters, an approach that can be broadly defined as meta-analysis. For example, Guo et al. (2006) used a meta-analytic approach to refine positions of QTLs influencing SCN resistance by aligning the 95% confidence intervals reported in several different studies. QTLs were placed in the same cluster if their confidence intervals overlapped, allowing the authors to confirm QTL identities and positions on LGs A2, B1, E, G, and J. Multiple population analysis allows a larger portion of the genetic variation for a trait to be sampled and analyzed, rather than relying on the limited amount of variation that exists in a single biparental population (Xu 1998; Li et al. 2005b). Some issues must be further resolved, however, such as the potential heterogeneity of error variances resulting when several experiments with different errors or biases are combined into a much larger dataset (Holland 2007). Nevertheless, multi-population mapping approaches could be useful for confirming soybean QTLs.

Association mapping, also referred to as association analysis or linkage disequilibrium mapping, is another technique that will improve MAS capabilities. In association mapping, markers are used to detect statistically significant associations between genotypes and phenotypes in a large germplasm set (Buntjer et al. 2005). The lines in such a set would represent substantially more meiotic recombinations than what would have occurred in traditional biparental mapping populations (Gupta et al. 2005). The extensive sampling of meiotic events means that the physical proximity of a marker to a locus associated with a trait will be reflected in the level of linkage disequilibrium between the loci (Mackay and Powell 2006). Linkage disequilibrium (LD) refers to non-random associations between loci, including those between markers and genes or QTLs. LD generally results from physical linkage between loci, so association mapping uses the extent of LD to establish the strength of marker-trait associations, typically focusing on haplotypes, or specific combinations of alleles at linked loci, rather than on a single marker locus. Since population structure in association mapping studies increases the likelihood of detecting false positives (Pritchard et al. 2000), the methodology has been modified in various ways to detect and account for population structure present in the set of surveyed genotypes (Liu and Zeng 2000; Christiansen et al. 2006; Holland 2007).

While association mapping has been widely tested in allogamous species such as maize (Gupta et al. 2005; Yu et al. 2005, 2006; Mackay and Powell 2006), its application to an inbreeding species like soybean is different because of the higher LD. Based on a genetic diversity study by Zhu et al. (2003), it is estimated that any random region of the soybean genome can be categorized into one of only three to four different haplotypes, confirming the high level of LD in soybean. This makes it more difficult to determine the precise location of a locus in soybean than it would be in maize. While studies have quantified the extent of LD in soybean (Zhu et al. 2003; Hyten et al. 2006, 2007a), the actual use of association mapping to detect QTLs in soybean is in the early stages. However, association mapping is a powerful tool that is certain to be used in the future because of the increasing total number of available SNPs, and the development of highly informative SNP genotyping panels designed specifically for QTL mapping (Cregan et al. 2008; Hyten et al. 2008).

Along with better statistical methods for QTL mapping, the development of more informative, preferably genic markers (i.e., located within the gene itself) has also increased MAS efficiency. Most QTL studies in soybean have been completed in one or two biparental populations, and the markers that are defined as significantly associated with the trait may not actually be physically close to the gene(s). Reasons for this include low marker polymorphism in certain genomic regions and the relatively high LD discussed earlier. Markers that are not close to the gene have limited value for MAS in different populations, as recombination and a potential lack of polymorphism may limit their predictive ability. It would therefore be better to have markers that detect polymorphisms that actually occur in the gene sequence itself ("perfect markers"), because these markers can easily be used for MAS to select for specific alleles in a wide range of populations. Such perfect markers have also been referred to as allele-specific markers to highlight their usefulness compared to population-specific markers, which must be evaluated for MAS on a population-by-population basis. As discussed previously, tight marker-trait associations ensure more effective selection, less linkage drag, and a greater value for breeding applications (Asíns 2002; Holland 2004).

Allele-specific markers have been developed for several cultivated species, including soybean (Beuselinck et al. 2006). Some of these markers detect nucleotide variation that results in functional differences, and this makes them very powerful allele-specific markers for MAS. In one example from tomato, genes conditioning differences in total soluble solid content were mapped in a population of segregating lines, and much of the variation was explained by DNA base pair differences within the *LIN5* cell wall invertase gene (Fridman et al. 2004). The variation could be resolved to a single nucleotide substitution referred to as a quantitative trait nucleotide, or QTN, which modified the properties of the enzyme. The gene-specific nature of this marker also allowed researchers to quickly analyze the same gene in potato, a close relative of tomato (Li et al. 2005a).

Bilyeu et al. (2005, 2006) characterized mutated genes that resulted in lower contents of linolenic acid in soybean seeds, and developed allele-specific primers that allowed more accurate selection for this trait in soybean lines (Beuselinck et al. 2006). Similarly, allele-specific primers have been defined for enzymes involved in variation for palmitic acid content in soybean seeds (Cardinal et al. 2007), and for the *SACPD-C* gene that influences the stearic acid content in seeds (Zhang et al. 2008). Although MAS for disease resistance has traditionally been the most effective application, new research reveals that resistance gene architecture can be complex. Allele-specific markers are available for the *rhg1* and *Rhg4* conditioning resistance to soybean cyst nematode (*Heterodera glycines*), but there is still much to understand about allelic differences in these two genes when they are derived from different sources, such as PI 88788 and the cultivar Peking (Concibido et al. 2004). There is also still much to understand about the highly complex genomic region encompassing the *Rps1* locus for resistance to *Phytophthora sojae*, as it was determined that *Rps1-k*-conditioned resistance was mediated by at least two genes separated by approximately 20 kb of intervening highly repetitive DNA (Gao and Bhattacharyya 2008). It is important to understand whether DNA markers are needed from both genes to track resistance, or if one marker is sufficient.

As traits are better understood and are resolved into component candidate genes, it will become easier to develop more effective markers. Numerous technological advances like expression profiling are facilitating the development of allele-specific or functional markers for soybean (Anderson and Lübberstedt 2003; Varshney et al. 2005; Zhu and Salmeron 2007). In addition, the availability of soybean genomic sequence data will provide new possibilities to test and identify candidate genes for QTLs (Salvi and Tuberosa 2005; Jackson et al. 2006). Despite the merits of allele-specific markers, however, they are not necessarily essential for efficient MAS. For example, SSR markers on LG O were predictive and in tight linkage disequilibrium with a QTL conditioning resistance to the southern root-knot nematode (RKN) when surveyed in the Southern soybean germplasm base (Ha et al. 2004). These SSR markers work extremely well for identifying RKN-resistant lines within that germplasm, but may be less useful if the objective would be to introgress RKN resistance into a different germplasm base. Even if the development of genic markers is not yet possible for complex traits such as yield, the availability of SNP markers tightly linked to beneficial alleles will improve the efficiency of MAS. Selection with tightly linked SNPs would facilitate identification of rare recombinants in which the potential for linkage drag is reduced.

Other promising strategies include gene expression profiling and the development of functional markers from ESTs to obtain SNPs directly related to gene functionality and phenotype or in regions of the genome sparsely populated with other markers (Varshney et al. 2005; Choi et al. 2007; Varshney et al. 2005). Microarray gene expression profiling is particularly powerful, as it can identify thousands of single feature polymorphisms (SFPs) between different genotypes, each of which can be converted into a functional marker for the trait of interest (Varshney et al. 2005). In an extension of this technique, the expression profiles of individuals in mapping populations can be analyzed, allowing researchers to map eQTLs or expression QTLs, and providing a method for dissecting the underlying gene identities of the QTLs. Luo et al. (2007) reported, however, that while genotyping with the use of Affymetrix gene chips was robust, the SFPs primarily represented polymorphisms in *cis*-acting expression regulators.

Reverse genetics tools such as TILLING (McCallum et al. 2000) and virus-induced gene silencing (Zhang and Ghabrial 2006) may be used to identify and confirm candidate genes. A technique that should be particularly useful for "allele mining" in a germplasm collection is an application of TILLING called "Ecotilling" (Varshney et al. 2005). This technique, originally developed by Comai et al. (2004), involves the use of primers designed for candidate genes of interest to screen a germplasm collection for multiple types of polymorphisms, and relies on a unique method to detect sequence variations in alleles within the germplasm compared to the sequences of known alleles. Ecotilling enables both SNP discovery and haplotyping to be performed at a much lower cost than more traditional methods that require large-scale sequencing (Varshney et al. 2005).

The availability of complete sequence data for the *G. max* genome will benefit many technologies for gene mapping and identification of candidate genes underlying QTLs, thus enhancing the efficiency of MAS in soybean breeding (Chapter 10 of this volume). A project to sequence 'Williams 82' using a whole-genome shotgun approach was begun at the Joint Genome Institute of the U.S. Department of Energy in 2006, and preliminary data for an assembly with 7.23x coverage of the genome were posted on the web in January, 2008 (*www.phytozome.net/soybean*). More than 98% of the known protein-coding genes are thought to be represented in this assembly, based on comparison with the soybean EST set. A final chromosome-scale assembly with 8x coverage was to be completed by the end of 2008, thus fostering other objectives outlined in the 2008-2012 strategic plan for soybean genomics research (*http://soybase.org/resources/soygec*). The genome sequence has already been exploited extensively to develop new SNP markers through targeted resequencing of the genomes of 17-20 diverse accessions. With continuing advances in sequencing technologies, this is becoming increasingly rapid and affordable (Metzker, 2005). It is expected that a minimum of 15,000 SNPs (on average one SNP every 50 kb) will have been identified by the end of 2008, and that this number will rise to 120,000 by the year 2013. These SNPs will initially be mapped in a variety of biparental populations, but the ultimate intention is to haplotype the entire soybean germplasm collection, which consists of more than 18,000 accessions. In addition, the resequencing projects will also identify additional SSR loci, of which approximately 2,000 are expected to be mapped during 2008. Alignment of linkage maps, BAC-based physical maps, and whole-genome shotgun sequences will provide information on the genetic architecture of QTL-containing regions, thereby helping to elucidate whether certain loci are pleiotropic or simply linked so tightly that recombination between them is extremely rare (Jackson et al. 2006).

For new markers to be exploited, genomics information must be made accessible to breeders in a user-friendly interface. The Soybean Breeders Toolbox, with its online databases, plays an important role in disseminating information from both genetics and molecular biology investigations of soybean. While markers linked to genes of importance will always receive the most attention, more precise knowledge about the locations of maturity genes and genes controlling deleterious traits is also likely to improve selection efficiency. A breeder's decision about whether or not to select for certain QTLs, particularly those with moderate or minor effects, may depend in part on whether a QTL with a major influence on another important trait is likely to have a deleterious allele in linkage disequilibrium with the allele being introgressed. One idea that evolved out of discussions among public-sector soybean breeders is to provide a "breeders tool kit" that would accompany the release of new lines and improved germplasm. This "tool kit" would consist of markers that tag one or more unique and useful genes or QTLs that have been introgressed into the new line, thus providing other breeders with a ready means to transfer the gene(s) quickly and efficiently into other genetic backgrounds.

The future of MAS depends on several key points that were summarized by Francia et al. (2005). These include access to cost-effective technologies that can process large numbers of samples, the ability to apply knowledge gained from other species related to soybean, flexibility to customize MAS to issues particular to soybean breeding, and being able to use expression data and resources such as candidate genes to further refine QTL identities and positions. We are truly moving into an era where the term "genomics-assisted breeding" rather than "marker-assisted selection" will describe the essence of our research programs (Varshney et al. 2005).

In summary, major improvements in technologies and techniques, together with a wealth of experience gained during two decades of soybean genetic mapping and MAS studies, has now made marker-assisted soybean breeding more efficient and more accessible than ever before. Increased marker availability together with the development of innovative mapping approaches that promote an improved understanding of the genetic basis of all important agronomic traits and allelic variation at the loci affecting those traits may one day offer breeders the ability to design improved genotypes *"in silico"*, as proposed by Peleman and van der Voort (2003). In the meantime, MAS already holds enormous promise for enhancing efforts to develop new soybean cultivars with improved nutritional value, disease and pest resistance, tolerance to abiotic stress, and seed yields.

# References

Abe J, Xu DH, Suzuki Y, Kanazawa A, Shimamoto Y (2003) Soybean germplasm pools in Asia revealed by SSRs. Theor Appl Genet 106: 445–453.

Akkaya MS, Bhagwat AA, Cregan PB (1992) Length polymorphisms of simple sequence repeat DNA in soybean. Genetics 132: 1131–1139.

Akkaya MS, Shoemaker RC, Specht JE, Bhagwat AA, Cregan PB (1995) Integration of simple sequence repeat DNA markers into a soybean linkage map. Crop Sci 35: 1439–1445.

Altshuler D, Brooks LD, Chakravarti A, Collins FS, Daly MJ, Donnely P (2005) A haplotype map of the human genome. Nature 437: 1299–1320.

Andersen JR, Lübberstedt T (2003) Functional markers in plants. Trends Plant Sci 8: 554–560.

Anderson JA, Chao S, Liu S (2007) Molecular breeding using a major QTL for Fusarium head blight resistance in wheat. Crop Sci 47(S3): S112–S119.

Apuya NR, Frazier BL, Keim P, Roth EJ, Lark KG (1988) Restriction fragment length polymorphisms as genetic markers in soybean, *Glycine max* (L) Merrill. Theor Appl Genet 75: 889–901.

Ashikari M, Sakakibara H, Lin SY, Yamamoto T, Takashi T, Nishimura A, Angeles ER, Quian Q, Kitano H, Matsuoka M (2005) Cytokinin oxidase regulates rice grain production. Science 309: 741–745.

Asíns MJ (2002) Present and future of quantitative trait locus analysis in plant breeding. Plant Breed 121: 281–291.

Beckman J, Soller M (1986) Restriction fragment length polymorphisms in plant genetic improvement. Oxford Surv Plant Mol Biol Cell Biol 3: 197–250.

Bernardo R (2008) Molecular markers and selection for complex traits in plants: Learning from the last 20 years. Crop Sci 48: 1649–1664.

Bernardo R, Yu J (2007) Prospects for genomewide selection for quantitative traits in maize. Crop Sci 47: 1082–1090.

Beuselinck PR, Sleper DA, Bilyeu KD (2006) An assessment of phenotype selection for linolenic acid using genetic markers. Crop Sci 46: 747–750.

Bilyeu K, Palavalli L, Sleper DA, Beuselinck P (2005) Mutations in soybean microsomal omega-3 fatty acid desaturase genes reduce linolenic acid concentration in soybean seeds. Crop Sci 45: 1830–1836.

Bilyeu K, Palavall L, Sleper DA, Beuselinck P (2006) Molecular genetic resources for development of 1% linolenic acid soybeans. Crop Sci 46: 1913–1918.

Buntjer JB, Sørensen AP, Peleman JD (2005) Haplotype diversity: the link between statistical and biological association. Trends Plant Sci 10: 466–471.

Brim CA, Schultz WM, Collins FI (1968) Maternal effect on fatty acid composition and oil content of soybean, *Glycine max* (L) Merrill. Crop Sci 8: 517–518.

Brown–Guedira GL, Thompson JA, Nelson RL, Warburton ML (2000) Evaluation of genetic diversity of soybean introductions and North American ancestors using RAPD and SSR markers. Crop Sci 40: 815–823.

Brummer EC, Graef GL, Orf J, Wilcox JR, Shoemaker RC (1997) Mapping QTL for seed proteins and oil content in eight soybean populations. Crop Sci 37: 370–378.

Burnham KD, Francis DM, Dorrance AE, Fioritto RJ, St Martin SK (2002) Genetic diversity patterns among *Phytophthora* resistant soybean plant introductions based on SSR markers. Crop Sci 42: 338–343.

Cardinal AJ, Burton JW, Camacho-Roger AM, Yang JH, Wilson RF, Dewey RE (2007) Molecular analysis of soybean lines with low palmitic acid content in the seed oil. Crop Sci 47: 304–310.

Cardon LR, Abecasis GR (2003) Using haplotype blocks to map human complex trait loci. Trends Genet 19: 135–140.

Carter TE, Jr, Nelson RL, Sneller CH, Cui Z (2004) Genetic diversity in soybean. In: HR Boerma, JE Specht (eds) Soybeans: Improvement, Production, and Uses. 3rd edn. ASA, CSSA, and SSSA Madison, WI, USA, pp 303–416.

Chantarangsu S, Cressey T, Mahasirimongkol S, Tawon Y, Ngo-Giang-Huong N, Jourdain G, Lallemant M, Chantratita W (2007) Comparison of the TaqMan and LightCycler systems in evaluation of CYP2B6 516G>T polymorphism. Mol Cell Probes 21: 408–411.

Cheesbrough TM (1989) Changes in the enzymes for fatty acid synthesis and desaturation during acclimation of developing soybean seeds to altered growth temperature. Plant Physiol 90: 760–764.

Chen CY, Gu C, Mensah C, Nelson RL, Wang D (2007) SSR marker diversity of soybean aphid resistance sources in North America. Genome 50: 1104–1111.

Choi I-Y, Hyten DL, Matukumalli LK, Song Q, Chaky JM, Quigley CV, Chase K, Lark KG, Reiter RS, Yoon MS, Huang E-Y, Yi S-I, Young ND, Shoemaker RC, van Tassell CP, Specht JE, Cregan PB (2007) A soybean transcript map: Gene distribution, haplotype and single-nucleotide polymorphism analysis. Genetics 176: 685–696.

Christiansen MJ, Feenstra B, Skovgaard IM, Andersen SB (2006) Genetic analysis of resistance to yellow rust in hexaploid wheat using a mixture model for multiple crosses. Theor Appl Genet 112: 581–591.

Collard BCY, Jahufer MZZ, Brouwer JB, Pang ECK (2005) An introduction to markers, quantitative trait loci (QTL) mapping and marker-assisted selected for crop improvement: The basic concepts. Euphytica 142: 169–196.

Comai L, Young K, Till BJ, Reynolds SH, Greene EA, Comodo CA, Enns LC, Johnson JE, Burnter C, Odden AR, Henikoff S (2004) Efficient discovery of DNA polymorphisms in natural populations by Ecotilling. Plant J 37: 778–786.

Concibido VC, La Vallee B, Mclaird P, Pineda N, Meyer J, Hummel L, Yang J, Wu K, Delannay X (2003) Introgression of a quantitative trait locus for yield from *Glycine soja* into commercial soybean cultivars. Theor Appl Genet 106: 575–582.

Concibido VC, Diers BW, Arelli PR (2004) A decade of QTL mapping for cyst nematode resistance in soybean. Crop Sci 44: 1121–1131.

Cregan PB (2008) Soybean molecular genetic diversity. In: G Stacey (ed) Genetics and Genomics of Soybean. Springer, New York, NY, USA, pp 17–34.

Cregan PB, Bhagwat AA, Akkaya MS, Rongwen J (1994) Microsatellite fingerprinting and mapping in soybean. Meth Mol Cell Biol 5: 49–61.

Cregan PB, Jarvik T, Bush AL, Shoemaker RC, Lark KG, Kahler AL, Kaya N, VanToai TT, Lohnes DG, Chung J, Specht JE (1999) An integrated genetic linkage map of the soybean genome. Crop Sci 39: 1464–1490.

Cregan PB, Hyten D, Song Q, Choi I-Y, Cannon SB, Farmer AD, May GD, Shoemaker RC, Specht JE (2008) SNP analysis for QTL discovery and whole genome analysis In: Plant & Anim Genome XVI Conf, San Diego, CA, USA: *www.intl-pagorg/16/abstracts*

Dekkers JCM, Hospital F (2002) The use of molecular genetics in the improvement of agricultural populations. Nat Rev Genet 3: 22–32.

Delannay X, Rodgers DM, Palmer RG (1983) Relative genetic contributions among ancestral lines to North American soybean cultivars. Crop Sci 23: 944–949.

Dreher K, Khairallah M, Ribaut J-M, Morris M (2003) Money matters (I): costs of field and laboratory precedures associated with conventional and marker-assisted maize breeding at CIMMYT. Mol Breed 11: 221–234.

Eathington SR, Crosbie TM, Edwards MD, Reiter RS, Bull JK (2007) Molecular markers in a commercial breeding program. Crop Sci 47(S3): S145–S163.

Flagel L, Christiansen JR, Gustus CD, Smith KP, Olhoft PM, Somers DA, Matthews PD (2005) Inexpensive, high throughput microplate format for plant nucleic acid extraction: Suitable for multiplex Southern analyses of transgenes. Crop Sci 45: 1985–1989.

Francia E, Tacconi G, Crosatti C, Barabaschi D, Bulgarelli D, Dall'Aglio E, Valè G (2005) Marker assisted selection in crop plants. Plant Cell Tiss Org Cult 82: 317–342.

Fridman E, Carrari F, Liu YS, Fernie AR, Zamir D (2004) Zooming in on a quantitative trait for tomato yield using interspecific introgressions. Science 305: 1786–1789.

Frisch M, Bohn M, Melchinger AE (1999a) Minimum sample size and optimal positioning of flanking markers in marker-assisted selection for transfer of a target gene. Crop Sci 39: 967–975.

Frisch M, Bohn M, Melchinger AE (1999b) Comparison of selection strategies for marker-assisted backcrossing of a gene. Crop Sci 39: 1295–1301.

Gao H, Bhattacharyya MK (2008) The soybean-Phytophthora resistance locus *Rps1*-k encompasses coiled coil-nucleotide binding-leucine rich repeat-like genes and repetitive sequences. BMC Plant Biol 8: 29.

Gizlice Z, Carter TE, Jr, Burton JW (1993) Genetic diversity in North American soybean: I Multivariate analysis of founding stock and relation to coefficient of parentage. Crop Sci 33: 614–620.

Gizlice Z, Carter TE, Jr, Gerig TM, Burton JW (1996) Genetic diversity patterns in North American public soybean cultivars based on coefficient of parentage. Crop Sci 36: 753–765.

Guo B, Sleper DA, Lu P, Shannon JG, Nguyen HT, Arelli PR (2006) QTLs associated with resistance to soybean cyst nematode in soybean: Meta-analysis of QTL locations. Crop Sci 46: 595–602.

Gupta PK, Rustgi S, Kulwal PL (2005) Linkage disequilibrium and association studies in higher plants: Present status and future prospects. Plant Mol Biol 57: 461–485.

Guzman PS, Diers BW, Neece DJ, St Martin SK, LeRoy AR, Grau CR, Hughes TJ, Nelson RL (2007) QTL associated with yield in three backcross-derived populations of soybean. Crop Sci 47: 111–122.

Ha B-K, Boerma HR (2008) High-throughput SNP genotyping by melting curve analysis for resistance to Southern root-knot nematode and frogeye leaf spot in soybean. J Crop Sci Biotechnol 11: 91–100.

Ha B-K, Bennett JB, Hussey RS, Finnerty SL, Boerma HR (2004) Pedigree analysis of a major QTL conditioning soybean resistance to Southern root-knot nematode. Crop Sci 44: 758–763.

Ha B-K, Hussey RS, Boerma HR (2007) Development of SNP assays for marker-assisted selection of two southern root-knot nematode resistance QTL in soybean. Plant Genome 2: S73–S82 [Publ in Crop Sci 47 (S2)].

Hauge BM, Wang ML, Parsons JD, Parnell LD (2006) Nucleic acid molecules and other molecules associated with cyst nematode resistance. US Patent 7,154,021.

Heffner EL, Sorrells ME, Jannink J-L (2009) Genomic selection for crop improvement. Crop Sci 49: 1–12 .

Hill CB, Li Y, Hartman GL (2006a) A single dominant gene for resistance to the soybean aphid in the soybean cultivar Dowling. Crop Sci 46: 1601–1605.

Hill CB, Li Y, Hartman GL (2006b) Soybean aphid resistance in soybean Jackson is controlled by a single dominant gene. Crop Sci 46: 1606–1608.

Holland JB (2004) Implementation of molecular markers for quantitative traits in breeding programs—challenges and opportunities. Proc 4th Int Crop Sci Congr, 26 Sep–1 Oct 2004, Brisbane, Australia: *www.cropscience.org.au*

Holland JB (2007) Genetic architecture of complex traits in plants. Curr Opin Plant Biol 10: 156–161.

Hospital F, Chevalent C, Mulsant P (1992) Using markers in gene introgression breeding programs. Genetics 132: 1199–1210.

Hospital F, Moreau L, Lacoudre F, Charcosset F, Gallais A (1997) More on the efficiency of marker-assisted selection. Theor Appl Genet 95: 1181–1189.

Huang N, Angeles ER, Domingo J, Magpantay G, Singh S, Zhang G, Kumaravadivel N, Bennett J, Khush GS (1997) Pyramiding of bacterial blight resistance genes in rice: Marker-assisted selection using RFLP and PCR. Theor Appl Genet 95: 313–320.

Hurley JD, Engle LJ, Davis JT, Welsh AM, Landers JE (2004) A simple, bead-based approach for multi-SNP molecular haplotyping. Nucl Acids Res 32: e186.

Hyten DL, Song Q, Zhu Y, Choi I-K, Nelson RL, Costa JM, Specht JE, Shoemaker RC, Cregan PB (2006) Impacts of genetic bottlenecks on soybean genome diversity. Proc Natl Acad Sci USA 103: 16666–16671.

Hyten DL, Choi I-Y, Song Q, Shoemaker RC, Nelson RL, Costa JM, Specht JE, Cregan PB (2007a) Highly variable patterns of linkage disequilibrium in multiple soybean populations. Genetics 175: 1937–1944.

Hyten DL, Hartman GL, Nelson RL, Frederick RD, Concibido VC, Narvel JM, Cregan PB (2007b) Map location of the *Rpp1* locus that confers resistance to soybean rust in soybean. Crop Sci 47: 837–840.

Hyten DL, Choi I-Y, Song Q, Specht JE, Carter T Jr, Shoemaker RC, Nelson RL, Cregan PB (2008) Soybean Consensus Linkage Map 40 and the development of a Universal 1,536 Soy Linkage Panel for QTL mapping. In: Plant & Anim Genome XVI Conf, San Diego, CA, USA:*www.intl-pagorg/16/abstracts*

Hyten DL, Smith JR, Frederick RD, Tucker ML, Song Q, Cregan PB (2009) Bulked segregant analysis using the GoldenGate assay to locate the *Rpp3* locus that confers resistance to soybean rust in soybean. Crop Sci 49: 265–271.

Ishii T, Yonezawa K (2007) Optimization of the marker-based procedures for pyramiding genes from multiple donor lines: I Schedule of crossing between the donor lines. Crop Sci 47: 537–546.

Jackson SA, Rokhsar D, Stacey G, Shoemaker RC, Schmutz J, Grimwood J (2006) Toward a reference sequence of the soybean genome: A multiagency effort. Crop Sci 46: 55–61.

Kabelka EA, Diers BW, Fehr WR, LeRoy AR, Baianu IC, You T, Neece DJ, Nelson RL (2004) Putative alleles for increased yield from soybean plant introductions. Crop Sci 44: 784–791.

Kamaya M, Kiguchi T (2003) Rapid DNA extraction method from soybean seeds. Breed Sci 53: 277–279.

Kang HW, Cho YG, Yoon UH, Eun MY (1998) A rapid DNA extraction method for RFLP and PCR analysis from a single dry seed. Plant Mol Biol Rep 16: 90.

Keim P, Olson TC, Shoemaker RC (1988) A rapid protocol for isolating soybean DNA. Soybean Genet Newsl 15: 150–152.

Keim P, Diers BW, Olson TC, Shoemaker RC (1990) RFLP mapping in soybean: Association between marker loci and variation in quantitative traits. Genetics 126: 735–742.

Kim K-S, Hill CB, Hartman GL, Mian MAR, Diers BW (2008) Discovery of soybean aphid biotypes. Crop Sci 48: 923–928.

Knapp SJ (1998) Marker-assisted selection as a strategy for increasing the probability of selecting superior genotypes. Crop Sci 38: 1164–1174.

Koebner RMD, Summers RW (2003) 21st century wheat breeding: Plot selection or plate detection? Trends Biotechnol 21: 59–63.

Lambert L, Tyler J (1999) Appraisal of insect-resistant soybeans. In: JA Webster , BR Wiseman (eds) Economic, Environmental, and Social Benefits of Insect Resistance in Field Crops. Thomas Say, Lanham, MD, USA, pp 131–148.

Lande R, Thompson R (1990) Efficiency of marker-assisted selection in the improvement of quantitative traits. Genetics 124: 743–756.

Lee GJ, Boerma HR, Villagarcia MR, Zhou X, Carter Jr TE, Li Z, Gibbes MO (2004a) A major QTL conditioning salt tolerance in S-100 soybean and descendent cultivars. Theor Appl Genet 109: 1610–1619.

Lee S-H, Walker DR, Boerma HR (2004b) Comparison of four flow cytometric SNP detection assays and their use in plant improvement. Theor Appl Genet 110: 167–174.

Li L, Strahwalk J, Hofferbert HR, Lubeck J, Tacke E, Junghans H, Wunder J, Gebhart C (2005a) DNA variation at the invertase locus *invGE/GF* is associated with tuber quality traits of potato breeding clones. Genetics 170: 813-821.

Li R, Lyons MA, Wittenburg H, Paigen B, Churchill GA (2005b) Combining data from multiple inbred line crosses improves the power and resolution of quantitative trait loci mapping. Genetics 169: 1699–1709.

Li Z, Qiu L, Thompson JA, Welsh MM, Nelson RL (2001) Molecular genetic analysis of US and Chinese soybean ancestral lines. Crop Sci 41: 1330–1336.

Liu P, Zhu J, Lou X, Lu U (2003) A method for marker-assisted selection based on QTLs with epistatic effects. Genetica 119: 75–86.

Liu Y, Zeng Z-B (2000) A general mixture model approach for mapping quantitative trait loci from diverse cross designs involving multiple inbred lines. Genet Res 75: 345–355.

Luo ZW, Potokina E, Druka A, Wise R, Waugh R, Kearsey MJ (2007) SFP genotyping from Affymetrix arrays is robust but largely detects *cis*-acting expression regulators. Genetics 176: 789–800.

Mackay I, Powell W (2006) Methods for linkage disequilibrium mapping in crops. Trends Plant Sci 12: 57–63.

McCallum CM, Comai L, Greene EA, Henikoff S (2000) Targeting Induced Local Lesions IN Genomes (TILLING) for plant functional genomics. Plant Physiol 123: 439–442.

Melchinger AE (1990) Use of molecular markers in breeding for oligogenic disease resistance. Plant Breed 104: 1–19.

Mensah C, DiFonzio C, Wang D (2008) Inheritance of soybean aphid resistance in PI 567541B and PI 567598B. Crop Sci 48: 1759–1763.

Metzker ML (2005) Emerging technologies in DNA sequencing. Genome Res 15: 1767–1776.

Mian MAR, Kang S-T, Beil SE, Hammond RB (2008) Genetic linkage mapping of the soybean aphid resistance gene in PI 243540. Theor Appl Genet 117: 955–962.

Missaoui AM, Phillips DV, Boerma HR (2007) DNA marker analysis of 'Davis' soybean and its descendants for the *Rcs3* gene conferring resistance to *Cercospora sojina*. Crop Sci 47: 1263–1270.

Monteros MJ, Ha B-K, Boerma HR (2007a) Development of a SNP assay to detect an Asian soybean rust resistance gene from 'Hyuuga' soybean. In: Annu Meet Am Soc Agron Abstracts, New Orleans, LA, USA.

Monteros MJ, Missaoui AM, Phillips DV, Walker DR, Boerma HR (2007b) Mapping and confirmation of the 'Hyuuga' red-brown lesion resistance gene for Asian soybean rust. Crop Sci 47: 829–836.

Monteros MJ, Burton JH, Boerma HR (2008) Molecular mapping and confirmation of QTLs associated with oleic acid content in N00-3350 soybean. Crop Sci 48: 2223–2234.

Moraes RMA de, Soares TCB, Colombo LR, Salla MFS, Barros JG, Piovesan ND, Barros EG, Moreira MA (2006) Assisted selection by specific DNA markers for genetic elimination of the Kunitz trypsin inhibitor and lectin in soybean seeds. Euphytica 149: 221–226.

Morgante M, Rafalski J, Biddle P, Tingey S, Olivieri AM (1994) Genetic mapping and variability of seven soybean simple sequence repeat loci. Genome 37: 763–769.

Morris M, Dreher K, Ribaut J-M, Khairallah M (2003) Money matters (II): Costs of maize inbred line conversion schemes at CIMMYT using conventional and marker-assisted selection. Mol Breed 11: 235–247.

Narvel JM, Jakkula LK, Phillips DV, Wang T, Lee SH, Boerma HR (2001a) Molecular mapping of *Rxp* conditioning bacterial pustule in soybean. J Hered 92: 267–270.

Narvel JM, Walker DR, Rector BG, All JN, Parrott WA, Boerma HR (2001b) A retrospective DNA marker assessment of the development of insect-resistant soybean. Crop Sci 41: 1931–1939.

Nelson RR (1978) Genetics of horizontal resistance to plant diseases. Annu Rev Phytopathol 16: 359–378.

Oliva ML, Shannon JG, Sleper DA, Ellersieck MR, Cardinal AJ, Paris RL, Lee JD (2006) Stability of fatty acid profile in soybean genotypes with modified seed oil composition. Crop Sci 46: 2069–2075.

Orf JH, Chase K, Adler FR, Mansur LM, Lark KG (1999) Genetics of soybean agronomic traits: II. Interactions between yield quantitative trait loci in soybean. Crop Sci 39: 1652–1657.

Orf JH, Diers BW, Boerma HR (2004) Genetic improvement: Conventional and molecular-based strategies. In: HR Boerma , JE Specht (eds) Soybeans: Improvement, Production, and Uses. 3rd edn. ASA, CSSA, SSSA, Madison, WI, USA, pp 417–450.

Pantalone VR, Walker DR, Dewey RE, Rajcan I (2004) DNA marker-assisted selection for improvement of soybean oil concentration and quality. In: RF Wilson, HT Stalker, EC Brummer (eds) Legume Crop Genomics. AOCS Press, Champaign, IL, USA, pp 283–311.

Peleman JD, van der Voort JR (2003) Breeding by design. Trends Plant Sci 8: 330–334.

Pritchard JK, Stephens M, Rosenberg NA, Donnelly P (2000) Association mapping in structured populations. Am J Hum Genet 67: 170–181.

Rafalski A, Morgante M (2004) Corn and humans: recombination and linkage disequilibrium in two genomes of similar size. Trends Genet 20: 103–111.

Reyna, N, Sneller CH (2001) Evaluation of marker-assisted introgression of yield QTL alleles into adapted soybean. Crop Sci 41: 1317–1321.

Ribaut JM, Hoisington D (1998) Marker-assisted selection: New tools and strategies. Trends Plant Sci 3: 236–239.

Ribaut JM, Betrán J (1999) Single large-scale marker-assisted selection (SLS-MAS). Mol Breed 5: 531–541.

Rongwen J, Akkaya MS, Bhagwat AA, Lavi U, Cregan PB (1995) The use of microsatellite DNA markers for soybean genotype identification. Theor Appl Genet 90: 43–48.

Saghai Maroof MA, Jeong SC, Gunduz I, Tucker DM, Buss GR, Tolin SA (2008) Pyramiding of soybean mosaic virus resistance genes by marker-assisted selection. Crop Sci 48: 517–526.

Salvi S, Tuberosa R (2005) To clone or not to clone plant QTLs: Present and future challenges. Trends Plant Sci 10: 297–304.

Sebolt AM, Shoemaker RC, Diers BW (2000) Analysis of a quantitative trait locus allele from wild soybean that increases seed protein concentration in soybean. Crop Sci 40: 1438–1444.

Sleper DA, Shannon JG (2003) Role of public and private soybean breeding programs in the development of soybean varieties using biotechnology. AgBioForum 6: 27–32.

Smalley MD, Fehr WR, Cianzio SR, Han F, Sebastian SA, Streit LG (2004) Quantitative trait loci for soybean seed yield in elite and plant introduction germplasm. Crop Sci 44: 436–442.

Song QJ, Marek LF, Shoemaker RC, Lark KG, Concibido VC, Delannay X, Specht JE, Cregan PB (2004) A new integrated genetic linkage map of the soybean. Theor Appl Genet 109: 122–128.

Staub JE, Serquen FC (1996) Genetic markers, map construction, and their application in plant breeding. HortScience 31: 729–741.

Stuber CW, Polacco M, Senior ML (1999) Synergy of empirical breeding, marker-assisted selection, and genomics to increase crop yield potential. Crop Sci 39: 1571–1583.

Syvänen A-C (2005) Toward genome-wide SNP genotyping. Nat Genet 37: S5–S10.

Tanksley SD, Rick CM (1980) Isozyme gene linkage map of the tomato: applications in genetics and breeding. Theor Appl Genet 57: 161–170.

Tanksley SD, Nelson JC (1996) Advanced backcross QTL analysis: a method for the simultaneous discovery and transfer of valuable QTLs from unadapted germplasm into elite breeding lines. Theor Appl Genet 92: 191–203.

Tanksley SD, McCouch SR (1997) Seed banks and molecular maps: Unlocking genetic potential from the wild. Science 277: 1063–1066.

Tanksley SD, Young ND, Paterson AH, Bonierbale MW (1989) RFLP mapping in plant breeding: New tools for an old science. Bio/Technology 7: 257–264.

Ude GN, Kenworthy WJ, Costa JM, Cregan PB, Alvernaz J (2003) Genetic diversity of soybean cultivars from China, Japan, North America, and North American ancestral lines determined by amplified fragment length polymorphism. Crop Sci 43: 1858–1867.

van Berloo R, Stam P (2001) Simultaneous marker-assisted selection for multiple traits in autogamous crops. Theor Appl Genet 102: 1107–1112.

Varshney RK, Graner A, Sorrells ME (2005) Genomics-assisted breeding for crop improvement. Trends Plant Sci 10: 621–630.

Visscher PM, Haley CS, Thompson R (1996) Marker-assisted introgression in backcross breeding programs. Genetics 144: 1923–1932.

Walker DR, Boerma HR, All JN, Parrott WA (2002) Combining *cry1Ac* with QTL alleles from PI 229358 to improve soybean resistance to lepidopteran pests. Mol Breed 9: 43–51.

Walker DR, Narvel JM, All JN, Boerma HR, Parrott WA (2004) A QTL that enhances and broadens Bt insect resistance in soybean. Theor Appl Genet 109: 1051–1057.

Walker DR, Scaboo AM, Pantalone VR, Wilcox JR, Boerma HR (2006) Genetic mapping of loci associated with seed phytic acid content in CX1834-1-2- soybean. Crop Sci 46: 390–397.

Wang D, Shi J, Carlson SR, Cregan PB, Ward RW, Diers BW (2003) A low-cost, high-throughput polyacrylamide gel electrophoresis system for genotyping with microsatellite DNA markers. Crop Sci 43: 1828–1832.

Wang L, Guan R, Zhangxiong L, Chang R, Qiu L (2006) Genetic diversity of Chinese cultivated soybean revealed by SSR markers. Crop Sci 46: 1032–1038.

Webb DM (1996) Soybean cyst nematode resistant soybeans and methods of breeding and identifying resistant plants. US Patent 5,491,081.

Webb DM (2003) Quantitative trait loci associated with soybean cyst nematode resistance and uses thereof. US Patent 6,538,175.

Wilson RF (2004) Seed composition. In: HR Boerma, JE Specht (eds) Soybeans: Improvement, Production, and Uses. 3rd edn. ASA, CSSA, and SSSA Madison, WI, USA, pp 621–677.

Xu S (1998) Mapping quantitative trait loci using multiple families of line crosses. Genetics 148: 517–524.

Xu Y, Crouch JH (2008) Marker-assisted selection in plant breeding: From publications to practice. Crop Sci 48: 391–407.

Yamanaka N, Hiroyuki S, Yang Z, Dong He X, Catelli LL, Binneck E, Arias CAA, Abdelnoor RV, Nepomuceno AL (2007) Genetic relationships between Chinese, Japanese, and Brazilian soybean gene pools revealed by simple sequence repeat (SSR) markers. Genet Mol Biol 30: 85–88.

Yates JL (2006) Use of diverse germplasm to improve peanut root-knot nematode resistance and seed protein content in soybean. PhD Dissert. Univ of Georgia, Athens, USA.

Yoon MS, Song QJ, Choi IY, Specht JE, Hyten DL, Cregan PB (2007) BARCSoySNP23: a panel of 23 selected SNPs for soybean cultivar identification. Theor Appl Genet 114: 885–899.

Young ND (1999) A cautiously optimistic vision for marker-assisted breeding. Mol Breed 5: 505–510.

Young ND, Mudge J, Ellis THN (2003) Legume genomes: More than peas in a pod. Curr Opin Plant Biol 6: 199–204.

Yu J, Arbelbide M, Bernardo R (2005) Power of in silico QTL mapping from phenotypic, pedigree, and marker data in a hybrid breeding program. Theor Appl Genet 110: 1061–1067.

Yu J, Pressoir G, Briggs WH, Bi IV, Yamasaki M, Doebley JF, McMullen MD, Gaut BS, Nielsen DM, Holland JB, Kresovich S, Buckler ES (2006) A unified mixed-model method for association mapping that accounts for multiple levels of relatedness. Nat Genet 38: 203–208.

Yuan J, Njiti VN, Meksem K, Iqbal MJ, Triwitayakorn K, Kassem MA, Davis GT, Schmidt ME, Lightfoot DA (2002) Quantitative trait loci in two soybean recombinant inbred line populations segregating for yield and disease resistance. Crop Sci 42: 271–277.

Zabala G, Vodkin LO (2007) A rearrangement resulting in small tandem repeats in the F3'5'H gene of white flower genotypes is associated with the soybean *W1* locus. Crop Sci 47(S2): S113–S124.

Zhang C, Ghabrial SA (2006) Development of Bean pod mottle virus-based vectors for stable protein expression and sequence-specific virus-induced gene silencing in soybean. Virology 344: 401–411.

Zhang G, Gu C, Wang D (2009) Molecular mapping of soybean aphid resistance genes in PI 567541B. Theor Appl Genet 118: 473–482.

Zhang P, Burton JW, Upchurch RG, Whittle E, Shanklin J, Dewey RE (2008) Mutations in a Δ9-stearoyl-ACP-desaturase gene are associated with enhanced stearic acid levels in soybean seeds. Crop Sci 48: 2305–2313.

Zhu S, Walker DR, Boerma HR, All JN, Parrott WA (2008) Effects of defoliating instect resistance QTLs and a *cry1Ac* transgene in soybean near-isogenic lines. Theor Appl Genet 116: 455–463.

Zhu T, Salmeron J (2007) High-definition genome profiling for genetic marker discovery. Trends Plant Sci 12: 196–202.

Zhu YL, Song QJ, Hyten DL, Van Tassell CP, Matukumalli LK, Grimm DR, Hyatt SM, Fickus EW, Young ND, Cregan PB (2003) Single-nucleotide polymorphisms in soybean. Genetics 163: 1123–1134.

# 7

# Map-based Cloning of Genes and QTL in Soybean

*Madan K. Bhattacharyya*

## ABSTRACT

Map-based cloning is a powerful strategy for isolation of genes by their functions. Despite the power of this method, only a few soybean genes have been isolated by applying this approach. With the availability of soybean genome sequences, highly sensitive and high volume marker technologies such as SNPs, and a large collection of recombinant inbred lines, mapping populations and improved transformation technology—it is expected that map-based cloning will become more attractive to the soybean geneticists or biologists. In this chapter, the steps involved in cloning soybean genes by their map positions are described. To avoid the lengthy transformation step for gene identification, generation of a collection of independent point mutations in the gene of interest can be considered. Map-based cloning approach in soybean is expected to continuously benefit from new technological inventions; e.g. less expensive high throughput sequencing, new marker technologies, etc. The key to the success of a map-based gene cloning approach in soybean will of course remain on phenotyping the trait encoded by the gene, and investigation of a large segregating material for mapping the gene in a narrow genetic interval. Possible pitfalls, one may encounter during map-based cloning of a soybean gene are also addressed.

**Keywords:** map-based cloning; positional cloning; high-density genetic map; high-resolution genetic map; physical map; recombinants; gene identification

---

Department of Agronomy, Iowa State University, Ames, IA 50011, USA; e-mail: *mbhattac@iastate.edu*

## 7.1 Introduction

Soybean is a very important crop agronomically. It is a good source of both protein (40%) and oil (20%). In addition to human consumption, soybean is a major protein source of animal feeds. Soybean is also becoming a major crop for biodiesel production. Despite the economic importance of this legume, the molecular basis of physiological processes controlling these important traits are still largely unknown. Several factors contributed to this slow progress. Until very recently, the genome sequence of soybean was unavailable. Transformation in soybean is lengthy and difficult to apply for large scale functional analyses of genes. Likewise, suitable active endogenous transposable elements for studying gene functions have not been reported. It is an ancient tetraploid and thus has many duplicated genes. The polyploid nature of soybean makes the genetic studies difficult.

In this chapter, the steps involved in map-based or positional cloning of soybean genes are addressed. Although limited applications of this approach in identifying soybean genes are documented (Searle et al. 2003; Ashfield et al. 2004; Gao et al. 2005), with the availability of soybean genome sequence this gene cloning strategy will be attractive to soybean geneticists. Inadequate transformation procedures required for gene-identification, however, will continue to obstruct the application of this powerful technology for forward genetic studies in soybean. Generation of many independent point mutations in the gene of interest will eliminate the problem faced during the gene-identification step. Pitfalls and future prospects of this gene cloning strategy in soybean are also discussed.

## 7.2 General Steps in Map-based Cloning of Soybean Genes

In map-based cloning approach, as the name implies, genes are identified or isolated based on their map positions on chromosomes. Thus, map-position is the basis for this gene isolation method. Although it is hard to prepare a general protocol for map-based cloning, here a general strategy for cloning soybean genes is discussed (Fig. 7-1).

### 7.2.1. Phenotyping Segregants

A reliable phenotype is considered to be a key to the rapid and successful map-based cloning of genes. As discussed later, an alternative strategy can be applied for traits that show poor penetrance or heritability.

Step 1: Phenotyping segregants

Step 2: Segregating materials: $F_{2:3}$ and/or RILs

Step 3: Genetic mapping

Step 4: High density and high-resolution mapping

Step 5: Physical mapping of the region containing the gene

Step 6: Isolation of the DNA fragment containing the gene

Step 7: Identification of the gene

**Figure 7-1** General steps involved in map-based cloning of a soybean gene.

## 7.2.1.1 Alternative Alleles of the Gene

To determine the genetic map position of the gene on chromosome, one must identify the alternative forms or alleles of the gene. This can easily be achieved by identifying natural variants of the gene. For example, an allele or gene that confers resistance against a serious disease can be identified by screening the available germplasm. One can also create alternate alleles by treating the line carrying the wild type allele with chemical mutagens, such as ethylmethane sulfonate (EMS). The mutant population is then screened to identify the EMS-induced mutants that carry the alternative allele. The mutant alleles usually show loss of function. For example, if we mutate a disease resistant cultivar carrying a disease resistant (*R*) gene, we expect to identify susceptible genotypes that carry susceptible alleles of the *R* gene.

Identification of several independent point mutations in the *R* gene through screening of EMS-induced mutant population can greatly expedite the gene identification process, which will be discussed later. Secondly, if the mutants can be created in the cultivar Williams 82, whose genome sequence is available, physical mapping and large insert library construction may be avoided and cloning of the gene can be accomplished rapidly. EMS-induced Williams 82 mutant populations are already available in the research community (Cooper et al. 2008). Although the mutant screening makes the process slower at the beginning, identification of the gene becomes certain and rapid.

Chemical mutagenesis can be applied for identifying alternate alleles mapped to the quantitative trait loci (QTL) (Mohan et al. 2007); however, looking for naturally available alleles is a better choice considering the high cost involved in phenotyping QTL.

### 7.2.1.2 Phenotyping Segregants

Success of the map-based gene cloning approach depends on accurate phenotyping of the segregants. We can be 99% certain about phenotypes if at least 16 progeny of individual $F_2$s are tested or scored (Mather 1957). A modified approach for genes with poor penetrance is to phenotype approximately eight progenies of each $F_2$ and select only the $F_2$ or $F_3$ homozygous recessive individuals for mapping experiments. For QTL, recombinant inbred lines (RILs) should be developed by selfing at least five generations through single-seed descent method.

## 7.2.2 Segregating Population for Mapping the Gene

Once the alternate alleles are identified, the second obvious step is to map the gene with molecular markers. A population of around 100 individuals segregating for wild type and mutant phenotypes may be generated for the initial map of the gene. Progeny of each individual segregant should be phenotyped. The segregating $F_{2:3}$ materials developed by crossing diverse parents are ideal for mapping single genes. One can also develop segregating materials by crossing available near-isogenic lines (NILs) that differ for the alleles of the gene under investigation. The introgressed region containing the gene is usually diverse enough for identifying and mapping linked molecular markers (Kasuga et al. 1997). One can also consider phenotyping available recombinant inbred lines (RILs) (Ashfield et al. 2003). For mapping QTL, RILs developed through several selfing generations are suitable because rigorous phenotyping requires an abundant seed supply. For high-resolution mapping, one requires large numbers of segregating progeny. Evaluation of a thousand $F_2$s or RILs could be a good starting point.

## 7.2.3 Genetic Mapping

After a protocol for phenotyping the trait is established and segregating materials are generated, the next critical step is to accurately map the position of the gene on the chromosome. Molecular mapping of the gene is ideal for achieving this goal.

### 7.2.3.1 Identification of Linked Molecular Markers

For many traits, NILs have been created in soybean and can be obtained from Dr. Randy Nelson, University of Illinois. A pair of NILs differing for the alleles of the locus to be cloned is suitable for isolating molecular markers. If NILs are not available, one can create two bulks of ~20 $F_{2:3}$s or RILs that are homozygous for either the wild type or the mutant allele. Alternatively, a bulk of ~20 homozygous recessive $F_2$s and a bulk of ~20 heterozygous and homozygous dominant $F_2$s can be created. These two bulks and/or a pair of NILs are used in isolating linked molecular markers (Fig. 7-2A; Michelmore et al. 1991). Simple sequence repeat (SSR) markers are ideal for creating the first molecular map of the region that contains the gene (Song et al. 2004). One should select >250 SSR markers that are polymorphic between the two parents used in generating the two bulks or NILs and are distributed evenly (one in every 10–15 cM) on all 20 soybean chromosomes. One can identify >250 appropriate SSR markers by investigating over more than 1,000 SSR markers currently available for mapping soybean genes (Song et al. 2004).

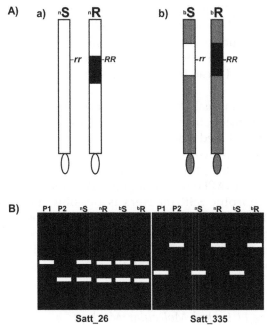

**Figure 7-2** Isolation of linked molecular markers.
**A)** The upper arm of a chromosome, to which an *R* gene of interest was mapped. **a)** Two near-isogenic lines in the cultivar Williams differing for the *R* gene. Note the introgressed region from the donor parent is in black color, and the gene is located at one end of this introgressed region. **b)** Two bulks differing for the *R* gene were developed by bulking 20 *rr* and 20 *RR* lines. Note that the region containing either *r* or *R* allele show only DNA from the parent, from which allele was descended, whereas rest of the upper arm shown with gray color is composed of bulk DNA from both parents. Ellipsoids are used to show the location of the centromeres.
**B)** Identification of an SSR marker, Satt_335 linked to the *R* gene. [n]S, susceptible NIL; [n]R, resistant NIL; [b]S, susceptible bulk pool of 20 *rr* $F_{2:3}$ families; [b]R, resistant bulk pool of 20 *RR* $F_{2:3}$ families. For Satt_26, no polymorphisms were observed between either NILs or bulk pools. For Satt_335, polymorphisms were observed between either NILs or bulk pools. Results suggested that Satt_335 putatively linked to the *R* gene.

From evaluation of >250 selected SSR markers using two NILs and/or bulks we should be able to identify at least one marker that shows a clear polymorphism between the two NILs and/or bulks (Fig. 7-2B). Such a polymorphic marker is presumably linked to the locus to be isolated. Once a candidate linked marker is identified (e.g., Satt_335 in Fig. 7-2B), additional markers linked to this marker should be investigated for their possible polymorphism between the two bulks and/or NILs. If we observe polymorphism between NILs and bulks (as shown for Satt_335 in Fig. 7-2B) for additional linked markers, then it will indicate that the gene is located in that genomic region.

If we will fail to identify SSR markers for a large genomic regions (>20 cM) that are polymorphic between the parents used for developing the bulks or NILs, then we will identify cleaved amplified polymorphic sequences (CAPS) or single nucleotide polymorphisms (SNPs) for such regions (Konieczny and Ausubel 1993; Henry 2001; Hyten et al. 2008). Once the region is known, CAPS can be developed to place the gene into a small genomic region.

CAPS or SNP can easily be developed because the soybean genome sequence is now available. In the future, the use of either SNP or SSR marker technology for map based cloning or any mapping experiments will be determined by the relative costs of conducting SNP or SSR analyses. Although amplified fragment length polymorphism (AFLP), random amplified polymorphic DNA (RAPD) or developing CAPS for the regions with no available polymorphic SSR markers is possible, most likely these technologies will not be cost effective once affordable SNP assays are available to the research community. Given that the soybean genome sequence is now available, one can apply any marker technology to identify linked markers and based on marker sequences the putative genomic region containing the gene can easily be identified.

### 7.2.3.2 *Mapping of Linked Molecular Markers*

Once several linked markers show polymorphism between NILs and/or bulk pools, markers of this candidate genomic region are investigated for segregation among a collection of RILs or $F_{2:3}$ families that have been characterized for phenotypes governed by the alternative alleles. The initial map of the molecular markers that show polymorphisms between the two bulks and/or NILs can easily be achieved by studying 50 homozygous recessive $F_2$ plants or 50 $F_{2:3}$ families that are completely classified for the alternate phenotypes of the trait. This mapping population will yield the smallest mapping distance of ~1 cM for a single recombination event between two linked codominant molecular markers such as SSR, CAPS, etc. A suitable mapping program, such as Map Manager QTX program (Manly et al. 2001) can be used to generate the molecular map of the genomic region containing the gene of interest. An example of scoring a few $F_2$ individuals for two linked SSR markers is shown in Figure 7-3.

## 7.2.4 High-Density and High-Resolution Mapping

Once the gene is mapped to an interval between two molecular markers (Fig. 7-3B), the next step is to develop a high-density and high-resolution genetic map of the region containing the gene. This step will allow the identification of molecular markers that are only a fraction cM away from the gene to be cloned. Two resources required simultaneously for this step are (i) recombinants and (ii) molecular markers.

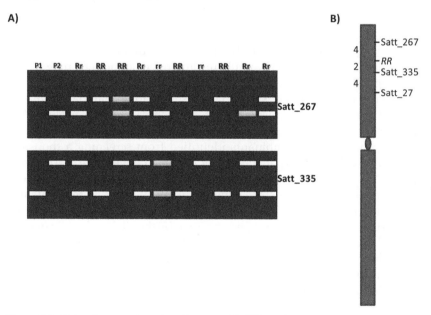

**Figure 7-3** Molecular mapping of an *R* gene with SSR markers.
**A)** Genotypes of two SSR markers, linked to an *R* gene, are shown for a sample of $F_2$ plants. Results of disease phenotypes from screening 25 $F_3$ progenies of each $F_2$ individual are shown at the top of the panel. P1, the resistant parent; P2, the susceptible parent; Rr, heterozygous $F_2$ plants, $F_3$ progenies of whose were segregating in the 3:1::Resistant:Susceptible ratio; RR, homozygous resistant; rr, homozygous susceptible. Recombinants are shown by blue DNA fragments.
**B)** Map position of the *R* gene is shown. The gene is mapped in between Satt_267 and Satt_335. Centi-Morgan (cM) distances between loci are shown on left side of the genetic map.

### 7.2.4.1 Identification of Recombinants and their Characterization

The two flanking SSR/CAPS markers encompassing the genomic region containing the gene or QTL (gene- or QTL-interval) should be used to screen a large segregating material to identify the recombinant plants that carry recombination breakpoints between the two molecular markers (Fig. 7-3A). It is preferable to have this interval relatively small (< 5 cM) so that we can keep the number of recombinants to a minimum. Now the question is how many segregants should be investigated for identifying recombination breakpoints in the gene- or QTL-interval.

One can roughly estimate the number of segregants to be studied by considering the physical distance covered by the two flanking markers. This can now easily be calculated by looking at the relationship between the genetic distance between two nearest flanking markers (hypothetical markers, Satt_267 and Satt_335 in Fig. 7-3) and the sequence covered by these two markers in the soybean genome sequence (*www.Phytozome.org*; *www.soybase.org*). A small physical distance/cM in the gene- or QTL-interval will indicate that the region is highly recombinogenic and screening ~1,000 segregants ($F_2$s or RILs) will be sufficient for generating a high-resolution map of the interval. Otherwise, a larger segregating material should be assayed for identifying recombinants, which can be accomplished in multiple steps of 1,000 segregants.

The number of recombinants identified from screening a thousand segregants will depend upon the genetic distance between the two markers used for identifying recombinants. For a genetic distance of 5 cM between the two molecular markers, we expect to identify about 50 recombinants from 1,000 segregants or RILs. These recombinants can then be progeny-tested for determining phenotypes governed by the alleles of the locus to be isolated. Although it is expected that the breakpoints among these recombinants are distributed randomly in the gene- or QTL-interval, we may find places that carry fewer breakpoints. Thus, a large collection of recombinants can be very useful in dividing the gene- or QTL-interval into smaller sections.

One can also use an alternative approach in isolating recombinants. If the phenotypes encoded by the alleles of the gene have high penetrance, homozygous recessive $F_2$ segregants can be used for developing the high-resolution map. In this scenario, one can score the segregants for the phenotypes governed by the alleles of the gene prior to scoring for molecular markers. Once the recombinants are identified, the phenotypes of individual recombinant can be confirmed by determining the phenotypes of its 16 progenies. For quantitative or oligogenic traits with low penetrance or for traits that require extensive evaluation, as described above, molecular marker-based isolation of recombinants for the interval containing the gene or the QTL can be considered first. Once the recombinants are identified, the progenies of the recombinants are evaluated rigorously (25 progenies/ recombinant; multiple testing, etc.) for a QTL. Once the gene to be cloned is narrowed down into a genomic region of < 100 kb, the recombinant isolation can be discontinued.

## 7.2.4.2 Identification of Molecular Markers

As the genome-sequence of the soybean cultivar Williams 82 is available, many different means can be used to generate molecular markers for the targeted region containing the gene or QTL of interest. For examples, (i) develop CAPS markers to saturate the gene- or QTL-interval using the available Williams 82 genome sequence; (ii) Williams 82 can be one of the parents, and the entire genome of the other parent can be shotgun sequenced to generate CAPS or SNP markers for the gene- or QTL-interval; (iii) develop a BAC library for the gene or the QTL containing parent (if it is not Williams 82) and identify and sequence the BACs carrying the gene or QTL for identifying CAPS or SNP markers.

CAPS markers are the marker type of choice if SNP assays are unavailable. The interval containing the gene or QTL can be identified in the soybean genome sequence by searching sequences of the two markers that flank the interval. Once the sequence is identified, approximately two kb unique DNA sequence should be amplified from both parents (if one of the parents is not Williams 82) and sequenced to generate CAPS markers (Fig. 7-4). CAPS should initially be generated to divide the gene- or QTL-interval into segments of ~100 kb DNA fragments as shown in Figure 7-4. One can also explore the possibility of generating SSR markers by searching SSR sequences in the gene- or QTL-interval and then designing primers to generate polymorphic SSR markers. Considering the large genome size of soybean, 1 cM genetic distance can be close to a megabase pair of DNA and may contain up to 100 genes. Thus, the smaller the gene- or QTL-interval, the quicker will be the process of identifying the gene. We need to continue isolating markers until every recombination breakpoint is flanked by polymorphic markers and the target gene is localized into an <100 kb region.

## 7.2.4.3 Generation of High-Density and High-Resolution Genetic Maps

Once many recombinants and molecular markers for the gene- or QTL-intervals are identified, the next step is to apply the markers and recombinants to map the gene or QTL of interest. This process is then expected to result in development of a precise molecular marker map with the nearest marker within 50 kb DNA from the gene or QTL of interest.

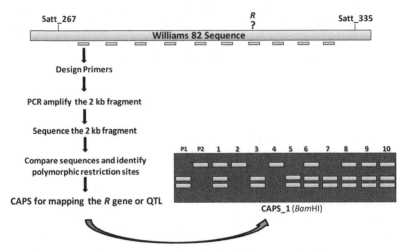

**Figure 7-4** Identification of CAPS markers for high density molecular map.
Steps to be followed in isolating CAPS between Williams 82 and the parent containing the alternative allele are presented on the left. Possible genomic locations for developing CAPS (at a ~100 kb interval) are shown with rectangles just below the William 82 sequence. CAPS_1 polmorphic for *Bam*HI and was used to show polymorphisms of ten $F_{2:3}$ families containing recombination breakpoints between Satt_267 and Satt_335.

The actual number of $F_2$ individuals to be investigated at this stage of mapping varies. For genes mapped to highly recombinogenic regions, ~1,000 $F_2$ plants are a good start, whereas >4,000-5,000 $F_2$s will be required for genes located in a recombination-poor region. It is expected that the currently available high-density global genetic map of the soybean genome (Song et al. 2004) together with the soybean genome sequence will provide the necessary information to determine the number of $F_2$s needed for high-resolution mapping experiments.

The available soybean genome sequence will also allow developing CAPS markers as stated earlier (Fig. 7-4). The gene- or QTL-interval can be divided into segments of roughly equal sizes for developing CAPS markers in every < 100 kb sequence. Polymorphic CAPS will be then used to map the gene or QTL with the help of recombinants that mapped to the gene- or QTL-interval.

Once the gene- or QTL-interval is reduced to about <100 kb, it is worthwhile to look for candidate genes that are located in that interval. Again, in the case of a QTL one may find it difficult to map all genes for QTL in such a small physical distance. To be safe, one should consider reinvestigating the phenotypes as well as genotypes of the recombinants that were used to map the gene or QTL into a small genomic (<100 kb) region.

## 7.2.5 Physical Mapping of the Region Containing the Gene

It is worth noting that the publicly available Williams 82 genome sequence may not cover some genomic regions. Williams 82 may not carry the region containing the gene or QTL. Physical mapping becomes essential for these cases. This step requires pulse-field gel electrophoresis (PFGE) of high molecular weight DNA digested with various rare cutting restriction enzymes. This study will clarify if the Williams 82 DNA sequence contains any missing sequence in the interval containing the gene or QTL. Secondly, it will also show if the genomic region containing the functional gene or QTL in wild type haplotype is comparable in size to that in Williams 82 (Fig. 7-5).

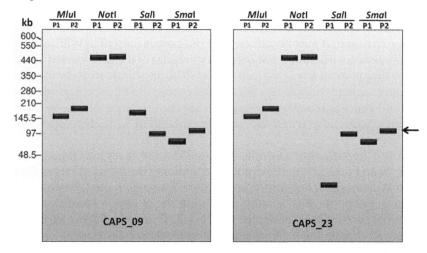

**Figure 7-5** Physical mapping of the genomic region carrying an *R* gene. Two DNA filters were prepared by transferring DNA digested with four restriction enzymes, *Mlu*I, *Not*I, *Sal*I and *Sma*I from gels following pulse field gel electrophoresis. P1, Williams 82; P2, the parent that carries the *R* gene. Arrow indicates the approximately 100 kb *Sma*I DNA fragment that hybridized to both flanking markers CAPS_09 and CAPS_23 in P2. The fragment is slightly smaller in Williams 82. Similar results were also observed for DNA digested with *Mlu*I. In both *Mlu*I and *Sma*I enzyme digested DNA, fragments were larger in P2 than those in Williams 82.

## 7.2.6 Isolation of the DNA Fragment Containing the Gene

Once physical map suggests that the gene or QTL to be isolated is within an approximately 100 kb fragment, one can consider isolating the DNA fragment containing the gene. Cloning of the DNA fragment becomes essential if the region containing the functional allele is absent in Williams 82.

There are several large insert libraries available in the soybean community (Marek and Shoemaker 1997; Danesh et al. 1998; Salimath et al. 1999; Tomkins et al. 1999; Meksem et al. 2000; Santra et al. 2003; Wu et al. 2004, 2008). Prior to constructing a library, one may consider investigating the possibility of using one of these libraries. If the gene or QTL is absent in these libraries, it will be appropriate to prepare a large insert library in *Escherchia coli*. Bacterial artificial chromosomes (BACs) are easier to create when compared to construction of yeast artificial chromosomes (YACs). The BAC library should be generated from sheared DNA rather than enzyme-digested DNA as the library developed from sheared DNA better represents the entire genome as compared to libraries from enzyme-digested high molecular weight DNA (Osoegawa et al. 2007).

If the size of the gene- or QTL-region in the wild type haplotype is comparable to that in Williams 82 haplotype, a library constructed in lambda phage or phagemid may be sufficient. If the gene is present in Williams 82, then library construction is not necessary because of the genome sequence of this line is available. One can apply PCR approach in isolating the alternative mutant or non-functional alleles from the P2 parent. However, one should be cautious about the possible absence of the gene in the genome sequence, especially when the gene is embedded in highly repetitive sequences.

In our hypothetical example (Figure 7-5), the *R* region (flanked by CAPS_09 and CAPS23) in P2 haplotype containing the *R* gene is larger than that in Williams 82. Therefore, a BAC library was constructed. Sequencing of two overlapping BAC clones containing the *R* region identified two additional genes that are absent in Williams 82 (Figure 7-6). Since Williams 82 does not contain the *R* gene, we hypothesize that one of the two genes is the *R* gene to be cloned.

### 7.2.7 Identification of the Gene

Once we narrow down the number of candidate genes to a manageable number by utilizing recombinants and molecular markers and sequencing BAC clones as described earlier, the final step is to identify the gene that governs the phenotype specific to the wild type or functional allele of the gene. The two most commonly used strategies used in this step are: (i) complementation of the mutant by transferring the candidate gene through stable transformation (Paz et al. 2006; Figure 7-7); (ii) characterization of the mutant alleles of the candidate gene. The first approach was applied in map-based cloning the *Rps1*-k that confers *Phytophthra* resistance in soybean (Gao et al. 2005). Stable transformants and their progeny were investigated for race reaction in cloning this *R* gene. The second approach requires a number of independent mutant alleles of the gene. The mutant alleles are created usually through treatment with EMS. The point mutations induced by EMS cause the loss of wild type phenotype among the mutants. In this approach, we PCR amplify the open reading frame sequences of all candidate genes from individual mutants and then determine which gene contains mutations among the mutant alleles. This can be achieved in a step by step fashion. We first sequence all candidate genes in one mutant. We then sequence the other mutant alleles for the candidate gene that shows a mutation in the first mutant allele. This approach was applied in map-based cloning the first soybean gene, *GmNARK* (*Glycine max* nodule autoregulation receptor kinase) that autoregulates the nodule formation (Searle et al. 2003). A transient complementation assay was applied in cloning the *Rpg1-b* gene that confers bacterial blight resistance in soybean (Mindrinos et al. 1994; Ashfield et al. 2004).

In addition to map-based cloning of these three soybean genes, there are additional map-based cloning projects that are in progress. For example, a strong candidate gene for soybean cyst nematode (SCN) resistance at the *rhg1* locus has been identified from map-based cloning effort (Ruben et al. 2006). Graham et al. (2002) reported the construction of a BAC contig that includes three genes, two of which encode disease resistance and the third one is involved in the nodulation process.

## 7.3 Pitfalls Encountered in Map-based Cloning and their Possible Solutions

Since it is very unlikely that any two map-based gene cloning projects will be identical, the steps described above will usually need to be modified to account for specific details for the trait or gene of interest. As stated earlier, having an easily scorable trait expedites the cloning of the target gene. If we have to work on a trait that is difficult to score, then we may consider conducting elaborate progeny testing to make sure that phenotypes of individual $F_2$s or RILs are accurate. Based on progeny testing, homozygous recessive plants may be selected for initial mapping of these traits. Once the gene interval is established using about 40–50 homozygous recessive $F_2$ plants, two markers that encompass the interval can be used in the high-density and high-resolution mapping experiments. The two markers can be used to identify the recombinants of the gene interval. These will be a small subset of a larger mapping population and can be progeny tested for phenotypes. At least 25–30 progeny may be tested to accurately distinguish the homozygous dominant from heterozygous individuals. If necessary, additional progeny testing can be considered for traits with poor penetrance.

In the hypothetical example presented in Figure 7-6, of the eight genes of haplotype P2 only six were found in Williams 82 that does not carry the functional allele of the gene. Therefore, the two P2-specific genes are candidates for the *R* gene. In this case, we have to study only two candidates for identifying the gene of interest. In most cases, multiple genes may be encountered for functional characterization and gene identification. We may encounter 15 or more genes in 100 kb fragment that contains the gene of interest. If we cannot develop a transient assay system to identify the gene, gathering additional recombinants and markers to reduce the number of candidate genes can be considered. Sequences of the candidate genes from both parents may also be considered as a good option. Once the candidate gene number is reduced to a manageable number, one can generate stable transformants to conduct complementation analyses.

In cloning *Rpg1*-b and *Rps1*-k, it was observed that the BAC clones for the regions containing these two genes were highly underrepresented in BAC libraries (Ashfield et al. 2003; Bhattacharyya et al. 2005). Several BAC libraries representing over 24 genome equivalents genomic DNA were screened to obtain two overlapping BAC clones that contain the *Rps1*-k gene (Bhattacharyya et al. 2005; Gao and Bhattacharyya 2008).

It is possible that a gene of interest may be located in a region that is very poor in recombination. For those genes, map-based cloning will require an enormous amount of resources. Coupled with this difficulty, if the gene does not show high penetrance and the phenotype is difficult to score, then one may have to reconsider the importance of map-based cloning of such genes.

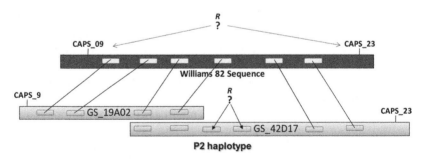

**Figure 7-6** Physical map of the *R* region.
In the hypothetical example, the *R* gene to be cloned is present only in parent P2. Blue rectangles represent putative genes. Physical mapping data (Fig. 7.5) showed that the P2 haplotype is bigger than the Williams 82 haplotype. Therefore, a BAC library for the P2 haplotype was constructed and two overlapping BACs, GS_19A02 and GS_42D17, carrying CAPS_09 and CAPS_23, respectively, were isolated. Sequencing of these two BACs revealed that there are two additional open reading frames in the *R* region of P2 haplotype as compared to that in Williams 82.

*Color image of this figure appears in the color plate section at the end of the book.*

## 7.4 Future of Map-based Cloning in Soybean

Map-based cloning approach is technology driven. Although there will be little changes in phenotyping various traits and creation of genetic materials, the cloning process will be expedited if RILs segregating for the traits under consideration are available. It is expected that the size of available RIL pools will only increase with time. Cloning of soybean *Rpg1-b* is a good example of the use of RILs available to the research community (Ashfield et al. 2003). In other words, the first step for any map-based cloning will be assessment of the RILs, other segregating materials and large insert libraries available to the research community.

The major progress one can expect is in the direction of sequencing any soybean lines to generate polymorphic markers for identifying the candidate genes from a small genomic region. The genome sequence of cultivar the Williams 82 is already available to the research community. DNA sequencing cost has been going down in an unprecedented manner. Now, one can sequence one soybean genome equivalent DNA for a cost of ~US$5,000 with the help of Solexa® technology (*http://cat.ucsf.edu/pdfs/ SS_DNAsequencing.pdf*), and the cost of sequencing is expected only to fall. Thus, sequencing the parents of crosses used in mapping experiments will become the preferred approach for rapidly creating SNPs, SSR or CAPS markers to narrow down the region containing the gene.

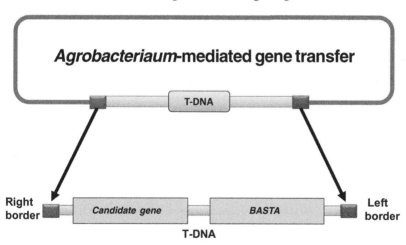

**Figure 7-7** Complementation analysis for identifying the gene of interest.
A candidate gene integrated into the T-DNA of *Agrobacterium tumefacians* in the binary vector, pTF101.1 was used to transform the soybean cultivar Williams 82 that does not carry the *R* gene. Stable transformants were taken to the next generation and tested for segregation of the complemented disease resistance phenotype. Of the two candidate genes (Fig. 7-6), one showed to confer disease resistance against the destructive soybean pathogen

Another advance, which will revolutionize high-density molecular mapping, is the availability of polymorphism information of the entire pool of USDA soybean germplasm (~ 20,000 lines) for a set of 50,000 SNPs (Hyten 2008). This means, we will have on average at least one SNP in every 100 kb DNA for any pair of soybean lines, even if only 11,500 of the 50,000 SNPs are polymorphic between the pair of lines for the genome of 1,150 mb DNA (Arumuganathan et al. 1991).

There will be continuous progress in the gene identification step. Stable soybean transformation is now feasible. For some traits, transient expression assay should be sufficient for gene identification. Targeting induced local lesions in genomes (TILLING) has now been documented for soybean. In this study, two soybean lines were treated with two chemical mutagens to generate four populations saturated with point mutations (Cooper et al. 2008). The soybean TILLING facility established at the Southern Illinois University is expected to play a significant role in the future map-based cloning of soybean genes (*http://www.soybeantilling.org/*), if the gene of interest is present in the cultivars used in TILLING experiments. Virus-induced gene silencing (VIGS) method is now available for soybean (Zhang and Ghabrial 2006). This transient gene-silencing assay should expedite the gene identification step at least for some of the genes that can be evaluated in soybean trifoliates. Many of the genes expressed in roots can be silenced through RNAi with the aid of *Agrobacterium rhizogenes*-mediated transformation for their functional analyses (Subramanian et al. 2005; Graham et al. 2007).

## References

Arumuganathan K, Earle ED (1991) Estimation of nuclear DNA content of plants by flow cytometry. *Plant Mol Biol Rep* 9: 229–241.

Ashfield T, Bocian A, Held D, Henk AD, Marek LF, Danesh D, Peñuela S, Meksem K, Lightfoot DA, Young ND, Shoemaker RC, Innes RW (2003) Genetic and physical localization of the soybean *Rpg1-b* disease resistance gene reveals a complex locus containing several tightly linked families of NBS-LRR genes. Mol Plant-Micr Interact 16: 817–826.

Ashfield T, Ong LE, Nobuta K, Schneider CM, Innes RW (2004) Convergent evolution of disease resistance gene specificity in two flowering plant families. Plant Cell 16: 309–318.

Bhattacharyya MK, Narayanan NN, Gao H, Santra DK, Salimath SS, Kasuga T, Liu Y, Espinosa B, Ellison L, Marek L, Shoemaker R, Gijzen M, Buzzell RI (2005) Identification of a large cluster of coiled coil–nucleotide binding site-leucine rich repeat-type genes from the *Rps1* region containing *Phytophthora* resistance genes in soybean. Theor Appl Genet 111: 75–86.

Cooper JL, Till BJ, Laport RG, Darlow MC, Kleffner JM, Jamai A, El-Mellouki T, Liu S, Ritchie R, Nielsen N, Bilyeu KD, Meksem K, Comai L, Henikoff S (2008) TILLING to detect induced mutations in soybean. BMC Plant Biol 8: 9.

Danesh D, Penuela S, Mudge J, Denny RL, Nordstrom H, Martinez JP, Young ND (1998) A bacterial artificial chromosome library for soybean and identification of clones near a major cyst nematode resistance gene. Theor Appl Genet 96: 196–202.

Gao H, Bhattacharyya MK (2008) The soybean-*Phytophthora* resistance locus *Rps1*-k encompasses coiled coil-nucleotide binding-leucine rich repeat-like genes and repetitive sequences. BMC Plant Biol 8: 29.

Gao H, Narayanan NN, Ellison L, Bhattacharyya MK (2005) Two classes of highly similar coiled coil-nucleotide binding-leucine rich repeat genes isolated from the *Rps1*-k locus encode *Phytophthora* resistance in soybean. Mol Plant-Micr Interact 18: 1035–1045.

Graham MA, Marek LF, Lohnes D, Shoemaker RC (2002) Organization, expression and evolution of a disease-resistant gene cluster in soybean. Genetics 164: 1961–1977.

Graham TL, Graham MY, Subramanian S, Yu O (2007) RNAi silencing of genes for elicitation or biosynthesis of 5-deoxyisoflavonoids suppresses race-specific resistance and hypersensitive cell death in *Phytophthora sojae* infected tissues. Plant Physiol 144: 728–740.

Henry RJ (2001) Plant Genotyping: The DNA Fingerprinting of Plants. CABI Publ, Oxford, UK.

Hyten D (2008) Whole Genome Analysis of the USDA Soybean Germplasm Collection and Applications for New Gene Discovery (50,000 Snps): *http://www.ars.usda.gov/ research/projects/projects.htm?accn_no=413473*

Kasuga T, Salimath SS, Shi J, Gijzen M, Buzzell RI, Bhattacharyya MK (1997) High resolution genetic and physical mapping of molecular markers linked to the *Phytophthora* resistance gene *Rps1*-k in soybean. Mol Plant-Micr Interact 10: 1035–1044.

Konieczny A, Ausubel FM (1993) A procedure for mapping Arabidopsis mutations using co-dominant ecotype-specific PCR-based markers. Plant J 4: 403–410.

Manly, KF, Cudmore Jr RH, Meer JM (2001) Map Manager QTX, cross-platform software for genetic mapping. Mamm Genom 12: 930–932.

Marek LF, Shoemaker RC (1997) BAC contig development by fingerprint analysis in soybean. Genome 40: 420–42.

Mather K (1957) The Measurement of Linkage in Heredity. John Willey, New York, USA.

Meksem K, Zobrist K, Ruben E, Hyten D, Quanzhou T, Zhang H-B, LightfootD A (2000) Two large-insert soybean genomic libraries constructed in a binary vector: Applications in chromosome walking and genomic wide physical mapping. Theor Appl Genet 101: 747–755.

Michelmore RW, Paran I, Kesseli RV (1991) Identification of markers linked to disease-resistance genes by bulked segregant analysis: a rapid method to detect markers in specific genomic regions by using segregating populations. Proc Natl Acad Sci USA 88: 9828–9832.

Mindrinos M, Katagiri F, Yu GL, Ausubel FM (1994) The *A. thaliana* disease resistance gene *RPS2* encodes a protein containing a nucleotide-binding site and leucine-rich repeats. Cell 78: 1089–1099.

Mohan S, Chest V, Chadwick RB, Wergedal JE, Srivastava AK (2007) Chemical mutagenesis induced two high bone density mouse mutants map to a concordant distal chromosome 4 locus. Bone 41: 860–868.

Osoegawa K, Vessere GM, Li Shu C, Hoskins RA, Abad JP, de Pablos B, Villasante A, de Jong PJ (2007) BAC clones generated from sheared DNA. Genomics 89: 291–299.

Paz MM, Martinez JC, Kalvig AB, Fonger TM, Wang K (2006) Improved cotyledonary node method using an alternative explant derived from mature seed for efficient *Agrobacterium*-mediated soybean transformation. Plant Cell Rep 25: 206–213.

Ruben E, Jamai A, Afzal J, Njiti VN, Triwitayakorn K, Iqbal MJ, Yaegashi S, Bashir R, Kazi S, Arelli P, Town CD, Ishihara H, Meksem K, Lightfoot DA (2006) Genomic analysis of the *rhg1* locus: Candidate genes that underlie soybean resistance to the cyst nematode. Mol Genet Genom 276: 503–516.

Salimath SS, Bhattacharyya MK (1999) Generation of a soybean BAC library, and identification of DNA sequences tightly linked to the *Rps1*-k disease resistance gene. Theor Appl Genet 98: 712–720.

Santra DK, Sandhu D, Tai T, Bhattacharyya MK (2003) Construction and characterization of a soybean yeast artificial chromosome library and identification of clones for the *Rps6* region. Funct Integr Genom 3: 153–159.

Searle IR, Men AE, Laniya TS, Buzas DM, Iturbe-Ormaetxe I, Carroll BJ, Gresshoff PM (2003) Long-distance signaling in nodulation directed by a CLAVATA1-like receptor kinase. Science 299: 109–112.

Song QJ, Marek LF, Shoemaker RC, Lark KG, Concibido VC, Delannay X, Specht JE, Cregan PB (2004) A new integrated genetic linkage map of the soybean. Theor Appl Genet 109: 122–128.

Subramanian S, Graham MY, Yu O, Graham TL (2005) RNA interference of soybean isoflavone synthase genes leads to silencing in tissues distal to the transformation site and to enhanced susceptibility to *Phytophthora sojae*. Plant Physiol 137: 1345–1353.

Tomkins JP, Mahalingam R, Smith H, Goicoechea JL, Knap HT, Wing RA (1999) A bacterial artificial chromosome library for soybean PI 437654 and identification of clones associated with cyst nematode resistance. Plant Mol Biol 41: 25–32.

Wu CC, Nimmakayala P, Santos FA, Springman R, Scheuring C, Meksem K, Lightfoot DA, Zhang HB (2004) Construction and characterization of a soybean bacterial artificial chromosome library and use of multiple complementary libraries for genome physical mapping.Theor Appl Genet 109: 1041–1050.

Wu X, Zhong G, Findley SD, Cregan P, Stacey G, Nguyen HT (2008) Genetic marker anchoring by six-dimensional pools for development of a soybean physical map. BMC Genom 9: 28.

Zhang C, Ghabrial SA (2006) Development of Bean pod mottle virus-based vectors for stable protein expression and sequence-specific virus-induced gene silencing in soybean. Virology 344: 401–411.

# Candidate Gene Analysis of Mutant Soybean Germplasm

*R. E. Dewey\* and P. Zhang*

## ABSTRACT

Biological information is often available from model systems or alternate species that provides a molecular genetic basis for individual phenotypes which also occur in soybean. When such information is available, a candidate gene approach can be utilized to determine if related genes provide the underlying cause of the phenotype for soybean. The candidate gene approach is a rapid method to confirm or exclude soybean genes associated with traits without chromosomal position information that would otherwise take intensive mapping efforts to discover without the gene knowledge provided from other species. This method has been particularly effective in defining soybean genes corresponding to disease resistance loci, oil quality traits, and flower and seed pigmentation.

**Keywords:** candidate gene analysis; mutant soybean, disease resistance; oil quality; carbohydrate mutant analysis

## 8.1 Introduction

A major goal for molecular genetic studies in soybean involves the identification and characterization of genes associated with important agronomic traits. In many cases, a trait of interest is manifest as a simple heritable mutation, either natural or induced, that gives rise to a desirable phenotype. Techniques such as map-based cloning have proven to be

Department of Crop Science, North Carolina State University, Raleigh, NC 27695, USA.
*Corresponding author: *ralph_dewey@ncsu.edu*

effective in identifying the specific genes responsible for such traits (reviewed in Chapter 7), yet are often time consuming and labor intensive. In cases where sufficient information is available from model systems concerning the molecular genetics or biochemistry of a specific metabolic pathway, and observed phenotypic outcomes that result from disruptions at a given step, reasonable predictions can be made regarding the gene or genes that may be involved when a similar phenotype is observed in soybean. This strategy, commonly referred to as a candidate gene approach, can serve as a relatively rapid and efficient means for characterizing gene mutations associated with traits of agronomic interest. In this chapter we will focus on how candidate gene approaches have been used to effectively define the molecular basis of important traits in soybean.

Candidate gene analysis for defining the molecular genetic basis of simply inherited phenotypic traits typically consists of four steps. First, one or more candidate genes are proposed according to cause-effect relationships that have been previously observed in the investigation of similar phenotypes in model species (such as *Arabidopsis thaliana*), or alternatively, based on rational predictions deduced from an understanding of metabolite flow through well characterized biochemical pathways. Second, molecular polymorphisms having the potential of altering gene function are identified in the candidate gene between genotypes exhibiting measurable variability at the investigated trait. Third, an association is established between the molecular polymorphism and phenotype in question in a population segregating for the trait. Fourth, the candidate gene is validated. In cases where a mutation in a gene is believed to be responsible for the trait in question as a consequence of gene dysfunction, the ultimate means of verification is to demonstrate functional complementation through transgenic expression of the wild type gene.

Certain characteristics of soybean present challenges for candidate gene analyses. Molecular evolution studies support a model predicting the modern soybean genome to be the product of both an ancient and more recent duplication event (Shoemaker et al. 1996; Schlueter et al. 2004). Therefore, for any given gene product, there are typically multiple independent functional isoforms (paralogs) residing in the genome. This can complicate the candidate gene approach, in some cases requiring knowledge of all functional isoforms of a gene in order to identify the casual gene (or alternatively conclude that the particular gene family is not involved in the trait in question). Furthermore, soybean transformation remains a relatively labor intensive, time consuming process that is not routinely conducted in most soybean research laboratories. Therefore, confirmation of candidate genes through genetic complementation is typically not used as a means of validation in soybean. However, when a candidate gene encodes an enzyme activity that can be assayed in vitro, comparisons of mutant versus wild type activities of the recombinant enzymes can provide strong supportive evidence that an observed polymorphism in a candidate gene is responsible for the observed trait. In some cases, mutations resulting in an in-frame stop codon or frame-shift mutation have been observed upstream of an essential functional domain of the encoded product. Under these circumstances, it is usually assumed that the mutation has rendered the gene nonfunctional.

Despite the above mentioned limitations, candidate gene analysis has proven to be a powerful tool in defining the molecular genetics of several important agronomic traits in soybean. Specific examples are highlighted below.

## 8.2 Disease Resistance Genes

One of the most important advancements in the field of plant molecular genetics has been the identification and characterization of disease resistance (*R*) genes (reviewed by Martin et al. 2003). Remarkably, from a wide diversity of plant species a small number of conserved structural motifs have been observed in *R* genes targeting a broad spectrum of pathogens (e.g., viruses, fungi, bacteria, insects and nematodes). The most abundant of the *R* gene classes identified to date is defined by the presence of an N-terminal nucleotide-binding domain (NBD) and a C-terminal leucine-rich repeat (LRR) motif. The NBD has been proposed to function in pathogen-dependent signal transduction cascades through a classical phosphorylation/ dephospholation mechanism, while the LRR is believed to be involved in ligand binding and pathogen recognition via protein-protein interactions (McHale et al. 2006). The conservation of these motifs, particularly the NBD has provided a means for retrieving candidate *R* genes from essentially any plant genome.

Using degenerate primers corresponding to highly conserved regions of the NBD of the tobacco *N* and *Arabidopsis RPS2* disease resistance genes, Kanazin et al. (1996) and Yu et al. (1996) identified several resistance gene analogs (RGAs) from soybean. In both studies, individual RGAs were genetically mapped to the vicinity of known resistance loci. Although this technique provided a rapid, powerful tool for identifying candidate *R* genes, subsequent analyses were hampered by the fact that RGAs reside throughout the genome as clustered arrays. For example, 16 RGAs were identified on a single 119 kb BAC clone corresponding to a region on soybean linkage group J where resistance to powdery mildew and Phytophthora stem and root rot had been mapped (Graham et al. 2002). Furthermore, 12 independent RGAs were found to map to a region of linkage group F where the *Rsv1* gene that confers resistance to soybean mosaic virus resides (Jeong et al. 2001).

Defining the specific RGA that represents a genuine *R* gene for a given pathogen requires either the transformation of each candidate RGA into a susceptible background followed by pathogen screening, or high-resolution mapping to eliminate all RGAs within a cluster that do not perfectly co-segregate with the resistance phenotype. Owing to the recalcitrance of soybean transformation, the latter strategy has been the preferred option for soybean researchers. Through fine-mapping, Hayes et al. (2004) were able to identify an RGA designated *3gG2* as the only candidate within the gene cluster that perfectly co-segregated with the major *Rsv1* soybean mosaic virus resistance allele. A PCR-based marker specific for the *3gG2* gene has been developed to facilitate marker-assisted breeding of *Rsv1* (Shi et al. 2008). Although difficulties in regard to the clustering of RGA genes, together with the non-facile nature of soybean transformation remain an impediment to progress in defining specific *R* genes through a candidate gene approach, it is likely that continued advances in soybean genome information (see Chapter 10) and high-resolution mapping will help alleviate some of these obstacles.

## 8.3 Lipid Metabolic Genes

The myriad uses for soybean oil in both food and industrial applications have led to an intense investigation of the genetics and physiology that underlie the regulation of lipid storage reserves within the soybean seed. Oil quality is primarily determined by the fatty acid composition of the storage triacylglycerols, and what defines the optimal combination and ratio of fatty acids in soybean oil differs depending on the specific end application. For example, oils containing minimal amounts of palmitic acid (16:0) are desired when used for human consumption because of the cholestrogenic nature of this fatty acid and its ability to promote arterial thrombosis formation (Hu et al. 1997; Hornstra and Kester 1997). Also, high levels of polyunsaturated fatty acids, particularly linolenic acid (18:3) reduce the oxidative stability of the oil and can lead to poor flavor quality. To meet the needs of the vegetable oil industry, plant breeders have developed an array of mutant germplasm lines (either natural or induced by mutagenesis) displaying a wide range of altered oil phenotypes. The availability of these lines, coupled with the wealth of information that has accrued regarding the biochemistry and molecular genetics of lipid metabolism in model plant systems, provides an ideal scenario for utilizing a candidate gene approach for mutant germplasm analysis.

A scheme representing the primary biochemical pathway responsible for triacylglycerol formation in oilseed species such as soybean is shown in Figure 8-1. The early steps of fatty acid formation occur within the plastids of the developing seed, whereas the subsequent attachment of the fatty acids to a glycerol backbone and the introduction of double bonds giving rise to polyunsaturated fatty acids take place in the cytosol/ER (Ohrogge and Jaworski 1997; Lung and Weselake 2006). The individual steps in the lipid biosynthetic pathway where researchers have been successful in identifying the specific genes responsible for a novel oil phenotype using a candidate gene approach are highlighted in Figure 8-1.

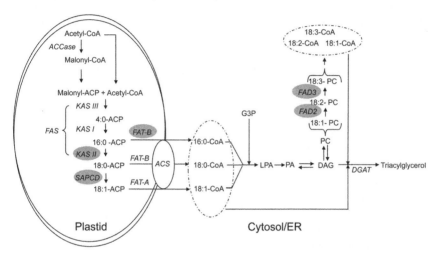

**Figure 8-1** Primary storage lipid biosynthetic pathway in developing seeds of soybean. Steps where candidate gene approaches have been used to identify causal gene mutations are highlighted in gray. ACP, acyl carrier protein; LPA, lysophosphatidic acid; PA, phosphatidic acid; DAG, diacylglyceol; PC, phosphatidylcholine; G3P, glycerol 3-phosphate; ACCase, acetyl-coA carboxylase; FAS, fatty acid synthase; KAS, 3-keto-acyl-ACP synthase; SACPD, Δ9-stearoyl-ACP-desaturase; FAT-B, 16:0 (18:0)-ACP thioesterase; FAT-A, 18:1-ACP thioesterase; ACS, acyl CoA synthetase; FAD2, ω-6 fatty acid desaturase; FAD3, ω-3 fatty acid desaturase; DGAT, diacylglycerol acyltransferase.

The fatty acid thioesterase (FAT) enzyme that catalyzes the cleavage of palmitic and stearic (18:0) acids from their respective acyl carrier protein (ACP) intermediates during fatty acid biosynthesis is designated FAT-B (in contrast to FAT-A enzymes that recognize oleic acid substrates). Because 16:0 fatty acids do not become further modified from the time they are released from the plastid to when they become incorporated into triacylglycerols, the FAT-B enzymatic reaction is pivotal in establishing the amount of 16:0 found in the storage oil. Therefore, reductions in FAT-B enzymatic activity would be predicted to lead to a decrease in the palmitic acid content of the storage triacylglycerols. In a candidate gene investigation of soybean germplasm possessing the $fap_{nc}$ locus that confers a reduced palmitic acid phenotype, Wilson et al. (2001) concluded from Southern blotting assays that a copy of a *FAT-B* gene was deleted in lines carrying $fap_{nc}$. Subsequently, it was determined that the soybean genome contains four unique functional isoforms encoding FAT-B enzymes, and that the isoform whose transcripts were most highly represented in soybean EST databases, *GmFATB1a*, was the gene specifically deleted in these lines (Cardinal et al. 2007). Molecular markers specific to *GmFATB1a* proved to be effective in following the *fap*$_{nc}$-mediated reduced 16:0 trait in segregating breeding populations (Cardinal et al. 2008).

Although the metabolic flow of fatty acid intermediates depicted in the biochemical pathway shown in Figure 8-1 would correctly lead to the prediction that enhancement of FAT-B enzyme activity would lead to an increase in 16:0 in the storage oil (as proven using transgenic modification of oil content), such "gain-of-function" mutations are rarely obtained through mutagenesis. Instead, recessive mutations yielding an elevated palmitic acid phenotype would more likely be caused by mutations in the 3-keto-acyl-ACP synthase activity that elongates 16:0-ACP to 18:0-ACP (KAS II, Fig. 8-1). During an investigation of the *fap*$_2$ locus associated with an enhancement of the 16:0 of the seed oil, it was revealed that the soybean genome expresses two major *KAS II* genes, designated *GmKAS IIA* and *GmKAS IIB* (Aghoram et al. 2006). Sequence analysis of lines containing *fap*$_2$ uncovered a point mutation within *GmKAS IIA* that introduced a premature stop codon within the reading frame, a mutation that would be expected to render the encoded product nonfunctional, thus providing a rational explanation for the enhanced 16:0 phenotype associated with the *fap*$_2$ locus.

Stearic acid levels are typically low in soybean (2–4% total fatty acid); therefore little attention has been paid toward reducing its content in the oil. Mutant germplasm has been characterized, however, in which 18:0 levels as high as 30% have been observed (Pantalone et al. 2002). The most straightforward means for obtaining an elevated 18:0 phenotype via a recessive mutation would involve altering the stearoyl-ACP-desaturase enzyme (SACPD) responsible for converting 18:0-ACP to 18:1-ACP (Fig. 8.1). An initial study of *SACPD* genes in soybean identified two specific isoforms, *SACPD*-A and *SACPD*-B, yet neither gene appeared to be altered in germplasm possessing the *fas*$^a$ locus that mediates a 30% 18:0 phenotype (Byfield et al. 2006). More recent studies revealed a third isoform (designated *SACPD-C*) that is specifically expressed during soybean seed development. A candidate gene analysis of germplasm containing the *fas*$^a$ locus showed that *SACPD-C* was completely deleted in this line (Zhang et al. 2008). Furthermore, an amino acid substitution mutation was found in the *SACPD-C* gene of soybean lines possessing *fas*$^{nc}$, an elevated 18:0 locus that is allelic to *fas*$^a$, yet less severe (9–15% 18:0). Interestingly, in enzyme assays conducted in vitro using recombinant proteins, the *fas*$^{nc}$-associated mutant SACPD-C enzyme displayed greater activity than the wild type enzyme. Given the strict genetic correlation that was observed between the mutant *SACPD-C* gene and the elevated 18:0 phenotype, however, it was speculated that the consequence of the observed amino acid substitution may be to reduce enzyme stability in vivo as opposed to reducing the inherent activity of the enzyme (Zhang et al. 2008).

The enzymes primarily responsible for the synthesis of polyunsaturated fatty acids during soybean seed development are ER-localized proteins termed FAD2 (18:1 to 18:2) and FAD3 (18:2 to 18:3) (Fig. 8-1). The genes encoding these enzymes represent the most obvious targets in mutant germplasm displaying reduced levels of 18:2 and/or 18:3. The soybean genome contains two distinct classes of *FAD2* genes. Genes designated *FAD2-1* are expressed specifically during seed development, while *FAD2-2* genes are constitutively expressed throughout the entire plant. Within these classes, two functional *FAD2-1* isoforms (*Fad2-1a* and *Fad2-1b*) have been characterized (Tang et al. 2005), and at least three *FAD2-2* paralogs have been identified (Schlueter et al. 2007). Candidate gene analysis of germplasm line M23, an elevated 18:1 (reduced 18:2 + 18:3) line developed by X-ray mutagenesis, revealed that the embryo-specific *Fad2-1a* isoform is deleted in both the M23 parental line and elevated 18:1 segregants in breeding populations derived from M23 (Alt et al. 2005; Sandhu et al. 2007).

Candidate gene strategies have been particularly effective in revealing the molecular basis of low 18:3 soybean germplasm. Similar to the other lipid biosynthetic enzymes described above, FAD3 desaturases in soybean are encoded by multiple functional isoforms. In a seminal report by Bilyeu et al. (2003), three unique *FAD3* paralogs were characterized: Gm*FAD3A*, Gm*FAD3B* and Gm*FAD3C*. In a series of studies, primarily by the same group, candidate gene strategies have led to a thorough understanding of the molecular genetics of linolenic accumulation in soybean oil. Mutations in all three soybean *FAD3* genes have been shown to be associated with lowering the 18:3 content of the seed. Breeders have developed multiple independent lines displaying a reduced 18:3 phenotype at a locus designated *fan*. In each case, null mutations (e.g., deletions, frame-shifts, premature stop codons and splice junction defects) were found in the Gm*FAD3A* gene (Bilyeu et al. 2003, 2005; Anai et al. 2005; Chappell and Bilyeu 2006, 2007). Potentially debilitating mutations were also found within the Gm*FAD3B* and Gm*FAD3C* genes in additional reduced 18:3 germplasm lines (Anai et al. 2005; Bilyeu et al. 2005, 2006). By pyramiding these individual mutations, soybean germplasm containing just 1% linolenic acid was developed (Bilyeu et al. 2006), a trait highly desired by the food industry because of its enhanced oxidative and flavor stability. The residual 18:3 that remains in the triple mutant lines is likely either the product of plastid-localized 18:2 desaturase activities, or the product of yet a fourth ER localized *FAD3* gene that has been reported (Anai et al. 2005).

Along with clarifying the genetic mechanisms responsible for altering soybean oil composition, characterizing the specific molecular lesions underlying these traits also provides information advantageous for the development of soybean cultivars with desirable oil profiles. In most of the studies described above, molecular markers were developed corresponding to the gene mutations that were revealed. Molecular markers based on causal mutations are "perfect" in that there is no chance that the trait of interest will become separated from the marker through recombination.

## 8.4 Pigmentation and Carbohydrate/Phytate Mutant Analysis

Similar to the lipid biosynthetic pathway discussed above, the biochemistry and molecular genetics of pigment forming flavonoid pathways in plants have been characterized to a level that makes candidate gene approaches for mutant analysis feasible. Previously published chromatographic data led Toda et al. (2002) to predict that the gray pubescence coloration phenotype conferred by the recessive *t* locus in soybean is a consequence of deficient flavonoid 3'-hydroxylase activity. This supposition was supported by the discovery of a frame-shift mutation within a gene encoding this activity that faithfully co-segregated with the phenotype. Using a similar rationale, it was proposed that the magenta flower color phenotype of soybeans possessing the recessive *wm* allele (versus the purple color mediated by *Wm*) could be explained by a mutation in a flavonol synthase gene (Takahashi et al. 2007). Candidate gene analysis of flavonol synthase genes from isogenic lines of the cultivar Harsoy revealed a debilitating frameshift mutation in the *wm*-associated allele. The inability of the truncated protein to perform its normal function was further validated through in vitro enzyme assays of recombinant mutant versus wild type enzymes (Takahashi et al. 2007). Finally, prior knowledge of the role and function of chalcone synthase genes in pigment formation was the basis for the discovery that various duplications and deletions of members of this multigene family defined the *I* locus that controls seed coat color in soybean (Todd and Vodkin 1996; Senda et al. 2002; Kasai et al. 2007).

A final example that well exemplifies the utility of the candidate gene approach involves the characterization of a mutant soybean line that displayed both a decreased raffinosaccharide and phytic acid seed phenotype. This line, designated LR33, was uncovered through a large scale screen of a mutagenized soybean population, followed by heritability studies that attributed the dual phenotype to a single, recessive locus (Sebastian et al. 2000). Although raffinosaccharides and phytate are products of distinct biochemical pathways, both can originate from myo-inositol 1-phosphate. This fact, combined with supportive in vivo labeling results, suggested a defect in a myo-inositol 1-phosphate synthase (MIPS) gene as the best candidate for mediating both phenotypes (Hitz et al. 2002). DNA sequence analysis revealed an amino acid substitution mutation in a highly conserved region of the enzyme. Subsequently, in vitro enzyme assays showed the specific activity of the recombinant MIPS protein originating from LR33 to be only ~10% of that observed with the wild type enzyme.

## 8.5 Conclusions

Candidate gene approaches have been used in soybean genetics for more than a decade and have become an increasingly popular means of determining the molecular genetic basis of agronomically useful traits. Although success in this area is typically reliant on previously established relationships between known metabolic pathways and their cognate genes, the continual progress that is being made in understanding these relationships in model systems will only further expand the opportunities for applying this information toward gene discovery in soybean. Furthermore, advances in the elucidation of the complete soybean genome (see Chapter 10) will serve to greatly alleviate the problem of having to identify the complete set of paralogs of a given gene that may serve as viable targets. An obstacle that is likely to remain for at least the near future, however, is the inability to readily transform the soybean plant. Without this ultimate means of validation through complementation, it is possible that situations will arise where a benign polymorphism in a proposed gene candidate will be mistakenly assigned to a phenotype dictated by the true casual gene when the two are closely linked. Despite this limitation, it is likely that candidate gene approaches will continue to play an increasingly important role in the molecular genetic analysis of important agronomic traits in soybean.

## References

Aghoram K, Wilson RF, Burton JW, Dewey RE (2006) A mutation in a 3-keto-acyl-ACP synthase II gene is associated with elevated palmitic acid levels in soybean seeds. Crop Sci 46: 2453–2459.

Alt JL, Fehr WR, Welke GA, Sandhu D (2005) Phenotypic and molecular analysis of oleate content in the mutant soybean line M23. Crop Sci 45: 1997–2000.

Anai T, Yamada T, Kinoshita T, Rahman SM, Takagi Y (2005) Identification of corresponding genes for three low-α-linolenic acid mutants and elucidation of their contribution to fatty acid biosynthesis in soybean seed. Plant Sci 168: 1615–1623.

Bilyeu KD, Palavalli L, Sleper DA, Beuselinck PR (2003) Three microsomal omega-3 fatty-acid desaturase genes contribute to soybean linolenic acid levels. Crop Sci 43: 1833–1838.

Bilyeu K, Palavalli L, Sleper D, Beuselinck P (2005) Mutations in soybean microsomal omega-3 fatty acid desaturase genes reduce linolenic acid concentration in soybean seeds. Crop Sci 45: 1830–1836.

Bilyeu K, Palavalli L, Sleper DA, Beuselinck P (2006) Molecular genetic resources for development of 1% linolenic acid soybeans. Crop Sci 46: 1913–1918.

Byfield, GE, Xue H, Upchurch RG (2006) Two genes from soybean encoding soluble Δ9 stearoyl-ACP desaturases. Crop Sci 46: 840–846.

Cardinal AJ, Burton JW, Camacho-Roger AM, Yang JH, Wilson RF, Dewey RE (2007) Molecular analysis of soybean lines with low palmitic acid content in the seed oil. Crop Sci 47: 304–310.

Cardinal AJ, Dewey RE, Burton JW (2008) Estimating the individual effects of the reduced palmitic acid $fap_{nc}$ and *fap1* alleles on agronomic traits in two soybean populations. Crop Sci 48: 633–639.

Chappell AS, Bilyeu KD (2006) A *GmFAD3A* mutation in the low linolenic acid soybean mutant C1640. Plant Breed 125: 535–536.

Chappell AS, Bilyeu KD (2007) The low linolenic acid soybean line PI361088B contains a novel *GmFAD3A* mutation. Crop Sci 47: 1705–1710.

Graham MA, Marek LF , Shoemaker RC (2002) Organization, expression and evolution of a disease resistance gene cluster in soybean. Genetics 162: 1961–1977.

Hayes AJ, Jeong SC, Gore MA, Yu YG, Buss GR, Tolin SA, Maroof MAS (2004) Recombination within a nucleotide-binding-site/leucine-rich-repeat gene cluster produces new variants conditioning resistance to soybean mosaic virus in soybeans. Genetics 166: 493–503.

Hitz WD, Carlson TJ, Kerr PS, Sebastian S (2002) Biochemical and molecular characterization of a mutation that confers a decreased raffinosaccharide and phytic acid phenotype on soybean seeds. Plant Physiol 128: 650–660.

Hornstra, G and Kester ADM (1997) Effect of the dietary fat type on arterial thrombosis tendency: systematic studies with a rat model. Atheroschlerosis 131: 25–33.

Hu, FB, Stamfer MJ, Manson JE, Rimm E, Colditz GA, Rosner BA, Hennekens CH, Willett WC (1997) Dietary fat intake and the risk of coronary heart disease in women. N. Engl J Med 337: 1494–1499.

Jeong SC, Hayes AJ, Biyashev RM, Maroof MAS (2001) Diversity and evolution of a non-TIR-NBS sequence family that clusters to a chromosomal "hotspot" for disease resistance genes in soybean. Theor Appl Genet 103: 406–414.

Kanazin V, Marek LF, Shoemaker RC (1996) Resistance gene analogs are conserved and clustered in soybean. Proc Natl Acad Sci USA 93: 11746–11750.

Kasai A, Kasai K, Yumoto S, Senda M (2007) Structural features of GmIRCHS, candidate of the I gene inhibiting seed coat pigmentation in soybean: implications for inducing endogenous RNA silencing of chalcone synthase genes. Plant Mol Biol 64: 467–479.

Lung SC, Weselake RJ (2006) Diacylglycerol acyltransferase: a key mediator of plant triacylglycerol synthesis. Lipids 41: 1073–1088.

Martin GB, Bogdanove AJ, Sessa G (2003) Understanding the functions of plant disease resistance proteins. Annu Rev Plant Biol 54: 23–61.

McHale L, Tan X, Koehl P, Michelmore RW (2006) Plant NBS-LRR proteins: adaptable guards. Genom Biol 7: 212.

Ohlrogge JB, Jaworski JG (1997) Regulation of fatty acid synthesis. Annu Rev Plant Mol Biol 48: 109–136.

Pantalone VR, Wilson RF, Novitzky WP, Burton JW (2002) Genetic regulation of elevated stearic acid concentration in soybean oil. J Am Oil Chem Soc 79: 549–553.

Sandhu D, Alt JL, Scherder CW, Fehr WR, Bhattacharyya MK (2007) Enhanced oleic acid content in the soybean mutant M23 is associated with the deletion in the Fad2-1a gene encoding a fatty acid desaturase. J Am Oil Chem Soc 84: 229–235.

Sebastian SA, Kerr PS, Pearlstein RW, Hitz WD (2000) Soybean germplasm with novel genes for improved digestibility. In: JK Drakely (ed) Soy in Animal Nutrition. Feder Anim Sci Soc, Savoy, IL, USA, pp 56–74.

Senda M, Jumonji A, Yumoto S, Ishikawa R, Harada T, Niizeki M, Akada S (2002) Analysis of the duplicated CHS1 gene related to the suppression of the seed coat pigmentation in yellow soybeans. Theor Appl Genet 104: 1086–1091.

Schlueter JA, Dixon P, Granger C, Grant D, Clark L, Doyle JJ, Shoemaker RC (2004) Mining EST databases to resolve evolutionary events in major crop species. Genome 47: 868–876.

Schlueter JA, Vaslenko-Sanders IF, Deshpande S, Yi J, Siegfried M, Roe BA, Schlueter SD, Scheffler BE, Shoemaker RC (2007) The FAD2 gene family of soybean: Insights into the structural and functional divergence of a paleopolyploid genome. Crop Sci 47: S17–S26.

Shi A, Chen P, Zheng C, Hou A, Zhang B (2008) A PCR-based marker for the Rsv1 locus conferring resistance to soybean mosaic virus Crop Sci 48: 262–268.

Shoemaker RC, Polzin K, Labate J, Specht J, Brummer EC, Olson T, Young N, Concibido V, Wilcox J, Tamulonis JP, Kochert G, Boerma HR (1996) Genome duplication in soybean (*Glycine* subgenus *soja*). Genetics 144: 329–338.

Takahashi R, Githiri SM, Hatayama K, Dubouzet EG, Shimada N, Aoki T, Ayabe S, Iwashina T, Toda K, Matsumura H (2007) A single-base deletion in soybean flavonol synthase gene is associated with magenta flower color. Plant Mol Biol 63: 125–135.

Tang GQ, Novitzky WP, Griffin HC, Huber SC, Dewey RE (2005) Oleate desaturase enzymes of soybean: evidence of regulation through differential stability and phosphorylation. Plant J 44: 433–446.

Toda K, Yang D, Yamanaka N, Watanabe S, Harada K, Takahashi R (2002) A single-base deletion in soybean flavonoid 3′-hydroxylase gene is associated with gray pubescence color. Plant Mol Biol 50: 187–196.

Todd JJ, Vodkin LO (1996) Duplications that suppress and deletions that restore expression from a chalcone synthase multigene family. Plant Cell 8: 687–699.

Wilson RF, Burton JW, Novitzky WP, Dewey RE (2001) Current and future innovations in soybean (*Glycine max* L. Merr.) oil composition. J Oleo Sci 50: 353–358.

Yu YG, Buss GR, Maroof MAS (1996) Isolation of a superfamily of candidate disease-resistance genes in soybean based on a conserved nucleotide-binding site. Proc Natl Acad Sci USA 93: 11751–11756.

Zhang P, Burton JW, Upchurch RG, Whittle E, Shanklin J, Dewey RE (2008) Mutations in a Δ9- stearoyl-ACP-desaturase gene are associated with enhanced stearic acid levels in soybean seeds. Crop Sci 48: 2305–2313.

# 9

# Functional Genomics— Transcriptomics in Soybean

*Sangeeta Dhaubhadel,\* Frédéric Marsolais, Jennifer Tedman-Jones* and *Mark Gijzen*

## ABSTRACT

Transcriptomics is a new area of research that exploits advances in technology and computing to perform large scale gene expression analyses. The development of a wide variety of tools such as microarray and tag-based platforms has facilitated the study of genes that are associated with a specific treatment or developmental state in soybean. The primary goal of a global gene expression analysis is to characterize transcriptional regulatory networks underlying a biological process or response which can be employed for soybean improvement through biotechnology.

**Keywords:** soybean; microarray; gene expression; seed; quantitative trait loci

## 9.1 Introduction

Functional genomics is a field of molecular biology that utilizes the vast wealth of genomics data to elucidate gene function and genetic interactions. It concentrates on the dynamic features of "omics" related research such as transcriptomics, proteomics, metabolomics and protein-protein interactions. The transcriptome is the total set of transcript (messenger RNA) in a given

Southern Crop Protection and Food Research Center, Agriculture and Agri-Food Canada, London, Ontario, Canada N5V 4T3.
\*Corresponding author: *sangeeta.dhaubhadel@agr.gc.ca*

species, or the specific subset of transcripts present in a particular cell type at a given condition. The transcriptome can vary with the developmental stages, tissue types or external environmental conditions. Even though the absolute activity of a gene is determined by the protein it encodes, measurement of mRNA levels has proven to be an important tool for investigating gene function. The availability of complete genome sequence information of several plant species and large sets of expressed sequence tags (ESTs) from many organisms have prompted the development of efficient methods to access the global gene expression patterns and genome-wide analysis of genetic variation. A wide variety of technologies such as cDNA and oligonucleotide microarrays, serial analysis of gene expression (SAGE), massively parallel signature sequencing (MPSS), and robust analysis of 5'-transcript ends (5' RATE) have been used in many organisms to study gene expression profiles and to elucidate the functional roles of several genes at once.

## 9.2 Soybean Transcriptomic Tools

The public soybean EST project (Shoemaker et al. 2002) marked the beginning of large scale gene expression analysis in soybean. This project generated over 80 cDNA libraries created from many different tissues and organ systems of soybean at various stages of development as well as plants that were challenged with several different biotic and abiotic stresses. Currently, 1,279,502 ESTs are available which represent 1,33,549 unique cDNA sequences and 70,880 tentative contig (TC) sequences that are available at DFCI Soybean Gene Index (*http://compbio.dfci.harvard.edu/tgi/cgi-bin/tgi/gimain.pl?gudb=soybean*). The database also categorizes the TC sequences into their functional roles based on gene ontology vocabularies. The main purpose of the database is to represent a non-redundant view of all soybean genes and available data on their expression patterns, functional and cellular roles, and evolutionary relationships. The shortcoming of expression profiling using EST analysis is its lower rate of novel transcript discovery because rare transcripts are usually not well represented in EST collections.

The most widely used method of gene expression profiling in soybean is microarray analysis. Microarray technology is a targeted expression profiling approach where the expression levels of those genes for which a clone or a sequence is available can be determined using spotted cDNAs or oligonucleotides (Schena et al. 1995; Stears et al. 2003). Two 18 k soybean cDNA arrays, each containing 18,432 single spotted PCR products from cDNAs of low redundancy Gm-c1021, Gm-c1083, and Gm-c1070 unigene cDNAs (18k-A array) and low redundancy cDNA sets Gm-r1088 and Gm-r1089 (18k-B) are available at *http://soybeangenomics.cropsci.uiuc.edu* (Vodkin et al. 2004). The 18k-A array provides highly representative mRNAs expressed in roots of seedlings and adult plants, flower buds, flowers, pods, and various stages of immature embryos and seed coats whereas the 18k-B array is derived from tissue culture embryos, cotyledons of young seedlings, and leaves and seedlings exposed to various biotic and abiotic stresses. These arrays have been used extensively by many soybean researchers to study the global gene expression profiles in developing embryos (Dhaubhadel et al. 2007), in response to pathogen attack (Moy et al. 2004; Zou et al. 2005; Zabala et al. 2006), under elevated $CO_2$ condition (Ainsworth et al. 2006), during somatic embryogenesis (Thibaud-Nissen et al. 2003), and to study copy number variation (CNV) by comparative genomic hybridization (CGH) (Gijzen et al. 2006). The discovery of the molecular basis of pink flower (*wp*) locus in soybean was also resolved using cDNA microarray (Zabala and Vodkin 2005). The other alternative to cDNA microarray is the oligo array. Soybean oligo array containing a total of 38,000 unique 70-mer long oligos are designed and printed on two sets of slides each with 19,200 spots by Illumina, Inc./ Invitrogen (San Diego, CA). The advantage of using these arrays is that the oligos were designed from most distant 3' region of cDNAs which enables the experimenter to determine individual expression profiles for members of the same gene family. Another oligo array platform is Affymetrix technology, which uses 25-mers containing 37,500 probe sets on the array. The array also contains transcripts from two soybean pathogens, *Phytophthora sojae* and *Heterodera glycines* (soybean cyst nematode). Since the GeneChip soybean array (Affymetrix) platform is comprised of oligos representing both soybean and two soybean pathogens, this platform has been generally used for studying plant pathogen interactions (Ithal et al. 2007a; Klink et al. 2007b). To provide an integrated view of gene expression in the interaction between soybean and soybean cyst nematode, the soybean Genomics and Microarray Database (*http://psi081.ba.ars.usda.gov/SGMD/default.htm*) was established in 1999. This database contains genomics, EST and gene expression data with tools that allow correlation of soybean ESTs with their gene expression profiles (Alkharouf and Matthews 2004). Table 9-1 provides a listing of publicly available soybean transcriptome databases.

Another approach to study large scale gene expression profiles is by generating sequence tags from a given RNA sample. Generation of sequence tags is independent of prior knowledge of the particular gene, so information on expression of unknown genes can also be obtained. However, this approach requires large scale sequencing and a reference genome for establishing gene identity. The SAGE (Velculescu et al. 1995), MPSS (Brenner et al. 2000) and 5' RATE (Gowda et al. 2007) are methods that utilize the sequence tag approach. In SAGE, a frequent cutter restriction enzyme usually *Nla*III, is used to generate large number of 15–21 bp tags each taken from a unique site (especially 3' untranslated regions) on a single transcript. These tags are long enough to identify transcripts and unique to the transcript under investigation. The number of copies of a particular tag is proportional to the expression level of the gene that the tag represents. Soybean SAGE libraries and data tables are available at soybean genomics initiative site (*http://soybean.ccgb.umn.edu/*). In MPSS, cDNA libraries are made on microbeads and 17–20 mer signature tags from these cDNAs are sequenced using a MPSS system (Brenner et al. 2000). Similar to the SAGE method, the abundance of the signature sequence reflects gene expression level in that particular sample in MPSS. However, MPSS tags are easier to connect with known genes and a large number of distinct signatures can be identified in a single analysis, thus providing a greater coverage of transcriptome than SAGE. MPSS is generally considered a high throughput method since it involves Lynx cloning, and sequencing is done in a miniaturized platform. A new method called 5' RATE was recently developed which involves 5' oligo capping of mRNAs, isolating *Nla*III tags followed by ditag formation and then pyrosequencing of the ditags. This method has been used in soybean, maize and rice for identifying the transcript diversity at 5' regions of the expressed genes (Gowda et al. 2007). A comparison of different tag based platforms for gene expression analysis is summarized in Table 9-2.

Prior to the development of microarrays and tag-based platforms for gene expression analysis, various differential display (Liang and Pardee 1992) or subtractive hybridization techniques were widely employed to identify responsive transcripts. These methods are still used today, often to complement or augment a microarray analysis. Alternative methods that actually isolate differential transcripts would not be constrained in this way. Even though the platforms with tag-based expression platforms are much more sensitive and are therefore better suited for routine detection of low level expressed and novel transcripts, microarrays have the advantage of being high-throughput technology for the analysis of large number of samples at the same time. Therefore, microarray technology is more commonly used in studying soybean gene expression compared to other technologies available. The transcriptomic studies in soybean using various platforms are listed in Table 9-3.

## 9.3 Gene Expression Studies Related to Soybean Seed Development

Seed development initiates with a double fertilization event that leads to the differentiation of the major compartment of the seeds such as the embryo, endosperm and seed coat. Each of these compartments has a distinct origin that play special roles in seed development (Le et al. 2007). A soybean seed is comprised of a well developed embryo surrounded by the seed coat. There are only traces of endosperm present in mature soybean seeds. As shown in

**Table 9-1** Soybean transcriptome database.

| Name | Description | URL |
|------|-------------|-----|
| DFCI gene index, Dana Farber Cancer Institute | Inventory of soybean ESTs, consensus (TCs), expression profiles | *http://compbio.dfci. harvard .edu /tgi/cgi-bin/tgi/pl?gudb= soybean* |
| TIGR Plant Transcript Assemblies, J. Craig Venter Institute | Transcript assembly (TAs) | *http://plantta.tigr.org/ cgi-bin/plantta_release.pl* |
| Legume Information System (LIS), National Center for Genome Resources | Consensus EST contigs, legume comparative genomics | *http://comparative-ligumesorg/* |
| Plant Genome DataBase, Iowa State University | EST assembly, comparative genomics | *http://www.plantgdb. org/search/misc/plantlis-tconstruction.php?my Species=Glycine%20max* |
| Soybean cDNA and oligo array, University of Illinois | Generation of cDNA and oligo array | *http://soybeangenomics. cropsci.uiuc.edu* |
| The Soybean SAGE page, Soybean Genomics Initiative | Contains SAGE tags from soybean cDNA libraries | *http://soybean.ccgb. umn.edu/* |
| Soybean genomics and microarray database, USDA | Soybean resistance to soybean cyst nematode | *http://psi081.ba.ars. usda.gov/SGMD/ default.htm* |
| Soybean Transcription Factor Database, Peking University | Expression pattern of 1891 soybean transc-ription factors | *http://planttfdb.cbi.pku. edu. cn/web/index. php?sp=gm* |
| Dissecting soybean resistance to *Phytophthora* by QTL analysis of host and pathogen expression profiles, Virginia Bioinformatics Institute | Soybean resistance to *Phytophthora sojae* | *http://soy.vbi.vt.edu/* |

**Table 9-2** Comparison of tag-based platforms for transcript analysis. (source: modified from Gowda et al. 2007).

| Features | 5′ RATE | EST/ FL-cDNA | SAGE | MPSS |
|----------|---------|--------------|------|------|
| Position of tags | 5′ | 3′ or 5′ | 3′ or 5′ | 3′ |
| Length of tag (bp) | >80 | 300–3000 | 5–26 | 17–21 |
| mRNA coverage | Complete | Partial | Complete | Complete |
| Abundance of transcripts covered | high and low | high | high and low | high and low |
| Detection system | Pyroseq-uencing | Sanger Sequencing | Sanger Sequencing | Bead based hybridization |
| Technical complexity | Simple | Complex | Complex | Complex |
| Relative cost | Low | Very high | Very high | Very high |
| Sensitivity (transcripts/ million) | na | na | 100 | 5 |

Table 9-3 Transcriptomic study in soybean.

| Description | Platforms | References |
|---|---|---|
| *Seed* | | |
| Embryo development and isoflavonoid biosynthesis | cDNA array | Dhaubhadel et al. 2007 |
| Somatic embryogenesis | cDNA array | Thibaud-Nissen et al. 2003 |
| Seed germination and seedling development | Oligo array | Gonzalez and Vodkin 2007 |
| *Biotic stress* | | |
| Expression profiling of soybean and soybean cyst nematode interaction | Affymetrix array | Klink et al. 2007a, b; Ithal et al. 2007a, b; Puthoff et al. 2007 |
| Soybean and soybean cyst nematode interaction | cDNA array | Alkharouf et al. 2006 |
| Soybean and soybean rust | cDNA array | Choi et al. 2008 |
| Soybean and Asian soybean rust | Affymetrix array | Panthee et al. 2007; van de Mortel et al. 2007 |
| Soybean and soybean mosaic virus interaction | cDNA array | Babu et al. 2008 |
| Elevated $CO_2$ and $O_3$ alter soybean resistance to Japanese beetles | Affymetrix array | Casteel et al. 2008 |
| Soybean and *Pseudomonas syringae* interaction | cDNA array | Zabala et al. 2006; Zou et al. 2005 |
| Soybean and *Fusarium solani* f. sp glycines interaction | cDNA array | Iqbal et al. 2005 |
| Soybean and *Phytophthora sojae* interaction | cDNA array | Moy et al. 2004 |
| Early stage of soybean and *Phytophthora sojae* interaction | Suppression subtractive hybridization | Chen et al. 2007 |
| Soybean and aphid interaction | cDNA array | Li et al. 2008b |
| *Abiotic stress* | | |
| Global gene expression of iron deficiency in near isogenic lines | cDNA array | O'Rourke et al. 2007 |
| Expression profiling of osmotic stresses | cDNA array | Irsigler et al 2007 |
| Soybean leaves induced by NaCl stress | Suppression subtractive hybridization | Liao et al. 2003 |
| Effect of elevated $CO_2$ on gene expression | cDNA array | Ainsworth et al. 2006 |
| *Nodulation* | | |
| Soybean nodulation by *Bradyrhizobium japonicum* | cDNA array | Brechenmacher et al. 2008 |
| Analysis of root nodule enhanced transcriptome | cDNA array | Lee et al. 2004 |
| miRNA in soybean and *Bradyrhizobium japonicum* interaction | 454 Life Sciences sequencing, small RNAs | Subramanian et al. 2008 |
| Analysis of syntenic region between soybean and *Medicago truncatula* | Genomic tiling oligo array | Li et al. 2008a |

| Description | Platforms | References |
|---|---|---|
| *Other* | | |
| Effect of transgene in global gene expression in soybean | Affymetrix array | Cheng et al. 2008 |
| Use of cDNA microarray for global gene expression | cDNA array | Vodkin et al. 2004 |
| Gene amplification of *Hps* locus in *Glycine max* | cDNA array | Gijzen at al. 2006 |
| Change in flower color from purple to pink in soybean | cDNA array | Zabala and Vodkin 2005 |
| Tissue-specific gene expression in soybean | cDNA array | Maguire et al. 2002 |

Figure 9-1, the process of embryogenesis consists of five developmental stages: globular, heart, cotyledon, maturation and dormancy (Walbot 1978). Several unique cellular, morphological and physiological events occur during these stages of development that are controlled by changes in gene expression patterns (Goldberg et al. 1994; Gehring et al. 2004). Accumulation of seed storage products such as, proteins, trypsin inhibitors, lectins and isoflavonoids take place during the maturation stage of embryo development. This process is triggered by a complex regulatory network that includes transcription and physiological reprogramming of many different pathways (Wobus and Weber 1999).

Gene expression and regulation during soybean seed development has been an area of interest for many decades. Using solution hybridization, Goldberg et al. (1981a, b) showed that most of the embryonic mRNAs in soybean continue to express throughout the seed development and are stored in the dormant seeds later. Recently, the development of new genomic resources such as availability of EST database and cDNA/oligonucleotide microarrays has provided opportunities to study global gene expression during seed development in soybean. A detailed study of large scale gene expression patterns during the early stage of embryo development was performed using somatic embryos in soybean (Thibaud-Nissen et al. 2003). In this study, globular somatic embryos were induced with auxin treatment and gene expression analysis was performed using a 9,280 cDNA microarray. The results revealed that formation of somatic globular embryos coincided with the accumulation of transcript for storage protein and gibberellic acid synthesis. The study also highlighted several aspects of the changes in gene expression that occur during reprogramming of the cotyledon during the process of induction. Most recently, Dhaubhadel et al. (2007) conducted a detailed transcriptome profiling of soybean embryos from early to late seed maturation stage using an 18k-A cDNA microarray (Vodkin et al. 2004). The experiment included five different developmental stages of embryos from two soybean cultivars that differed in the level of total isoflavonoids in the seed that are grown in two different locations for

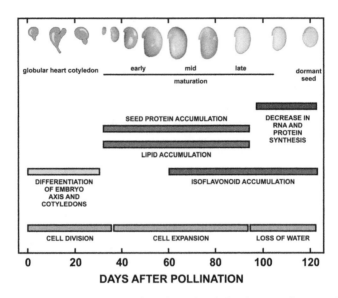

**Figure 9-1** A diagram representing physiological and developmental events that occur during soybean seed development (modified from Le et al. 2007).

*Color image of this figure appears in the color plate section at the end of the book.*

two consecutive years. This study examined the temporal changes of gene expression during embryo development at the time when seeds go through global changes in metabolism. The results revealed that synthesis and transport of seed storage products in soybean are transcriptionally regulated from early developmental stage to maturity and that environmental effects on gene expression of developing seeds are enormous and exceed cultivar specific differences. Comparison of gene expression pattern between two soybean cultivars that differed in isoflavonoid content pointed out the major role of *chalcone synthase* (*CHS*) *7* and *CHS8* genes in isoflavonoid biosynthesis in the seed. Analysis of expression profiles of transcription factor genes during embryo development and their correlation with key genes in isoflavonoid biosynthesis has assisted in identifying some candidate transcription factors that may regulate the biosynthesis of isoflavonoids in soybean seeds (Yi and Dhaubhadel, unpublished results). Identification of key gene(s) and gene regulators for isoflavonoid biosynthesis will help in metabolic engineering of isoflavonoids in soybean and other plants. An abundance of transcripts for some of the unknown genes at certain stages of seed development was also observed in soybean which offers opportunities for identifying functions of those genes using reverse genetic tools. The role of specific elements of the glyoxylate pathway was discovered in soybean cotyledon during seedling development using a set of 19,200 70-mer oligo array (Gonzalez and Vodkin 2007). During seedling development, cotyledons supply the nutrient needs of the young plants and undergo a functional transition to a photosynthetic tissue. The results highlighted the changes in expression profile of the metabolic pathway genes involved in various biological processes during development. Several other genes with known or unknown functions were also found to participate significantly during the transition between soybean seedling development and the cotyledonary function.

Gene expression during seed coat development was investigated in a separate study using the 18k-A platform (Qutob and Gijzen, unpublished results). The results indicated that seed coat tissues express a greater variety of transcripts than the embryo. This is not surprising because the seed coat is a composite organ containing distinct layers with highly differentiated cells. The specialized cell layers of the seed coat change dramatically during seed growth and maturation. The seed coat provides a source and conduit for nutrients to the developing embryo; whereas at seed maturity it provides a protective covering and controls germination by regulating water uptake. The complexity of the seed coat transcriptome, in comparison to the embryo, illustrates the importance of this organ in overall seed development.

Taken together, these gene expression studies have provided a great deal of information on how the genes involved in different metabolic pathways are regulated during seed development which will allow us to engineer value-added products or to eliminate antinutritional factors from the seeds that will ultimately be beneficial for human health and nutrition.

## 9.4 Transcriptomics and QTL Mapping in Soybean

Most of the economically important traits in plant species are quantitative in nature and are controlled by more than one gene. These genes can be mapped on chromosomes by calculating the correlation between quantitative traits and allelic states at linked genetic markers (Thoday 1961). The regions on the genome that these traits are mapped to are called quantitative trait loci (QTL) where functionally different alleles segregate resulting in measurable effects on quantitative traits of interest (Salvi and Tuberosa 2005). Such regions are generally very large and may contain thousands of putative genes. Fine-mapping within the QTL reduces the number of candidate genes to hundreds. The validation process can be very time-consuming and expensive to systematically reduce the list of putative loci. The strategy of combining QTL mapping and fine-mapping with global gene expression data has yielded many meaningful discoveries that have helped in identifying interesting and novel candidate genes for traits of interest whose expression differs between the parental lines (Jansen and Nap 2001; Wayne and McIntyre 2002; Jordan et al. 2007). This strategy can also be used to generate an integrated view of statistical and physiological measures of epistasis (Demuth and Wade 2006). This approach was termed "genetical genomics", "expression genetics" or "expression QTL (eQTL)" by different researchers (Jansen and Nap 2001; Schadt et al. 2003; Varshney et al. 2005). New bioinformatic tools have been developed to merge visualization of gene expression data with QTL mapping to identify the eQTLs (Fischer et al. 2003). Since eQTL analysis uses segregating populations, it allows us to determine if the expression of the gene of interest is regulated in *cis* (gene will be located physically within the eQTL) or *trans* (gene will be located outside the corresponding eQTL) which in turn helps to localize these components on the genetic map (Jansen and Nap 2001; Jansen 2003).

The combination of transcriptomic study and QTL analysis holds enormous potential to identify the genes involved in controlling agronomic traits in soybean. With this approach, the candidate gene can be used as a potential source for developing a perfect marker for selecting phenotype in marker-assisted breeding (Varshney et al. 2007). Identification of candidate genes or regions controlling a particular phenotype using genetical genomics approach has been emerging recently in economically important plants such as cotton (Zhang et al. 2008), wheat (Jordan et al. 2007) and maize (Shi et al. 2007). A lot of effort has been devoted to QTL mapping or gene expression analysis in soybean for several economically important traits such as seed composition, yield, oil quality, isoflavone content, disease resistance etc. Global gene expression analysis in developing soybean embryos from two soybean cultivars that differed in the level of total isoflavonoid content suggested that *CHS7* and *CHS8* genes are critical for isoflavonoid synthesis and accumulation in soybean seeds (Dhaubhadel et al. 2007). This conclusion was supported by two independent QTL analysis studies that identified several QTLs that lie in the same linkage group as *CHS* genes (Kassem et al. 2004; Primomo et al. 2005). Researchers at the Virginia Bioinformatics Institute are currently using a genomics approach to dissect the mechanism of soybean partial resistance to *P. sojae* (*http://soy.vbi.vt.edu/*). The Affymetrix gene chip used in the study included 38,000 soybean genes and 15,800 *P. sojae* genes. The experiment combined gene expression profiling with conventional and eQTL mapping in 300 recombinant inbred lines. The results should lead to new models to account for partial resistance of soybean to *P. sojae*.

## 9.5 Transcriptional Responses to Biotic and Abiotic Stresses

The application of microarrays for studying gene expression patterns in soybean plants subject to biotic and abiotic stresses is now becoming routine, as evidenced by the numerous references summarized in Table 9-3. It is true that many of these studies are largely descriptive, in that they portray particular genes or groups of genes that respond to the treatments, with speculation on the reasons behind the observed changes. Certainly we appear to be in a stage where our capacity to generate data exceeds our ability to fully analyze and interpret it. Despite these caveats, it remains imperative to do the work because the data is extremely valuable in many ways; it provides insight, it confirms past hypotheses or leads to new ones, and it is durable and adaptable over time. There are also creative ways of designing transcriptome experiments to address specific lines of inquiry, rather than as a general profiling tool. The interrogation of gene expression patterns in soybean after pathogen infection or attack has been performed for a variety of parasitic organisms, oomycete, fungal, bacterial, and invertebrate. Infection or attack generally causes massive changes in gene expression as the plant activates its immune, defense, and repair pathways. As more studies are completed in this area, commonalities and differences in host responses are becoming apparent.

Root rot caused by the oomycete *P. sojae* is a widespread disease that plagues soybean production. The pathogen initiates infection as a biotroph and is dependent on living host cells for establishment, but rapidly switches to a highly destructive necrotrophic phase as it proliferates in the root cortex and stele. Soybean responses to *P. sojae* infection have been studied for many years, so there is a good background of literature to provide context for transcriptome analyses. Sampling of transcripts by generation of ESTs also preceded microarray-based experiments and provided critical databases that were relied upon for the isolation of key genes such as *IFS* (Steele et al. 1999; Qutob et al. 2000). Results from a time-course study of soybean gene expression after *P. sojae* infection demonstrated that a plethora of defense responses are activated in the host (Moy et al. 2004). This occurred despite the fact that the interaction was susceptible and characterized by fast-spreading lesions. The phenylpropanoid pathway, isoflavonoid phytoalexin biosynthesis, pathogenesis-related proteins, and active oxygen-generating systems clearly emerged as categories of transcripts that were strongly induced during infection. It was notable that salicylic acid mediated defense pathways were activated whereas those linked to jasmonic acid signaling were inactive or suppressed (Moy et al. 2004). It was also evident that a general down-regulation in photosynthetic processes accompanied the activation of defense pathways in soybean.

A transcriptome analysis of resistant and susceptible interactions of soybean with the bacterial pathogen *Pseudomonas syringae* provided a rich data set that detailed hypersensitive response (HR) and related processes during defense (Zou et al. 2005). The timing and magnitude of defenses were important features that distinguished resistant from susceptible interactions, rather than the identity of the transcripts themselves, confirming long-held concepts in this area. The down-regulation of photosynthesis was noted, especially during resistant interactions characterized by an HR. This was followed-up by functional studies, indicating that photosystem II operating efficiency was severely impaired during HR (Zou et al. 2005). The authors speculate on possible links between declining photosynthetic ability, disintegrating chloroplasts, and the generation of reactive oxygen molecules that drive the HR. The data was also analyzed in-depth with regard to phenylpropanoid branch-pathway regulation during pathogen challenge (Zabala et al. 2006). The results show that soybean preferentially activates the isoflavonoid and lignin branches, and likely the flavone and salicylic acid branches during defense; whereas flux through the anthocyanin, proanthocyanidin, and flavonol branches is restricted by down-regulation of flavanone 3-hydroxylase. Also, by measuring the abundance of eight different *CHS* transcripts, the investigators show that nearly all copies of *CHS* are coordinately induced, the exception being *CHS4*.

Soybean responses to fungal infections have also been subject to transcript expression analysis. Root infection by *Fusarium solani* f. sp. *glycines* was studied for susceptible and resistance interactions using a macroarray of 191 cDNAs derived from ESTs (Iqbal et al. 2005). Many defense-associated transcripts were identified that were more rapidly induced in resistant plants. Responses to the soybean rust pathogen, *Phakopsora pachyrhizi*, have been examined by a variety of research groups (Panthee et al. 2007; Van De Mortel et al. 2007; Choi et al. 2008). In a comparison of susceptible and *Rpp2*-mediated resistance types, a biphasic response to the pathogen was observed over a seven- day infection time course (Van De Mortel et al. 2007). Within the first 12 hours after infection, both susceptible and resistant soybean plants responded to the pathogen. This was followed by a refractory period at 24 hours when expression levels for many genes returned to base-line levels. Subsequently, resistant plants initiated a second response that was more rapid and intense than that observed in susceptible plants. These results suggest that the pathogen is actively suppressing initial plant immune responses while it establishes infection, even in the resistant type. Effective containment of the pathogen by *Rpp2* appears to depend on a second wave of defense responses that is detectable 72 hours after infection. Soybean responses to *Phakopsora pachyrhizi* were also compared for susceptible and *Rpp1*-mediated resistance, by isolating differentially expressed transcripts and constructing a custom cDNA micrarray (Choi et al. 2008). This time course only extended over 48 hours, so biphasic responses may have been missed. But this may not be relevant because the *Rpp1* gene presents a different resistance phenotype than *Rpp2*, resulting in no visible lesions whatsoever. Thus, *Rpp1* confers a more absolute immunity to infection than other known *Rpp* genes and likely operates at earlier infection stages. In the analysis of *Rpp1*-mediated resistance, many defense-associated genes were found to be preferentially expressed in *Rpp1* plants compared to susceptible plants, and the authors emphasized a potential role for peroxidase and lipoxygenase genes in immunity (Choi et al. 2008).

Soybean responses to viruses and nematodes offer yet another perspective of the plant's defense and immune transcriptome. Soybean mosaic virus (SMV) is a prevalent disease that is fostered by aphid transmission. A time-course study of SMV infection in a susceptible cultivar indicated that host defense responses to SMV appear to be unduly delayed, perhaps contributing to susceptibility (Babu et al. 2008). A general down-regulation of photosynthetic-associated transcripts was observed for SMV infected plants, confirming a trend that has been noted in other biotic-interaction studies. Expression profiling of host transcripts during the cyst nematode interaction has been performed using a variety of platforms and approaches. Time course studies using cDNA or oligonucleotide arrays have been compared for mRNA samples isolated from whole roots, from compatible and incompatible interactions, and from laser capture micro-dissected syncytial cells at the infection interface (Alkharouf and Matthews 2004; Alkharouf et al. 2006). These studies have illustrated differences in soybean responses to compatible and incompatible races of the nematode, even at early time points of infection prior to syncytial formation. Transcript analysis of syncytial cells themselves show that cell-specific expression patterns may be obscured by whole root sampling (Klink et al. 2007a). Potentially important syncytial cell responses to cyst nematode feeding include activation of lignin biosynthesis and suppression of JA mediated defenses (Ithal et al. 2007a, b). Defense suppression is likely mediated by secreted nematode effectors, but this has yet to be experimentally demonstrated.

As a final point, to complete this summary of transcriptome studies of soybean-biotic interactions, it is necessary to mention that the simultaneous monitoring of host and pathogen gene expression is made possible by using arrays that have targets designed for each of the interacting organisms. For example, in addition to the 37,600 *Glycine max* targets on the Affymetrix Soybean Genome Array, the chip contains probe sets for 15,800 *P. sojae* and 7,500 *H. glycines* transcripts. The data shown in Figure 9-2 illustrates how gene expression can be monitored during a time course of *P. sojae* infection of soybean hypocotyls. Although pathogen transcripts are difficult to detect during the early stages of infection, due to low hybridization signals, the "interaction transcriptome" clearly emerges as the disease progresses and the pathogen builds up biomass.

Abiotic environmental stresses, such as extreme of temperatures, drought, soil mineral composition and high salinity, represent major constraints for growth and productivity of plants including soybean. Using the subtractive hybridization technique, a purple acid phosphatase-like gene, *GmPAP3* was identified that may have a role in adaptation of soybean to salt stress (Liao et al. 2003). It is speculated that *GmPAP3* may function through reactive oxygen species formation and scavenging or signaling pathways that leads to stress response. A cDNA microarray study was

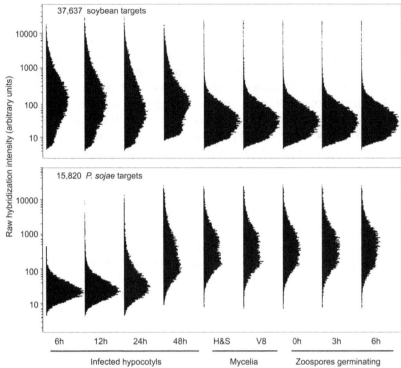

**Figure 9-2** Comprehensive microarray gene expression data for soybean and *P. sojae*. The distribution of raw hybridization intensities for 37,637 soybean and 15,820 *P. sojae* targets are shown. The nine treatments are as follows: four different time points of infection corresponding to 6, 12, 24, and 48 hours after inoculation of soybean hypocotyls with mycelia plugs of *P. sojae*; H&S, *P. sojae* mycelia grown on synthetic media; V8, *P. sojae* mycelia grown on rich media; and *P. sojae* germinating zoospores at 0, 3, and 6 hours after encystment. No normalization steps were performed on this data, in order to illustrate the range of signal intensities depending on the occurrence or ratio of *P. sojae* and soybean mRNA in the various treatments. Values represent the mean of three independent experiments (platform: Affymetrix GeneChip® Soybean Genome Array).

performed to identify the genes and physiological processes associated with iron deficiency response during the stress and recovery from stress in soybean (O'Rourke et al. 2007). The analysis identified four candidate genes during recovery that may have a role in long term effects of iron deficiency in soybean. Many of the genes that were differentially expressed in response to iron stress were also involved in other known stress induced pathways. The findings have opened up a new area to explore and investigate the functions of these gene products and their role in iron metabolism in soybean. A similar approach was used to investigate the expression profiles of the genes that may have a role in drought tolerance in soybean (Irsigler et al. 2007). Soybean leaves were exposed to either osmotic stress or to endoplasmic reticulum

(ER) stress inducers and changes in gene expression was monitored. The results demonstrated that the stimulus-specific positive changes exceeded the shared responses by osmotic and ER stress on soybean leaves and has provided a foundation for further investigation on connecting the gene expression data to the physiological relevance within the plant.

## 9.6 Functional Genomics Studies Related to Nodulation in Soybean

Soybean is an important crop model for studying the interaction with nitrogen fixing rhizobia leading to root nodule formation, along with the model legumes *Medicago truncatula*, *Lotus japonicus* and other legume crops including pea (Stacey et al. 2006; Sprent and James 2008). The tropical legumes soybean and common bean are characterized by determinate, desmodioid nodules that are formed by cell expansion after an initial stage of cell division. *L. japonicus* also forms determinate nodules, however most temperate legumes, including *M. truncatula*, form indeterminate nodules, maintaining an active meristem in young developing primordia. Another key distinction is that the tropical desmodioid legumes export the products of nitrogen fixation primarily as ureides, whereas amides are exported by the temperate legumes, including *L. japonicus*. There is no doubt that understanding the molecular basis of the phenotypic differences in nodulation in these different model systems will be a future challenge of comparative functional genomics.

Transcript profiling has been used to identify soybean genes that are differentially regulated during nodule development and function. These studies have successfully identified nodule associated genes, beyond the well-characterized nodulins, and revealed functional processes associated with nodule formation. An early study analyzed different developmental stages of nodule formation after inoculation with *Bradyrhizobium japonicum* USDA 110 by EST sequencing (Lee et al. 2004). Genes which appeared to be up-regulated during nodule development were spotted on a cDNA microarray comprising 382 clones. The cDNA array was used to analyze control roots and nodules at different developmental stages, and differentially regulated genes were grouped into clusters depending on expression patterns. The results revealed up-regulation of several enzymes and transporters involved in nodule metabolic processes. In a more recent study using the 18k-A and -B arrays, transcripts were profiled in roots of a supernodulating (*GmNARK*) mutant inoculated with *B. japonicum* at three developmental stages (Brechenmacher et al. 2008). This approach was found not to be effective for identifying differentially regulated genes at early time points (up to 72 hours), most likely due to tissue dilution within the sample. Results uncovered widespread changes in primary metabolism, including activation of most amino acid, carbohydrate, glycolysis, nucleic acid, TCA-cycle related genes and associated transporters, whereas genes of the glyoxylate pathway were repressed. In addition, the majority of defense-related and cell wall metabolic genes were repressed. The data also revealed complex regulatory functions, for example the repression of most ethylene responsive factors and WRKY transcription factors, along with ethylene and JA biosynthetic genes, and induction of cytokinin- and indole acetic acid catabolic genes. These data are consistent with repression of defense-related genes. Soybean transcript profiles were compared with those observed in five other studies performed with *M. truncatula* and *L. japonicus*. Putative orthologous genes were identified based on highest BLAST score. Very few putative orthologs were characterized as differentially expressed in nodules by multiple studies. However, all of these studies identified similar changes across different functional categories.

One of the most exciting new developments in transcript profiling of soybean is the large-scale identification of microRNAs (miRNAs) involved in mRNA silencing and their implication in nodule development (Subramanian et al. 2008). Two small RNA libraries were generated from control and *B. japonicum* inoculated roots, three hours post-inoculation. After extensive sequencing, bioinformatic analysis was performed to identify miRNAs based on several criteria, including predicted formation of hairpin precursors, and homology to known plant miRNAs. A total number of 55 families of miRNAs were identified, of which 35 were novel. Several miRNAs were validated by Northern blot analysis, and some were shown to be differentially expressed in nodulated roots. In a complementary *in silico* study of miRNAs in higher plants, 19 conserved families were predicted in soybean, *L. japonicus* and *M. truncatula* (Sunkar and Jagadeeswaran 2008), closely matching the empirical determination by Subramanian et al. (2008).

Another recent pioneering study addressed transcript profiling in the context of comparative functional genomics. A genomic oligonucleotide tiling array was used to analyze transcripts derived from a highly syntenic region of ca. 1 Mb in soybean and *M. truncatula*, encompassing the soybean *rhg1* and *Rhg4* loci for soybean cyst nematode resistance (Li et al. 2008). Tiling microarray profiling of six different organs confirmed expression for ca. 80% of 220 annotated genes. In addition, more than half of the transcriptionally active regions identified were found to lie in non-exonic regions, or antisense to predicted exons, extending similar findings in *Arabidopsis thaliana* and rice. Analysis of transcript levels revealed that ca. 30% of the annotated genes were differentially expressed in the six different organs. A similar number of genes (11 and 10) were preferentially expressed in nodules between *M. truncatula* and soybean, respectively. However, a single gene was represented in both gene sets. Analysis of transcript levels for 78 collinear gene pairs showed that their pattern of expression in different tissues was not conserved, suggesting a substantial divergence of their regulatory sequences during evolution. These observations are consistent with the relative lack of correlation in nodule expression pattern noted between putative orthologs from soybean, *M. truncatula* and *L. japonicus* (Brechenmacher et al. 2008).

## 9.7 Applications of Functional Genomics in Genomics-assisted Breeding

So far, there are a few examples of contributions from functional genomics to genomics-assisted improvement of soybean. One major limitation has been the lack of mutants, but the development of TILLING resources will enable the isolation of directed mutations (Cooper et al. 2008; Lightfoot 2008). As noted earlier, many transcript profiling studies performed in soybean are largely descriptive. However, the fundamental aim of this approach is to characterize transcriptional regulatory networks associated with a biological process or response. In the case of a biological response, analysis of early time points identifies primary target genes. In *A. thaliana*, the use of loss-of-function mutants has proven effective to identify master transcriptional regulators, whose transcript levels themselves may not be altered by the response, and putative target genes (Tepperman et al. 2001). In addition, subsets of transcriptional regulators are rapidly induced, regulating downstream targets (Tepperman et al. 2001; Rashotte et al. 2006). Bioinformatic analysis of promoter motifs in regulated genes may assist the identification of transcriptional regulators. Even for partially redundant transcriptional regulators, minimal sets of direct target genes can be identified through the proper use of mutant backgrounds (Zentella et al. 2007). In turn, the identification of master regulators facilitates studies of upstream signaling components. For a developmentally regulated process, temporal patterns of expression can be analyzed, along with promoter motifs, to select candidate transcriptional regulators whose expression should precede those of target genes.

Transcript profiling, therefore, provides functional candidate genes, particularly transcriptional regulators, which can be tested and used for crop improvement through biotechnology. There may be opportunities to use a biotechnological approach for complex traits conditioned by multiple genes, where variation is limited in currently available germplasm. Lightfoot (2008) listed some of the major biological questions to be addressed by genomics for soybean improvement. One of them concerns the limitation of seed isoflavonoid content, which does not exceed 6 mg/kg. This trait might be improved through the genetic manipulation of transcription factors that regulate the expression of isoflavonoid enzymes (Grotewold 2008). Another important question related to yield concerns the negative relationship between protein and oil content. A detailed knowledge of the regulatory networks controlling protein and oil accumulation could enable transgenic approaches to improve seed composition. Recently, a systematic study of families of soybean transcription factors has been initiated followed by transgenic expression in *A. thaliana* to evaluate the function of selected family members expressed during seed development or inducible by abiotic stress. Transgenic expression of soybean Dof transcription factors, GmDof4 and GmDof11, resulted in increased seed weight, and increased content of fatty acids in seeds, through up-regulation of two distinct fatty acid biosynthetic genes (Wang et al. 2007). The two proteins were found to down-regulate the expression of a seed storage protein gene, consistent with the negative relationship between protein and oil content in soybean seed. Transgenic expression of two WRKY transcription factors, GmWRKY21 and GmWRKY54 led to cold, and salt and drought resistance, respectively (Zhou et al. 2008).

By and large, most functional genomic efforts directed to soybean improvement in the near future are expected to focus on QTL cloning, taking advantage of existing natural genetic variation, since this can serve as a basis for conventional breeding. In rice, three years after the publication of the full genome sequence, achievements are already extremely impressive, with multiple QTL genes cloned, conditioning yield and abiotic stress resistance (Collins et al. 2008; Leung 2008; Sakamoto and Matsuoka 2008; Takeda and Matsuoka 2008). With the availability of the full soybean genome sequence, major advances in QTL cloning are expected. As already discussed, functional genomics may assist in the isolation of QTL genes by fine-mapping and positional cloning. Where multiple candidate genes are present in a QTL interval, in-depth expression studies and transcript profiling may help to identify and characterize the genes involved, especially when the QTL allele results in a change in transcript levels. eQTL analysis is currently very expensive due to the high number of hybridizations required, and cannot be performed on a routine basis. However, in a transcript profiling study of near-isogenic lines (NILs) of *A. thaliana*, changes in transcript levels were found to be exclusively associated with introgressed regions (Juenger et al. 2006). The use of NILs may, therefore, be far more advantageous, leading to fewer candidate genes. After cloning of a QTL gene, its function may be tested in mutants or overexpressing lines and transcript profiling used to provide insight into the mechanism underlying improvement of the agronomic trait. This knowledge may favor acceptance of the trait by stakeholders and consumers, especially when it originates from wild germplasm.

## 9.8 Impact of Sequencing and Microarray Technology

The recent advances in genetic analysis technologies have transformed all the life sciences. Soybean research has certainly benefited, as has been well-illustrated in the various chapters within this book. Despite that automated DNA sequencers and the first large EST studies on animals emerged in the early 1990s, many scientists were dismissive of high-throughput genetic approaches that were not hypothesis driven. The whole-genome shotgun sequencing of *Haemophilus influenzae* and *Mycoplasma genitalium* in 1995 changed everything; the revolution in genomics was fully unleashed (Fleischmann et al. 1995; Fraser et al. 1995). Yet, by the end of the 1990s, genomic resources for soybean research had still not developed; there were no public EST databases, no microarray platforms, and only a limited amount of sequence information for a selected number of genes existed. Certainly the accomplishments of the last ten years represent a revolution for soybean genetics, and most recently the crowning achievement was the successful completion of the soybean genome sequence.

The deep-sampling of soybean ESTs from dozens of different cDNA libraries was enabled by automated, Sanger-based sequencing technologies. In turn, EST databases provided unigene sets necessary for construction of representative microarrays. The microarrays were primarily designed for studying gene expression, and this is where they have found most use so far. But there are a plethora of other uses for microarrays. Genetic mapping of single nucleotide polymorphisms (SNP) and CGH, insertion-deletions (INDEL), or CNV represent powerful uses for microarrays that have yet to be fully applied towards soybean genetics. Other uses for arrays include: promoter analysis, diagnostics, and as an aid to improve genome annotation such as by studying exon expression or alternative splicing of transcripts. Furthermore, not just when and where the genes are expressed in the tissue but also how the expression of those genes are regulated can be studied by the use of these arrays. Cataloging the relationship between *cis*-regulating elements of genes and their transcription factors may provide a useful platform to dissect how different transcription factors regulate gene expression under different developmental stages and environmental stresses. In fact, a variety of array platforms with different target sets or configurations can be optimized and produced for particular uses, as is evident for the human, yeast, or mouse array resources that are now commercially available. So we remain in the early stages of applying array technology to soybean genetic studies. Certainly, arrays dedicated for genome mapping and fingerprinting in soybean, such as those based on scoring polymorphic SNP, INDEL, and CNV markers would find wide application especially as the cost of the technology declines. Likewise, massively parallel, flow-cell based DNA sequencing technologies offer powerful alternatives to array-based platforms for transcriptional profiling, genotyping, mapping, and gene expression studies.

The flow-cell based DNA sequencing technologies offer many opportunities, especially now that a draft genome sequence for soybean is available. Overall, the generation of DNA sequence data has simply produced a demand for more of it. As the cost of sequencing declines even further, procuring DNA sequence data as a tool for research will continue to penetrate all kinds of investigations. This will carry on a trend that has been evident for the last 20 years. Thus, in the future one can contemplate the saturation sequencing of the soybean transcriptome, and the sequencing of various cultivars or lines of soybean, or other *Glycine* species, to study comparative genomics within soybean and its close relatives. The availability of a collection of representative genome sequences will facilitate functional genomic studies and lead to a deeper understanding of the origins and natural phylogeny of cultivated soybeans. This will help to define clearly the associations between genes and traits, so that selection and breeding relies less on chance.

## 9.9 Conclusion

Further development and systematic use of soybean functional genomics is a requirement to understand soybean gene function and to develop strategies for increasing production, improving quality and overcoming various stresses. For these reasons, functional genomics research in soybean should focus on a genome wide approach, bringing together all the strategies that utilize sequence variation, forward and reverse genetics, and breeding applications. The identification of genes corresponding to a target trait and characterization of the physiological function of those genes will certainly contribute to the improvement of soybean plant leading to new varieties and product development with improved agronomical characteristics and better nutritional properties.

## References

Ainsworth EA, Rogers A, Vodkin LO, Walter A, Schurr U (2006) The effects of elevated $CO_2$ on soybean gene expression: An analysis of growing and mature leaves. Plant Physiol 142: 135–147.

Alkharouf NW, Klink VP, Chouikha IB, Beard HS, MacDonald MH, Meyer S, Knap HT, Khan R, Matthews BF (2006) Timecourse microarray analyses reveal global changes in gene expression of susceptible *Glycine max* (soybean) roots during infection by *Heterodera glycines* (soybean cyst nematode). Planta 224: 838–852.

Alkharouf NW, Matthews BF (2004) SGMD: The soybean genomics and microarray database. Nucl Acids Res 32: D398–D400.

Babu M, Gagarinova AG, Brandle JE, Wang A (2008) Association of the transcriptional response of soybean plants with soybean mosaic virus systemic infection. J Gen Virol 89: 1069–1080.

Brechenmacher L, Kim MY, Benitez M, Li M, Joshi T, Calla B, Lee MP, Libault M, Vodkin LO, Xu D, Lee SH, Clough SJ, Stacey G (2008) Transcription profiling of soybean nodulation by *Bradyrhizobium japonicum*. Mol Plant-Micr Interact 21: 631–645.

Brenner S, Johnson M, Bridgham J, Golda G, Lloyd DH, Johnson D, Luo S, McCurdy S, Foy M, Ewan M, Roth R, George D, Eletr S, Albrecht G, Vermaas E, Williams SR, Moon K, Burcham T, Pallas M, DuBridge RB, Kirchner J, Fearon K, Mao J, Corcoran K (2000) Gene expression analysis by massively parallel signature sequencing (MPSS) on microbead arrays. Nat Biotechnol 18: 630–634.

Choi JJ, Alkharouf NW, Schneider KT, Matthews BF, Frederick RD (2008) Expression patterns in soybean resistant to Phakopsora pachyrhizi reveal the importance of peroxidases and lipoxygenases. Funct Integr Genom: 1–19.

Collins NC, Tardieu F, Tuberosa R (2008) Quantitative trait loci and crop performance under abiotic stress: Where do we stand? Plant Physiol 147: 469–486.

Cooper JL, Till BJ, Laport RG, Darlow MC, Kleffner JM, Jamai A, El-Mellouki T, Liu S, Ritchie R, Nielsen N, Bilyeu KD, Meksem K, Comai L, Henikoff S (2008) TILLING to detect induced mutations in soybean. BMC Plant Biol 8: 9.

Demuth JP, Wade MJ (2006) Experimental methods for measuring gene interactions. Annu Rev Ecol Evol System 37: 289–316.

Dhaubhadel S, Gijzen M, Moy P, Farhangkhoee M (2007) Transcriptome analysis reveals a critical role of *CHS7* and *CHS8* genes for isoflavonoid synthesis in soybean seeds. Plant Physiol 143: 326–338.

Fischer G, Ibrahim SM, Brockmannn GA, Pahnke J, Bartocci E, Thiesen HJ, Serrano-Fernandez P, Moller S (2003) Expression view: Visualization of quantitative trait loci and gene expression data in Ecsembl. Genom Biol 4: R477.

Fleischmann RD, Adams MD, White O, Clayton RA, Kirkness EF, Kerlavage AR, Bult CJ, Tomb JF, Dougherty BA, Merrick JM, et al. (1995) Whole-genome random sequencing and assembly of *Haemophilus influenzae* Rd. Science 269: 496–512.

Fraser CM, Gocayne JD, White O, Adams MD, Clayton RA, Fleischmann RD, Bult CJ, Kerlavage AR, Sutton G, Kelley JM, Fritchman RD, Weidman JF, Small KV, Sandusky M, Fuhrmann J, Nguyen D, Utterback TR, Saudek DM, Phillips CA, Merrick JM, Tomb JF, Dougherty BA, Bott KF, Hu PC, Lucier TS (1995) The minimal gene complement of *Mycoplasma genitalium*. Science 270: 397–404.

Gehring M, Choi Y, Fischer RL (2004) Imprinting and seed development. Plant Cell 16: S203–S213.

Gijzen M, Kuflu K, Moy P (2006) Gene amplification of *Hps* locus in *Glycine max*. BMC Plant Biol 6: 6.

Goldberg RB, Hoschek G, Ditta GS, Breidenback RW (1981a) Developmental regulation of cloned superabundant embryo mRNAs in soybean seeds. Dev Biol 83: 218–231.

Goldberg RB, Hoschek G, Tam SH, Ditta GS, Breidenback RW (1981b) Abundance, diversity, and regulation of mRNA sequence sets in soybean embryogenesis. Dev Biol 83: 201–217.

Goldberg RB, Depaiva G, Yadegari R (1994) Plant embryogenesis-zygote to seed. Science 266: 605–614.

Gonzalez DO, Vodkin L (2007) Specific elements of the glyoxylate pathway play a significant role in the functional transition of the soybean cotyledon during seedling development. BMC Genom 8: 468.

Gowda M, Li H, Wang G-L (2007) Robust analysis of 5'-transcript ends: A high-throughput protocol for characterisation of sequence diversity of transcription start sites. Nat Protocol 2: 1622–1632.

Grotewold E (2008) Transcription factors for predictive plant metabolic engineering: Are we there yet? Curr Opin Biotechnol 19: 138–144.

Iqbal MJ, Yaegashi S, Ahsan R, Shopinski KL, Lightfoot DA (2005) Root response to *Fusarium solani* f. sp. *glycines*: Temporal accumulation of transcripts in partially resistant and susceptible soybean. Theor Appl Genet 110: 1429–1438.

Irsigler A, Costa M, Zhang P, Reis P, Dewey R, Boston R, Fontes E (2007) Expression profiling on soybean leaves reveals integration of ER- and osmotic-stress pathways. BMC Genom 8: 431.

Ithal N, Recknor J, Nettleton D, Hearne L, Maier T, Baum TJ, Mitchum MG (2007a) Parallel genome–wide expression profiling of host and pathogen during soybean cyst nematode infection of soybean. Mol Plant-Micr Interact 20: 293–305.

Ithal N, Recknor J, Nettleton D, Maier T, Baum TJ, Mitchum MG (2007b) Developmental transcript profiling of cyst nematode feeding cells in soybean roots. Mol Plant-Micr Interact 20: 510–525.

Jansen RC (2003) Studying complex biological systems using multifactorial pertubation. Nat Rev Genet 4: 145–151.

Jansen RC, Nap JP (2001) Genetical genomics: The added value from segregation. Trends Genet 17: 388–391.

Jordan MC, Somers DJ, Banks TW (2007) Identifying regions of the wheat genome controlling seed development by mapping expression quantitative loci. Plant Biotechnol J 5: 442–453.

Juenger TE, Wayne T, Boles S, Symonds VV, McKay J, Coughlan SJ (2006) Natural genetic variation in whole-genome expression in *Arabidopsis thaliana*: The impact of physiological QTL introgression. Mol Ecol 15: 1351–1365.

Kassem MA, Meksem K, Iqbal MJ, Njiti V, Banz WJ, Winters TA, Lightfoot DA (2004) Definition of soybean genomic regions that control seed phytoestrogen amounts. J Biomed Biotechnol 1: 52–60.

Klink V, Overall C, Alkharouf N, MacDonald M, Matthews B (2007a) Laser capture microdissection (LCM) and comparative microarray expression analysis of syncytial cells isolated from incompatible and compatible soybean (*Glycine max*) roots infected by the soybean cyst nematode (*Heterodera glycines*). Planta 226: 1389–1409.

Klink V, Overall C, Alkharouf N, MacDonald M, Matthews B (2007b) A time-course comparative microarray analysis of an incompatible and compatible response by *Glycine max* (soybean) to *Heterodera glycines* (soybean cyst nematode) infection. Planta 226: 1423–1447.

Le BH, Wagmaister JA, Kawashima T, Bui AQ, Harada JJ, Goldberg RB (2007) Using genomics to study legume seed development. Plant Physiol 144: 562–574.

Lee H, Hur CG, Oh CJ, Kim HB, Park SY, An CS (2004) Analysis of the root nodule-enhanced transcriptome in soybean. Mol Cell 18: 53–62.

Leung H (2008) Stressed genomics-bringing relief to rice fields. Curr Opin Plant Biol 11: 201–208.

Li L, He H, Zhang J, Wang X, Bai S, Stolc V, Tongprasit W, Young ND, Yu O, Deng XW (2008) Transcriptional analysis of highly syntenic regions between *Medicago truncatula* and *Glycine max* using tiling microarrays. Genom Biol 9: R57.

Liang P, Pardee AB (1992) Differential display of eukaryotic messenger RNA by means of the polymerase chain reaction. Science 275: 967–971.

Liao H, Wong F-L, Phang T-H, Cheung M-Y, Li W-YF, Shao G, Yan X, Lam H-M (2003) GmPAP3, a novel purple acid phosphatase-like gene in soybean induced by NaCl stress but not phosphorus deficiency. Gene 318: 103–111.

Lightfoot DA (2008) Soybean Genomics: Developments through the Use of Cultivar "Forrest". Int J Plant Genom 2008: doi:10.1155/2008/793158

Moy P, Qutob D, Chapman P, Atkinson I, Gijzen M (2004) Patterns of gene expression upon infection of soybean plants by *Phytophthora sojae*. Mol Plant-Micr Interact 17: 1051–1062.

O'Rourke J, Charlson D, Gonzalez D, Vodkin L, Graham M, Cianzio S, Grusak M, Shoemaker R (2007) Microarray analysis of iron deficiency chlorosis in near-isogenic soybean lines. BMC Genom 8: 476.

Panthee D, Yuan J, Wright D, Marois J, Mailhot D, Stewart C (2007) Gene expression analysis in soybean in response to the causal agent of Asian soybean rust ( *Phakopsora pachyrhizi* Sydow) in an early growth stage. Funct Integr Genom 7: 291–301.

Primomo VS, Poysa V, Ablett GR, Jackson CJ, Rajcan I (2005) Agronomic performance of recombinant inbred line populations segregating for isoflavone content in soybean seeds. Crop Sci 45: 2203–2211.

Puthoff DP, Ehrenfried ML, Vinyard BT, Tucker ML (2007) GeneChip profiling of transcriptional responses to soybean cyst nematode, *Heterodera glycines*, colonization of soybean roots. J Exp Bot 58: 3407–3418.

Qutob D, Hraber PT, Sobral BWS, Gijzen M (2000) Comparative analysis of expressed sequences in *Phytophthora sojae*. Plant Physiol 123: 243–253.

Rashotte AM, Mason MG, Hutchison CE, Ferreira FJ, Schaller GE, Kieber JJ (2006) A subset of Arabidopsis AP2 transcription factors mediates cytokinin responses in concert with a two-component pathway. Proc Natl Acad Sci USA 103: 11081–11085.

Sakamoto T, Matsuoka M (2008) Identifying and exploiting grain yield genes in rice. Curr Opin Plant Biol 11: 209–214.

Salvi S, Tuberosa R (2005) To clone or not to clone plant QTLs: Present and future challenges. Trends Plant Sci 10: 297–304.

Schadt EE, Monks SA, Drake TA, Lusis AJ, Che N, Colinayo V, Ruff TG, Milligan SB, Lamb JR, Cavet G, Linsley PS, Mao M, Stoughton RB, Friend SH (2003) Genetics of gene expression surveyed in maize, mouse and man. Nature 422: 297–302.

Schena M, Shalon D, Davis RW, Brown PO (1995) Quantitative monitoring of gene-expression patterns with a complementary-DNA microarray. Science 270: 467–470.

Shi C, Uzarowska A, Ouzunova M, Landbeck M, Wenzel G (2007) Identification of candidate genes associated with cell wall digestibility and eQTLP (expression qualtitative trait loci) analysis in a Flint x Flint maize recombinant inbred line population. BMC Genom 8: 22.

Shoemaker R, Keim P, Vodkin LO, Retzel E, Clifton SW, Waterson R, Smoller D, Coryell V, Khanna A, Erpelding J, Gai X, Brendel V, Raph-Schmidt C, Shoop EG, Vielweber CJ, Schmatz M, Pape D, Bowers Y, Theising B, Martin J, Dante M, Wylie T, Granger C (2002) A compilation of soybean ESTs: Generation and analysis. Genome 45: 329–338.

Sprent JI, James EK (2008) Legume-rhizobial symbiosis: An anorexic model? New Phytol 179: 3–5.

Stacey G, Libault M, Brechenmacher L, Wan J, May GD (2006) Genetics and functional genomics of legume nodulation. Curr Opin Plant Biol 9: 110–121.

Stears RL, Martinsky T, Schena M (2003) Trends in microarray analysis. Nat Med 9: 140–145.

Steele CL, Gijzen M, Qutob D, Dixon RA (1999) Molecular characterization of the enzyme catalyzing the aryl migration reaction of isoflavonoid biosynthesis in soybean. Arch Biochem Biophys 367: 146–150.

Subramanian S, Fu Y, Sunkar R, Barbazuk WB, Zhu JK, Yu O (2008) Novel and nodulation-regulated microRNAs in soybean roots. BMC Genom 9: 160.

Sunkar R, Jagadeeswaran G (2008) In silico identification of conserved microRNAs in large number of diverse plant species. BMC Plant Biol 8: 37.

Takeda S, Matsuoka M (2008) Genetic approaches to crop improvement: Responding to environmental and population changes. Nat Rev Genet 9: 444–457.

Tepperman JM, Zhu T, Chang HS, Wang X, Quail PH (2001) Multiple transcription-factor genes are early targets of phytochrome A signaling. Proc Natl Acad Sci USA 98: 9437–9442.

Thibaud-Nissen F, Shealy R, Khanna A, Vodkin LO (2003) Clustering of microarray data reveals transcript patterns associated with somatic embryogenesis in soybean. Plant Physiol 132: 118–136.

Thoday JM (1961) Location of polygenes. Nature 191: 368–370.

Van De Mortel M, Recknor JC, Graham MA, Nettleton D, Dittman JD, Nelson RT, Godoy CV, Abdelnoor RV, Almeida AMR, Baum TJ, Whitham SA (2007) Distinct biphasic mRNA changes in response to Asian soybean rust infection. Mol Plant-Micr Interact 20: 887–899.

Varshney RK, Graner A, Sorrells ME (2005) Genomics-assisted breeding for crop improvement. Trends Plant Sci 10: 621–630.

Varshney RK, Langridge P, Graner A (2007) Application of genomics to molecular breeding of wheat and barley. Adv Genet 58: 121–155.

Velculescu VE, Zhang L, Vogelstein B, Kinzler KW (1995) Serial analysis of gene expression. Science 270: 484–487.

Vodkin LO, Khanna A, Shealy R, Clough SJ, Gonzalez DO, Philip R, Zabala G, Thibaud-Nissen F, Sidarous M, Strömvik MV, Shoop E, Schmidt C, Retzel E, Erpelding J, Shoemaker RC, Rodriguez-Huete AM, Polacco JC, Coryell V, Keim P, Gong G, Liu L, Pardinas J, Schweitzer P (2004) Microarrays for global expression constructed with a low redundancy set of 27,500 sequenced cDNAs representing an array of developmental stages and physiological conditions of the soybean plant. BMC Genom 5.

Walbot V (1978) Control mechanisms for plant embryogeny. In: M Clutter (ed) Dormancy and Developmental Arrest. Academic Press, New York, USA, pp 113–166.

Wang HW, Zhang B, Hao YJ, Huang J, Tian AG, Liao Y, Zhang JS, Chen SY (2007) The soybean Dof-type transcription factor genes, GmDof4 and GmDof11, enhance lipid content in the seeds of transgenic Arabidopsis plants. Plant J 52: 716–729.

Wayne ML, McIntyre LM (2002) Combining mapping and arraying: An approach to candidate gene identification. Proc Natl Acad Sci USA 99: 14903–14906.

Wobus U, Weber H (1999) Seed maturation: genetic programs and control signals. Curr Opin Plant Biol 2: 33–38.

Zabala G, Vodkin LO (2005) The *wp* mutation of *Glycine max* carries a gene-fragment rich transposon of the CACTA superfamily. Plant Cell 17: 2619–2632.

Zabala G, Zou J, Tuteja J, Gonzalez DO, Clough SJ, Vodkin LO (2006) Transcriptome changes in the phenylpropanoid pathway of *Glycine max* in response to *Pseudomonas syringae* infection. BMC Plant Biol 6.

Zentella R, Zhang ZL, Park M, Thomas SG, Endo A, Murase K, Fleet CM, Jikumaru Y, Nambara E, Kamiya Y, Sun TP (2007) Global analysis of della direct targets in early gibberellin signaling in Arabidopsis. Plant Cell 19: 3037–3057.

Zhang HB, Li Y, Wang BC, P. W. (2008) Recent advances in cotton genomics. Int J Plant Genom 2008: doi: 10.1155/2008/742304.

Zhou QY, Tian AG, Zou HF, Xie ZM, Lei G, Huang J, Wang CM, Wang HW, Zhang JS, Chen SY (2008) Soybean WRKY-type transcription factor genes, GmWRKY13, GmWRKY21, and GmWRKY54, confer differential tolerance to abiotic stresses in transgenic Arabidopsis plants. Plant Biotechnol J 6: 486–503.

Zou J, Rodriguez-Zas S, Aldea M, Li M, Zhu J, Gonzalez DO, Vodkin LO, DeLucia E, Clough SJ (2005) Expression profiling soybean response to *Pseudomonas syringae* reveals new defense-related genes and rapid HR-specific down regulation of photosynthesis. Mol Plant-Micr Interact 18: 1161–1174.

# 10

# The Draft Soybean Genome Sequence

*Jeremy Schmutz,*[1,3*] *Steven Cannon,*[2] *Therese Mitros,*[3]
*Will Nelson,*[4] *Shengquiang Shu,*[3] *David Goodstein*[3] *and
Dan Rokhsar*[3]

## ABSTRACT

In this chapter, we describe the soybean genome sequencing project, which utilized the whole-genome shotgun sequencing method combined with a map-based chromosome scale assembly. We give an overview of the annotation results and gene content for the 7x sequence coverage assembly. We also describe the integration of the shotgun-sequenced scaffolds with the genetic and physical map and the resulting chromosome-scale soybean assembly.

**Keywords:** duplication; gene annotation; map integration; synteny; whole-genome shotgun sequencing

## 10.1 Introduction

The soybean (*Glycine max*) genome sequencing project began as an interagency project with the DOE's Joint Genome Institute providing the production sequencing throughput with the NSF and USDA funded groups providing genomic resources and soybean expertise to the project (Jackson

[1]HudsonAlpha Genome Sequencing Center, 601 Genome Way, Huntsville, AL 35806, USA.
[2]USDA-ARS Corn Insects and Crop Genetics Research Unit, Ames, IA 50011, USA.
[3]Joint Genome Institute, 2800 Mitchell Drive, Walnut Creek, CA 94598, USA.
[4]Arizona Genomics Computational Laboratory, BIO5 Institute, 1657 E. Helen Street, The University of Arizona, Tucson, AZ 85721, USA.
*Corresponding author: *jschmutz@hudsonalpha.org*

et al. 2006). The goal of the project is to produce an accurately sequenced, assembled, and annotated representation of the soybean genome that will form the basis for worldwide soybean genomics research. There are several major steps involved in producing this genome sequence including: sequencing, whole-genome shotgun (WGS) assembly, chromosome-scale assembly, and annotation. Here, we describe the 7x sequence coverage WGS assembly, the preliminary annotation of this assembly, and the efforts toward creating an ordered and oriented chromosome-scale representation of the soybean genome sequence.

## 10.2 Whole-genome Shotgun Sequencing of a Complex Plant Genome

The WGS genome sequencing method relies on the ability to assemble the majority of a genome sequence using paired-end sequences from clones with a range of insert sizes. End-sequences from small and mid-size inserts provide the majority of the sequence, and end-sequence from large inserts adds long-range contiguity across large repetitive sequence regions. For mammals, the activity of recent transposons is typically low, and polyploidy events are unusual. This allows a WGS strategy to approach the coverage and accuracy of a BAC-by-BAC based mapping and sequencing project (Venter et al. 2001). Plants, however, can have very complex duplicated genome structures due to polyploidy events and recent transposon activity (sometimes with very large transposons). These factors suggest that adopting a WGS strategy for a complex plant genome like soybean would not be the most efficient pathway to producing a near complete genome sequence, and that instead, a BAC-by-BAC strategy would produce a more complete and accurate genome sequence. However, beyond the benefit of streamlined data production, a WGS strategy allows us to identify and sort through the repetitive sequence of the plant species before the assembly and target regions of the genome that are able to be accurately assembled. In practice, this means that common isolated near-identical (99% or above) genomic repeats are accurately assembled using paired end backfill techniques, whereas collections of tandem repeats and transposon induced arrays are not assembled in the WGS sequence. These are the same regions that are not accurately assembled in BAC by BAC draft projects such as the ongoing maize sequencing project and, for the most part, do not appear in the finished genomic sequence of either BAC-by-BAC or WGS projects. These regions also tend to not be stably cloned in *E. coli*, and therefore may not reflect the genome sequence accurately even if sequenced to completion from the BAC. Furthermore, a WGS approach samples the majority of the genome, allowing assembly of representative sequence for the regions of the genome that are never sequenced in BAC-by-BAC approach. Such regions include near-centromeric and telomeric repeats, and rDNA structures. On the other hand, the WGS strategy has limited ability to distinguish near-identical regions that may result from polyploidy or large segmental duplications. In such duplicated regions, WGS assemblies tend to produce a single blended copy rather than two correctly-assembled regions. However, again, these are the same regions that are also generally not resolved in a BAC-by-BAC approach, which usually requires multiple coverage in finished clone sequences to correctly distinguish larger duplications. The soybean genome also presents an additional difficulty to a WGS sequencing effort, as its genomic history includes a relatively recent polyploidy event. Before we began to assemble the sequence of soybean, we were unsure of the impact of this duplicated sequence on the WGS assembly.

## 10.3 Sequencing and Whole Genome Assembly

For the soybean genome, as for most other JGI-sequenced WGS genomes, we constructed three differently sized subclone libraries from DNA isolated from the Williams 82 cultivar by the Gary Stacey laboratory at the University of Missouri. The largest library we made was a fosmid library, with insert sizes between 28 and 40 kb. The fosmid-length inserts help to add overall contiguity to the genomic assembly and to bridge over larger repetitive elements. The two small subclone libraries we sized to be about 3 kb and about 8 kb. However, as soybean is a very large genome, we eventually created three different 8 kb sized libraries and three different fosmid libraries during the course of the project, in order to produce enough subclones to complete the target coverage depths (Table 10-1). All of these libraries were checked for quality and organelle contamination before generating paired-end sequence.

**Table 10-1** Libraries included in the soybean genome assembly and their respective assembled sequence coverage levels in the 7x preliminary soybean release.

| Library | Insert Size | Read Number | Assembled Sequence Coverage (×) |
|---------|-------------|-------------|----------------------------------|
| 3kb | 3,287 | 5,204,177 | 2.82 |
| 8kb (1) | 6,547 | 2,610,150 | 1.34 |
| 8kb (2) | 6,806 | 3,261,934 | 1.76 |
| 8kb (3) | 8,106 | 604,128 | 0.28 |
| Fosmid (1) | 35,461 | 484,799 | 0.23 |
| Fosmid (2) | 36,675 | 991,002 | 0.47 |
| Fosmid (3) | 37,448 | 305,275 | 0.14 |
| GM_WBa | 113,793 | 59,286 | 0.03 |
| GM_WBb | 133,661 | 121,680 | 0.09 |
| GM_WBc | 135,211 | 175,191 | 0.07 |
| **TOTAL** | | 13,817,622 | 7.23 |

We performed several intermediate assemblies using the Arachne assembler (Batzoglou et al. 2002). These assemblies included an approximate 4x coverage assembly, from which we estimated the number of sequencing reactions for each library to take the genome to an 8.5x desired coverage depth. We also paused the production sequencing process at the projected 6x coverage stage and completed another intermediate assembly with Arachne. We included in this assembly the BAC end sequence (BES) data generated by the soybean resources effort (Shoemaker et al. 2008), and additional BAC end sequence data generated by the JGI consortium. All of the BAC end sequence data is from the three publicly available soybean libraries, GM_WBa, GM_WBb, and GM_Wbc, with average insert sizes ranging from 113 kb to 135 kb. For a complex plant genome such as soybean, these BES data are extremely valuable as they allow us to achieve long-range contiguity in a genomic assembly. These BES also provide solidly anchored clones, so that any soybean researcher may acquire a genomic region of interest from the soybean genome. While the target sequence depth of the 6x assembly was intended to be 6x, our assembly actually contained an average of 7.23x coverage; we named this the 7x preliminary soybean release.

We screened the sequence output from the Arachne assembler for organelles, bacterial contamination, and small wholly-repetitive scaffolds. Of the 4,262 scaffolds (see assembly statistics in Table 10-2), we released 3,118 scaffolds as belonging to the soybean genome. These scaffolds collectively contain 25,834 sequence contigs and cover an estimated 993.5 megabases of sequence. We assembled 967.4 megabases, meaning that 2.6% of the 7x assembly is gaps sized from paired-end sequences. This release had a scaffold N50 of 47, with an L50 of 6.5 megabases. The contig sequences had an N50 of 1,943 with an L50 of 147.5 kilobases. These scaffold and contig sizes appeared consistent with a 7x genome sequence coverage and indicated overall good contiguity of the genome.

We performed a simple analysis on the 7x assembly to evaluate the completeness of the genomic sequence surrounding the transcripts. We clustered 14,810 available 3' end expression sequence tags, and from each of the 6,203 clusters we selected one read to use in the analysis. We placed these reads using blat (Kent 2002), requiring greater than 95% sequence identity and greater than 95% of query read length coverage. Of the 6,203 reads, 5,645 (91%) met the placement criteria. Of the remaining unplaced reads, 392 (6.32%) were found with partial coverage, and only 166 (2.68%) were entirely absent from the assembly. We followed up on the missing gene segments by searching the WGS data set. Of the 166 missing gene segments, only one was found in the WGS reads, indicating that the majority of the missing gene segments werenot present in the (pre-assembly) genome sequence, and instead were likely the result of contamination in the EST libraries. As the genome coverage and assembly accuracy increases, we expect to be able to find more complete versions of those gene segments that are only partially found in the 7x assembly.

This 7x soybean genome assembly, although preliminary, allowed us to evaluate the accuracy and completeness of the sequencing project, to create a preliminary gene analysis, and to move forward with work building a chromosome-scale soybean genomic representation.

**Table 10-2** Summary statistics of the output of the whole genome shotgun assembly, before breaking and constructing chromosome scale pieces. The table shows total contigs and total assembled basepairs for each set of scaffolds greater than the given size.

| Size | # | Contigs | Scaffold Size | Basepairs | %Basepairs |
|---|---|---|---|---|---|
| 5,000,000 | 64 | 10,124 | 591,926,032 | 587,605,388 | 99.27 |
| 2,500,000 | 111 | 12,873 | 756,646,961 | 751,670,694 | 99.34 |
| 1,000,000 | 194 | 15,366 | 894,763,142 | 889,167,252 | 99.37 |
| 500,000 | 249 | 16,465 | 934,295,162 | 927,815,889 | 99.31 |
| 250,000 | 292 | 17,546 | 950,004,922 | 942,041,880 | 99.16 |
| 100,000 | 368 | 18,923 | 962,641,918 | 951,116,174 | 98.80 |
| 50,000 | 439 | 19,623 | 967,493,271 | 954,172,409 | 98.62 |
| 25,000 | 559 | 20,378 | 972,479,545 | 956,358,413 | 98.34 |
| 10,000 | 889 | 21,355 | 977,117,651 | 960,461,820 | 98.30 |
| 5,000 | 1,787 | 23,555 | 983,239,271 | 965,409,031 | 98.19 |
| 2,500 | 2,994 | 25,894 | 987,884,399 | 969,135,187 | 98.10 |
| 1,000 | 3,363 | 26,341 | 988,480,365 | 969,681,045 | 98.10 |
| 0 | 4,262 | 27,240 | 989,039,777 | 970,240,457 | 98.10 |

## 10.4 Annotating the Soybean Draft Genome

We produced preliminary gene calls and annotation on the 7x assembly. This preliminary annotation and assembly ("Gyma0") is available for download at *ftp://ftp.jgi-psf.org/pub/JGI_data/Glycine_max/*version0.1. The version 0.1 annotation of protein-coding genes was produced using a combination of homology modeling and alignments with publicly available ESTs and cDNAs from soybean, *Medicago*, and *Lotus*. The 0.1 annotation is intended as a preliminary release to facilitate early use of the soybean genome until the full chromosome assembly is released. It has not been manually curated and so contains errors as well as loci that remain unannotated. The next release, to accompany the chromosome scale assembly, will be improved through more thorough integration of information on repeats, homology, and expressed sequence.

To begin the annotation process, we masked the genome based on overrepresented 16-mers. The 16-mers that occur 20 or more times in the assembly represent just 5% of all 16-mers, but cover 41% of the assembled genome sequence. Putative gene loci were identified using homology and transcription evidence. Regions with homology were found by BLASTX of the soft-masked scaffolds to the *Arabidopsis* and rice proteomes using a significance cutoff of 1e-5 (Altshcul et al. 1997). Additionally, transcript assemblies from *Glycine*, *Lotus*, and *Medicago* species were downloaded from TIGR (*http://plantta.tigr.org/*) and aligned to the genome using blat (Kent 2002). Blat output was filtered to retain each top hit to the genome, as well as any other hit within 97% coverage of the best hit. Possible pseudogenous matches were filtered out by disallowing secondary matches with only a single exon if the best hit has multiple exons. Putative loci were defined by these peptide and EST assembly hits and joined if overlapping. Each region with flanking sequence was submitted with its best template to genomescan (Yeh et al. 2001) and FGENESH+ (Salamov and Solovyev 2000). The model sets from these programs were then submitted to PASA (Haas et al. 2003) to find models consistent with *Glycine*, *Lotus*, and *Medicago* ESTs. An additional run of PASA with *Glycine max* ESTs only was performed to predict alternative transcripts.

A preliminary gene set was chosen from the models generated as described above. This set includes all PASA-validated transcripts. In regions without a PASA-validated transcript, the model with the best blast hit to an *Arabidopsis* peptide was selected. Models at loci not validated by PASA and without an *Arabidopsis* hit were retained only if they overlapped a *Fabaceae* EST. In regions not containing a model by these criteria but where a PASA transcript assembly predicts a peptide of greater than 1 exon and greater than 50 amino acids, the longest open-reading frame from the EST assembly was used to create a model. The preliminary annotation set, Glyma0.1, consists of 58,556 genes comprised of 62,199 transcripts. These models can be viewed in the GBrowse environment at *http://www.phytozome.net/cgi-bin/gbrowse/soybean* (Fig. 10-1).

The models were analyzed for completeness against transcriptional evidence and other fully annotated proteomes. Of the predicted genes, 37,101, or 63.4%, are supported by EST evidence. An examination of the preliminary gene set shows a similar gene complement and gene size distribution to other fully sequenced angiosperm genomes. The median peptide length of the soybean annotation is somewhat shorter than that of other angiosperm annotations, though the median exon number is consistent with other angiosperm proteomes and the median intron size is slightly larger (Table 10-3). The peptide lengths of the soybean annotation have slightly higher representation of peptides under 100 amino acids than *Arabidopsis* and rice, the genomes that have undergone several rounds of annotation improvement, suggesting that there remain some spurious or truncated models in the gene set (Fig. 10-2b). The distribution of exons per gene is quite similar to other sequenced plants, with the most common number of coding exons per gene being one (Fig. 10-2a). Soybean shares the same peak

**Table 10-3** Median values for peptide length (in amino acids), number of exons per transcript, and intron length (in nucleotides) for longest at locus genes in the named proteomes.

|  | Peptide Length | Exons/ transcript | Intron Length |
|---|---|---|---|
| Soybean | 245 | 3 | 250 |
| Poplar | 262 | 3 | 162 |
| Grape | 297 | 3 | 238 |
| *Arabidopsis* | 354 | 4 | 93 |
| Rice | 307 | 3 | 155 |
| Sorghum | 329 | 3 | 144 |
| *Selaginella* | 317 | 4 | 52 |
| *Physcomitrella* | 277 | 3 | 208 |

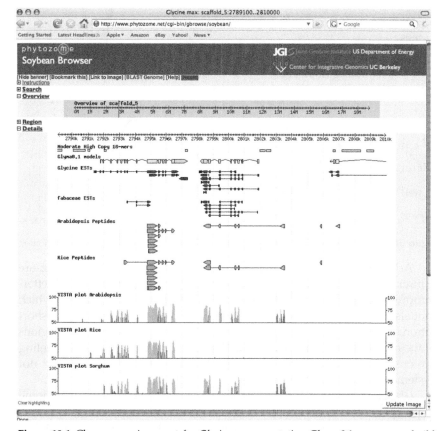

**Figure 10-1** Gbrowse environment for *Glycine max* annotation Glyma0.1 on genome build Glyma0 available at *http://www.phytozome.net/cgi-bin/gbrowse/soybean*

*Color image of this figure appears in the color plate section at the end of the book.*

in intron size of about 85 base pairs as the other angiosperm proteomes (*Arabidopsis*, rice, poplar, sorghum, and grape) (Fig. 10-2c). The soybean, grapevine, and *Medicago* annotations all contain an over-representation in intron sizes below a biologically plausible intron size of 50 nucleotides. This is most likely an artifact of gene calling algorithms that have inserted introns to avoid stop codons—and then have continued to extend the peptide. Despite an over-representation of extremely short introns, soybean has many more introns greater than 300 bp than *Arabidopsis*.

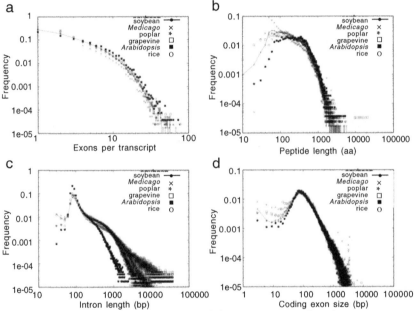

**Figure 10-2** Comparison of gene structure statistics across annotated angiosperm genomes.

Additional sequence coverage in the current assembly, as well as more transcript evidence, will help improve annotation of these models by better-informing the intron/exon boundaries and by helping to distinguish which models are spurious or pseudogenous. For both the short peptide and short intron models, the number of models falling into the over-represented regions of the distribution are relatively small, suggesting that errors in modeling are creating a small number of truncated or spurious models, and not systematic modeling errors affecting all models.

   *Glycine max* is known to have undergone two rounds of whole-genome duplication since its divergence from the Rosid clade represented by grape, *Arabidopsis*, and poplar (Schlueter et al. 2004; Pfeil et al. 2005; Shoemaker et al. 2006). To characterize these duplications we ran a self BLAST of longest at locus peptides from the *Glycine max* annotation to itself. Syntenic segments were found by locating blocks of genes for which there were no more than

10 intervening genes between BLAST hits with strength of 1e-18 or better. The age of these segments can be described by the 4DTv distance—the percentage of transversions in the 3rd codon position at all four-fold degenerate sites. The synonymous segment sizes are quite large given that this annotation is not on a chromosome-scale assembly, with the 10 largest segments containing over 200 loci each. The timing of the duplications along the soybean lineage can be seen in Figure 10-3. The most recent genome duplication has a peak at 0.051 4DTv and occurs on the soybean lineage only. The duplication event at 0.18 4DTv is shared with *Medicago* and is more recent than the divergence between other sequenced Rosid genomes. There are still small segments that contain the 4DTv signature of the most ancient soybean duplication at 0.4 4DTv, which is shared with *Arabidopsis*. This duplication likely represents an event that happened in the common ancestor of all Rosids. The peak is likely the same duplication as the peaks of soybean-poplar and soybean-grape at 0.3—poplar has been shown to have a slower-ticking 4DTv clock (Tuskan et al. 2006), and this comparison to grapevine shows a similar rate.

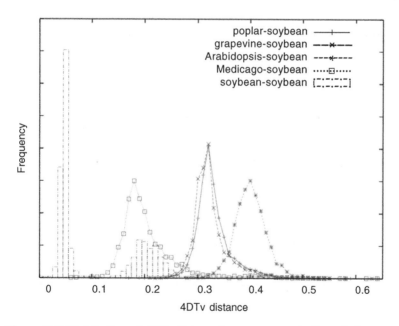

**Figure 10-3** Distribution of 4DTv distance between syntenically orthologous genes. Segments were found by locating blocks of BLAST hits with significance 1e-18 or better with less than 10 intervening genes between such hits. The 4DTv distance between orthologous genes on these segments is reported.

These characteristic 4DTv distances helped us define 32,133, or 55% of *Glycine max* genes that were retained in 15,026 paralogous pairs since the most recent duplication. This means that prior to this duplication the ancestor would have had 41,499 genes, assuming no losses or duplications among those pairs. The duplication at 0.18 4DTv, while more broken up and with greater losses, still represents 15,927 genes (7,072 paralogs). Prior to this duplication the ancestor would have had only 32,594 genes, a number close to that of other sequenced angiosperm annotations.

The preliminary gene set has a good representation of known domains. It contains 32,786 genes that have been assigned PFAM domains (Finn et al. 2006), 35,357 that have been assigned PANTHER classifications (Mi 2007), and 16,543 that have been assigned to a KOG (Tatusov et al. 2003). In total, 44,142 genes were assigned a domain or gene annotation (Table 10-4). This is comparable with the percentage of annotated genes in other organisms. Additionally, the numbers of distinct PFAM, KOG, and PANTHER annotations are relatively constant across proteomes. PFAM over and under-representation was calculated using Fisher's exact test. We do find some PFAM annotations that are significantly ($p \leq .05$) over- or under-represented relative to soybean in at least two genomes (poplar, *Arabidopsis*, grape, rice, and sorghum). Comparisons with poplar, another Eurosid I that has an independent recent whole-genome duplication (Table 10-5) show that the majority of over-represented genes in the *Glycine max* draft annotation come from known transposable elements which will be filtered out of annotations on the 8x assembly. However, the poplar annotation also has an over-representation of known TE-related domains in the hAT family dimerization domain and BED zinc-finger domain. Other interesting findings include the over-representation of stress up-regulated Nod 19, a domain implicated in nodule development.

The preliminary annotation, although already a valuable resource for comparative genomics, will of course be refined over time. The presence of many retroelements in the model set will be reduced by building a database of LTR-retrotransposons and other transcriptional elements for use in masking the genome and filtering the model set. Additionally, more ESTs

Table 10-4 Gene counts, percentage of genes assigned PFAM, PANTHER, or KOG annotation, and numbers of distinct PFAM, PANTHER, and KOG assigned per proteome.

|  | Soybean | Poplar | Grape | *Arabidopsis* | Rice |
|---|---|---|---|---|---|
| Genes | 58,556 | 45,488 | 30,433 | 26,811 | 41,078 |
| % of genes with annotation | 75.4 | 72.2 | 73.4 | 80.3 | 57.6 |
| KOG | 3,024 | 3,116 | 2,988 | 3,243 | 3,107 |
| PFAM | 2,624 | 3,013 | 2,613 | 2,675 | 2,612 |
| PANTHER | 3,558 | 3,822 | 2,936 | 2,725 | 2,805 |

**Table 10-5** Results from Fisher's exact test of gene counts with PFAM annotations. Results displayed are domains with p ≤ .05 to at least two angiosperm genomes. Transposable element related domains are highlighted.

| Under-represented PFAM domains relative to *Populus trichocarpa* | | | | | | |
|---|---|---|---|---|---|---|
| PFAM | Description | Soybean genes | % soybean | Poplar genes | % poplar | P-value |
| PF05699 | hAT family dimerisation domain | 20 | 0.03 | 79 | 0.17 | 3.80E-09 |
| PF01453 | D-mannose binding lectin | 119 | 0.20 | 207 | 0.46 | 9.57E-08 |
| PF08263 | Leucine rich repeat N-terminal domain | 353 | 0.60 | 456 | 1.00 | 9.44E-07 |
| PF04578 | Protein of unknown function, DUF594 | 5 | 0.01 | 36 | 0.08 | 8.54E-06 |
| PF00646 | F-box domain | 213 | 0.36 | 292 | 0.64 | 2.37E-05 |
| PF07714 | Protein tyrosine kinase | 1,240 | 2.12 | 1,295 | 2.85 | 4.75E-05 |
| PF02892 | BED zinc finger | 4 | 0.01 | 30 | 0.07 | 6.81E-05 |
| PF00685 | Sulfotransferase domain | 6 | 0.01 | 34 | 0.07 | 9.06E-05 |
| PF07567 | Protein of unknown function, DUF1544 | 23 | 0.04 | 62 | 0.14 | 9.64E-05 |
| PF00954 | S-locus glycoprotein family | 78 | 0.13 | 131 | 0.29 | 2.00E-04 |
| PF08488 | Wall-associated kinase | 0 | 0.00 | 17 | 0.04 | 2.58E-04 |
| PF03107 | C1 domain | 9 | 0.02 | 37 | 0.08 | 2.84E-04 |
| PF08276 | PAN-like domain | 88 | 0.15 | 134 | 0.29 | 2.31E-03 |
| PF00060 | Ligand-gated ion channel | 19 | 0.03 | 48 | 0.11 | 3.98E-03 |
| PF01397 | Terpene synthase, N-terminal domain | 16 | 0.03 | 43 | 0.09 | 3.98E-03 |
| PF03936 | Terpene synthase family, metal binding domain | 21 | 0.04 | 47 | 0.10 | 1.32E-02 |
| PF00560 | Leucine Rich Repeat | 1,214 | 2.07 | 1,205 | 2.65 | 1.48E-02 |
| PF00280 | Potato inhibitor I family | 7 | 0.01 | 26 | 0.06 | 1.48E-02 |
| PF07645 | Calcium binding EGF domain | 14 | 0.02 | 35 | 0.08 | 3.62E-02 |

| Over-represented PFAM domains relative to *Populus trichocarpa* | | | | | | |
|---|---|---|---|---|---|---|
| PFAM | Description | Soybean genes | % soybean | Poplar genes | % poplar | P-value |
| PF07727 | Reverse transcriptase (RNA-dependent DNA polymerase) | 1,865 | 3.18 | 0 | 0.00 | 0.00E+00 |
| PF08330 | Protein of unknown function (DUF1723) | 317 | 0.54 | 18 | 0.04 | 4.46E-59 |
| PF05970 | Eukaryotic protein of unknown function (DUF889) | 196 | 0.33 | 3 | 0.01 | 6.61E-45 |
| PF00665 | Integrase core domain | 162 | 0.28 | 7 | 0.02 | 1.00E-31 |
| PF00078 | Reverse transcriptase (RNA-dependent DNA polymerase) | 172 | 0.29 | 12 | 0.03 | 1.35E-29 |
| PF02370 | M protein repeat | 24 | 0.04 | 0 | 0.00 | 9.68E-05 |
| PF03732 | Retrotransposon gag protein | 133 | 0.23 | 58 | 0.13 | 1.39E-03 |
| PF07712 | Stress up-regulated Nod 19 | 23 | 0.04 | 2 | 0.00 | 6.77E-03 |
| PF01535 | PPR repeat | 894 | 1.53 | 626 | 1.38 | 1.32E-02 |

and cDNAs have been sequenced and will be used in the locus finding and model verification steps. Having a more accurate masking of the genome as well as more transcriptional evidence will allow better discrimination between spurious, partial gene models and real genes and will reduce transposable element contamination.

## 10.5 Building a Chromosome Scale Soybean Representation

Assembly of the soybean WGS can be thought of as occurring in two major phases: first, assembly of sequence reads into large, mostly-contiguous "scaffold" sequences; and second, ordering, orienting, and validation of the scaffolds to produce chromosome "pseudomolecule" sequences. The soybean chromosome-scale genome assembly makes use of the rich resources developed by the legume research community, including: extensive genetic map resources; a large, manually curated physical map; additional clone pairs not used in the primary assembly; genetic sequence landmarks such as centromeres and telomeres; and synteny comparisons within soybean and between soybean and *Medicago truncatula*. The following analyses were performed on a preliminary version of the 8x soybean genome sequence assembly.

A first approximate ordering and orienting (O&O) was made by positioning scaffolds by average cM value of marker positions in scaffolds, with marker positions determined by top e-PCR (Schuler 1997) match, then the top BLAST hit; and then orienting by comparing cM values of first and last markers in the scaffold. The genetic map for this stage included 5,503 markers from five mapping populations, and was constructed by Cregan, Hyten, and Song (Choi et al. 2007 and unpublished). About two thirds of these were SNP markers and the remainders were SSRs and RFLPs. Of the 5,503 markers in that map, 4,697 could be located on the 8x scaffolds.

This first-pass genetic/sequence integration identified 23 scaffolds that needed to be broken to align properly with the genetic map. These were scaffolds where at least two markers at the scaffold end came from a different chromosome from the rest of the markers on that scaffold. Scaffolds were recalculated using this information and were then reordered and oriented. Additional scaffold breaks were identified at later stages (described below).

The type of map-based O&O should be expected to have several kinds of problems. First, although the consensus map is a remarkably large, valuable resource, the genetic resolution is low because the mapping populations are small. The map was constructed from five mapping populations, each having approximately 100 individuals. Second, in pericentromeric regions, which constitute roughly half of the sequence space, there is very little recombination. Fully 37% of the mapped scaffolds have maximum cM separations of less than 2 cM. Also, in 33% of scaffolds, the distance between average cM values in those scaffolds is less than 2 cM. Third, markers are rare in the large pericentromeres, presumably because unique sequences suitable for marker design are relatively rare in these highly repetitive regions. This means many large scaffolds have too few markers to determine an orientation, and some scaffolds have no markers in the consensus map. Fourth, outlying markers at the scaffold ends may distort the calculation of orientation. For example, a scaffold might have markers with cM values 9, 8.7, 10, 10.1, 10.3, 8.5 (in order along the scaffold). A simple automated ordering procedure might conclude that this scaffold should be flipped, based on the first and last positions (9 and 8.5), whereas a close look at the markers would suggest that a positive orientation is more likely. To address the deficiencies in the first-pass ordering on the consensus map, we used a combination of additional genetic and physical map resources, synteny comparisons, and genomic landmarks.

### 10.5.1 A Special-Purpose, High-Resolution Map

A higher resolution map was developed in a mapping population of 444 progeny of a *G. max* by *G. soja* cross (Cregan, Hyten, Song, Cannon, Shoemaker, unpub). This map was produced using the Illumina GoldenGate platform, with 1,536 markers. Of the designed markers, 1,254 were usable, meaning they contained the predicted SNPs, were polymorphic, and had unique matches in the assembly scaffolds. These markers were also integrated with selected markers from the consensus map to generate a combined map of 1,791 markers. Besides the value from the higher mapping resolution, the *G. max* by *G. soja* map was helpful because the markers were designed from the 7x scaffold ends and from unmapped scaffolds, and were also therefore enriched for problematic areas in the 8x assembly. The markers were picked from a comparison of one run of Solexa reads of *G. soja* genomic DNA, and were selected to specifically validate scaffold ends, pull in more scaffolds, and improve orientations. Markers were selected from 491,115 reads (each 36 nucleotides long) that mapped to non-chloroplast, non-mitochondrial soybean genomic DNA.

The high-resolution, assembly-focused map accomplished most of the planned objectives. The map added markers to 249 scaffolds and facilitated placing of 24 previously unmapped scaffolds totaling 14 Mb. It also provided orientation information for 151 scaffolds (indicating orientation when there was at least 1 cM difference between the most widely separated markers on a scaffold). Finally, it helped validate scaffold integrities by adding markers to regions previously without markers and to scaffold ends. The last function is important because the Arachne scaffolds are more often problematic at their ends than their centers because a scaffold end, by definition, is not extendable by the assembler using existing clone pairs. Frequently, this is due to repetitive sequence near the end of the scaffold. Since good markers may not be designed from non-unique (repetitive) sequence, these regions also tend to be marker-poor. The *G. max* by *G. soja* map, with nearly half a million mapped reads to choose from, enabled selection of markers from many of these scaffold ends. On the basis of these new markers and physical map information (described next), we identified an additional 13 scaffold breaks.

## 10.5.2 Physical Map

The soybean cv. Williams physical map, a product of the NSF SoyMap project (Shoemaker et al. 2008), comprises, as of the June 2008 version, 1,745 contigs, incorporating 141,617 BACs from primarily three libraries: GM_WBa (35,145), GM_WBb (61,379), GM_WBc (37,658), as well as 7,435 from a minimal tile of an existing FPC map of the Forrest strain. The Williams map was constructed using the SNaPshot restriction enzyme fingerprinting technique (Luo et al. 2003) and assembled using the FPC software (FingerPrinted Contigs, Soderlund et al. 2000). Because the fingerprinting and FPC assembly process differs from WGS assembly, a physical map can provide additional information not captured by the WGS sequence, even if the WGS includes BES from the same BACs. For example, some BES may be highly repetitive and not informative for WGS, even though the overall restriction fingerprint of the BAC is sufficiently unique for FPC assembly. Similarly, since restriction fingerprinting captures information about a whole BAC, it can permit assembly of BACs that contain repetitive regions difficult to traverse with WGS; however, longer repetitive regions, such large tandem arrays of ribosomal DNA or centromeric repeats, may also not be spanable in the FPC map as they are likely to produce degenerate fingerprints with too few distinct bands to assemble.

The FPC software program contains tools that were developed during the course of the soybean project (Nelson and Soderlund, 2009) to facilitate and analyze the alignment of draft sequence to an FPC map. The 4x and 7x preliminary assemblies were aligned to the FPC map, providing valuable confirmation of the overall soundness of the assemblies (both map and draft). Detailed editing was not attempted on these early assemblies, but they were used rather to improve the FPC map in order to be maximally useful in editing the final assembly. Using the 4x assembly, many (~500) FPC contigs were automatically merged based on combined fingerprint and draft overlap evidence. A more detailed analysis of discrepancies was carried out with the 7x assembly, encompassing ~185 regions of inconsistent alignment, with most errors settled in favor of the draft contigs; 179 FPC contigs were broken apart at this stage. For the 8x assembly the detailed analysis was again carried out, indicating 51 potential scaffold edits, including 32 scaffold breaks (misassemblies) and 19 scaffold merges (i.e., scaffold pairs bridged by an FPC contig). An example is shown in Figure 10-4, where three different segments of Scaffold 22 have been aligned to FPC Contig 325 (these numbers may not correspond to the final sequence

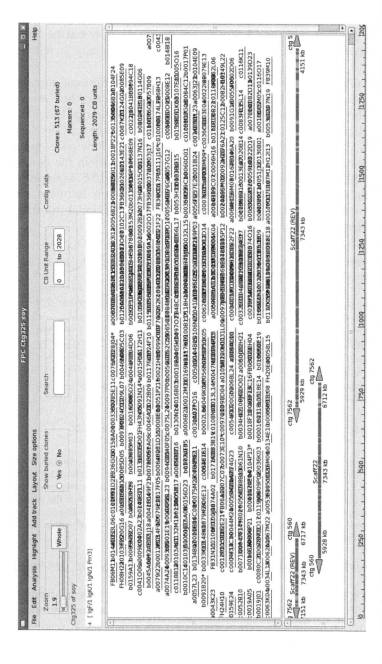

**Figure 10-4** An FPC representation of scaffold 22 that shows a potential scaffold break (inverted segment) between 5,928 kb and 6,712 kb. Color image of this figure appears in the color plate section at the end of the book.

and FPC map). This broken alignment suggests that the middle segment of Scaffold 22, from 5,928 kb to 6,712 kb, should be reversed. In analyzing such alignment discrepancies, the first step is to reassemble the FPC contig at more stringent cutoff values (1e-60 was used here) and verify that the sequence breakpoints are still indicated; otherwise, the problem may lie within the FPC contig.

Also related to physical map FPC data is using "additional clone pair" information. Not all clone pairs are included in the Arachne assembly because they fall outside of the expected BAC clone size range, but may still be considered and used to provide supporting evidence for O&O. Using blast and stringent match parameters, these "extra" clone-pairs spanned and validated approximately 26% of the scaffold pairs in the assembly. These additional associations are illustrated in Figure 10-5. In some cases, large numbers of clone pairs established the association. In others, single clones span the gap, in which case we required additional information before determining scaffold O&O.

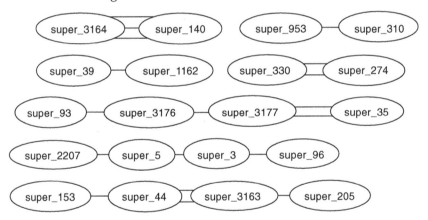

**Figure 10-5** BAC end and scaffold relationships. This shows a small portion of the links established between scaffolds by BAC-end pairs. The "super_##" are preliminary scaffolds from the 8x assembly. Lines between the scaffolds each represent one spanning clone pair, either from the Arachne assembly process (but not included as sequence in the assembly), or from comparisons using BLAST. BLAST parameters were ≥ 99% identity; ≥ 90 of query sequence matched; top matches only; and each BAC end matched within 150 kb of its scaffold end.

### 10.5.3 Synteny Comparisons and Applications in the Assembly

Even using the higher-resolution map as well as physical map information, many scaffolds were either too small or the genomic region had too little recombination to allow precise placements or orientations. To resolve such cases, we used synteny comparisons of the soybean genome to itself and to draft assemblies of the *Medicago truncatula* genome. The *Glycine* genome underwent a whole-genome duplication (WGD) at approximately 10–15 MYA (Blanc and Wolfe 2004; Schlueter et al. 2004; Pfeil et al. 2005). Another duplication is estimated to have occurred early in the legumes, at ~44–64 MYA (Schlueter et al. 2004; Pfeil et al. 2005; Cannon et al. 2006). Because of these two duplications, any given region usually matches at least one, and usually at least three other regions.

Identifying ancient relationships in genomic DNA is challenging due to gene loss and rearrangements. Identifying ancient duplications (such as the WGD near the origin of the legumes) is particularly challenging in a genome self-comparison where there have been multiple WGD episodes. First, ongoing transposition results in very high levels of self-similarity throughout the genome, obscuring the record of an ancient WGD. Second, each WGD episode weakens selection on any given gene. In the absence of gene losses or duplications, two rounds of WGD would produce four paralogs where there had initially been a single gene. All but one of these may be lost over time, leaving any single homoeologous region quite "gappy." Comparisons of two regions that have each, at random, lost 75% of their genes, would leave just 6% of the genes in common between the two regions.

To address the problem of repetitive DNA, we masked all but predicted genes and compared translations of the remaining genic nucleotide sequences using the promer program from the MUMmer package (Kurtz et al. 2004). Further, we considered only the top matches between any two chromosomes in a comparison. To address the problem of stochastic gene loss, we evaluated all potential syntenic relationships visually, as human pattern perception remains more sensitive than existing synteny detection programs. A sample of the MUMmer output is shown in Figure 10-6, showing a comparison of parts of chromosomes 1 and 2, which is 1 of 400 such comparisons in a whole-genome synteny comparison of soybean vs. itself. Several features in Figure 10-6 are typical of patterns seen throughout the Gm/Gm and Gm/Mt comparisons. In euchromatic arms, the synteny tends to be clear and unambiguous, which is evident in the downward sloping diagonal in Figure 10-6. Here, 2 scaffolds are spanned on the vertical axis (chromosome 2, scaffolds super_252 and super_100) and 5 are spanned on the horizontal axis (chromosome 1, starting with super_106). No large gaps are apparent across any of these scaffold boundaries, as a gap would appear as a horizontal or vertical shift at a boundary. Further, no scaffolds appear to be misorientated. A misorientation would appear as a slope sign change, with the inversion breakpoints occurring precisely at scaffold boundaries.

In contrast to the clear synteny in the lower left corner, the extended, broken synteny in the upper right identifies a region of pericentromeric expansion in the horizontal axis (chromosome 1). Here, as with the first synteny block, the synteny feature provides support for placement and orientation of all scaffolds in that region. In fact, the upper-right scaffold is

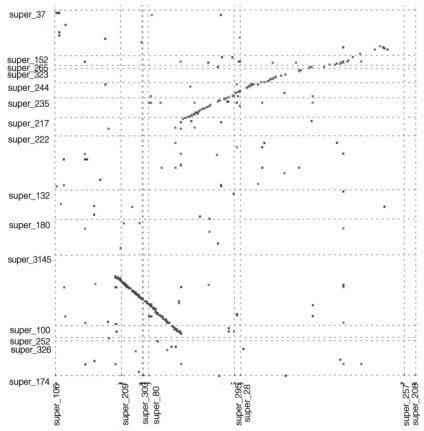

**Figure 10-6** Synteny comparisons between parts of two chromosomes. The horizontal axis is approximately one quarter of chromosome 1 (formerly D1A), and the vertical axis is approximately one quarter of chromosome 2 (formerly D1B). Each dot is the top "promer" match between predicted genes on two chromosomes. Thus, a given gene may have up to one top match per chromosome pair, or up to a total of 20 "top matches." Scaffold boundaries are gray dotted lines.

*Color image of this figure appears in the color plate section at the end of the book.*

particularly valuable, as it supports placement of 7 scaffolds on chromosome 2 and 3 in chromosome 1.

### 10.5.4 Integrating All Available Information

Several features of the assembly are shown together in Table 10-6. The high-resolution map generally provided a good guide to placement of the scaffolds. For the two example chromosomes (1/D1A and the first half of 2/D1B), scaffold-average cM values increase for each scaffold, and orientations are available in most cases. However, cM values are missing for some scaffolds.

Where no cM value is shown (e.g., super_295), placement was made via the physical map and/or additional clone-pair data. Other placements and orientations are suggested by cM values from the high-resolution map cM values where available, and confirmed with synteny analyses. Synteny results are shown in the last five columns, where numbers refer to chromosomes that show synteny with the indicated scaffold. The fifth column shows synteny with *Medicago*. Of interest in the synteny results is

**Table 10-6** Ordering and orienting of draft scaffolds in chromosomes 1 and 2. Scaffolds are numbered super_##. Scaffolds in bold are shown in Fig. 10-2. Orderings are generally supported by the average cM values for each scaffold. Where cM values are missing (e.g., super_295), scaffold placement was instead made using BAC end associations. The maximum separation of cM values for a scaffold, e.g., 20 for super_106 (first in Gm01), are influenced by recombination rates and by scaffold size. The five right-hand columns indicate where synteny exists with another chromosome—either in soybean or *Medicago*. For example, synteny exists between the first scaffolds in Gm01 and Gm02, and between most of the scaffolds shown in Gm02 and Gm01. Similarly, most of the scaffolds in both Gm01 and Gm02 show synteny with Mt chromosome 5.

| #LG | Scaffold (8 x draft) | Orient | Ave cM max-soja | cM sepa-ration | Scaffold size (kb) | Centro-or telo-mere | *Glycine-Glycine* synteny (with chromosome #) | | | Mt syn |
|---|---|---|---|---|---|---|---|---|---|---|
| Gm01 | **super_106** | + | 9.5 | 20.0 | 2,577 | telo | 2 | | | 5 |
| Gm01 | **super_209** | + | 26.5 | 8.8 | 844 | | 2 | 18 | 17 | 5 |
| Gm01 | **super_738** | ? | | | 12 | | 2 | 18 | 17 | |
| Gm01 | **super_300** | + | 31.0 | 0.9 | 213 | | 2 | 18 | 17 | 5 |
| Gm01 | **super_80** | + | 42.2 | 13.6 | 3,393 | | 2 | | 17 | 5 |
| Gm01 | **super_295** | – | | | 221 | | 2 | | | 5 |
| Gm01 | **super_28** | – | 47.6 | 1.2 | 6,439 | | | | | 5 |
| Gm01 | **super_257** | ? | 50.7 | 0.0 | 461 | | | | | |
| Gm01 | super_208 | – | | | 855 | | | | | |
| Gm01 | super_2207 | ? | | | 2 | | | | | |
| Gm01 | **super_5** | + | 49.2 | 0.5 | 12,003 | centro | | | | 5 |
| Gm01 | **super_3** | – | | 50.3 | 3.9 | 13,440 | 7 | 3 | 18 | |
| Gm01 | **super_96** | – | 55.8 | 2.2 | 2,844 | | 7 | 3 | 18 | |
| Gm01 | **super_7** | + | 89.2 | 55.7 | 11,132 | telo | | 3 | 18 | |
| Gm02 | **super_174** | – | 3.0 | 0.1 | 1,339 | telo | 10 | | | |
| Gm02 | **super_326** | + | 4.9 | 0.3 | 140 | | 10 | | | |
| Gm02 | **super_252** | + | 6.2 | 1.5 | 470 | | 1 | 10 | | 5 |
| Gm02 | **super_100** | + | 11.8 | 11.5 | 2,744 | | 1 | 16 | | 5 |
| Gm02 | **super_3145** | – | 21.6 | 45.9 | 1,341 | | 1 | 16 | 11 19 | 5 |
| Gm02 | **super_180** | + | 32.7 | 2.4 | 1,156 | | 16 | 11 | 19 | |
| Gm02 | **super_132** | + | 37.7 | 4.2 | 2,057 | | 16 | | | |
| Gm02 | **super_222** | + | 41.4 | 2.0 | 713 | | 1 | | | |
| Gm02 | **super_217** | + | 46.5 | 2.8 | 774 | | 1 | | | 5 |
| Gm02 | **super_235** | + | 49.2 | 2.1 | 586 | | 1 | | | 5 |
| Gm02 | **super_244** | – | 53.6 | 0.0 | 521 | | 1 | | | 5 |
| Gm02 | **super_323** | – | 54.1 | 0.0 | 145 | | 1 | | | |
| Gm02 | **super_265** | – | 55.5 | 1.0 | 382 | | 1 | | | |
| Gm02 | **super_152** | – | 59.3 | 3.6 | 1,737 | | 1 | 7 | | |
| Gm02 | super_37 | + | 76.0 | 0.0 | 5,391 | | | 7 | | |

(Gm02 is continued by 12 more scaffolds)

that, in the euchromatic chromosome arms, the reference chromosome often matches three other scaffolds, as would be expected following two rounds of duplications (e.g., Gm01/02/18/17.) Synteny generally helps provide validation of scaffold placements across each chromosome arm, and sometimes across entire chromosomes. Also of interest is that the duplicated relationship of chromosomes 1 and 2 is reflected in the relationship of both of these chromosomes to *Medicago* chromosome 5.

Genomic landmarks were also helpful. Eight of the chromosomes have telomeric repeats (tttaggg or ccctaaa) in both of the distal scaffolds and 11 other chromsomes have telomeric repeats found on a single arm. Also, internal scaffolds in 19 of 20 chromosomes contain a large block of characteristic 91- or 92-base pair pericentromeric repeats (Vahedian et al 1995; Lin et al. 2005).

One can perform a visual examination of the genetic and physical (sequence) integration by viewing a plot of genetic vs. physical distances (Fig. 10-7). The plots of chromosomes 1 and 2 in this figure correspond to the information in Figure 10-6 and Table 10-6, and are typical of plots for the other chromosomes. All show similar patterns of markedly diminished recombination (flat central slope) in broad pericentromeric regions, and consistent recombination (rising slopes) at chromosome ends. Misassemblies would be visible as negative slopes or as dramatic discontinuities.

Although the plots of suppressed recombination near chromosome centers are a feature of euchromatic genomes (e.g., Copenhaver et al. 1998; Fengler et al. 2007), it is striking that these regions are so large in soybean, and that that WGS sequencing in soybean has captured so much of the pericentromeric genomic sequence. This portion of the genome, being

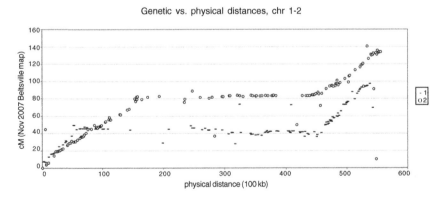

**Figure 10-7** Genetic and physical distances on chromosomes 1 and 2 (formerly D1A and D1B). Markers are from the 2007 Beltsville composite map (Cregan, Hyten and Song unpub.), and physical distances are determined as the top e-PCR locations of markers relative to draft 8x pseudomolecule sequence assemblies.

transposon-rich, should generally be more difficult to assemble. Furthermore, the presence of centromeric-characteristic repeats at distinct locations in 19 of 20 chromosomes, as well as telomeric sequences at most pseudomolecule ends, suggests near end-to-end sequencing of this rather large, challenging genome. Further work will be required to better characterize the assembly and the genome sequence, but initial assessments indicate a strong assembly and one that should be valuable to many researchers.

## 10.6 Summary

As the soybean genome project nears its end in 2008, we anticipate releasing a final draft assembly of the soybean genome sequence that will have been sequenced to an 8x sequence depth. The final assembly will be merged with the genetic and physical maps to produce corrected, chromosome scale sequences. These chromosome sequences will then be annotated with the most current available resource data sets and made available to soybean researchers via web browsers at *http://www.phytozome.net* and *http://soybase.org*. The final draft sequence will serve as the basis for soybean genomics work for many years to come.

## References

Altschul SF, Madden TL, Schaffer AA, Zhang J, Zhang Z, Miller W, Lipman DJ (1997) Gapped BLAST and PSI-BLAST: A new generation of protein database search programs. Nucl Acids Res 25: 3389–3402.

Batzoglou S, Jaffe DB, Stanley K, Butler J, Gnerre S, Mauceli E, Berger B, Mesirov JP, Lander ES (2002) ARACHNE: A whole-genome shotgun assembler. Genome Res 12: 177–189.

Blanc G, Wolfe KH (2004) Widespread paleopolyploidy in model plant species inferred from age distributions of duplicate genes. Plant Cell 16: 667–1678.

Cannon SB, Sterck L, Rombauts S, Sato S, Cheung F, Gouzy JP, Wang X, Mudge J, Vasdewani J, Scheix T, Spannagl M, Nicholson C, Humphray SJ, Schoof H, Mayer KFX, Rogers J, Quetier F, Oldroyd GE, Debelle F, Cook DR, Retzel EF, Roe BA, Town CD, Tabata S, Van de Peer Y, Young ND (2006) Legume genome evolution viewed through the *Medicago truncatula* and *Lotus japonicus* genomes. Proc Natl Acad Sci USA 103(40): 14959–14964.

Fengler K, Allen SM, Lin B, Rafalski A (2007) Distribution of genes, recombination, and repetitive elements in the maize genome. Crop Sci (The Plant Genome) 47: S83–S95.

Choi I-Y Hyten DL, Matukumalli LK, Song QJ, Chaky JM, Quigley CV, Chase K, Lark KG, Reiter RS, Yoon M-S, Hwang E-Y, Yi S-In, Young ND, Shoemaker RC, van Tassell CP, Specht JE, Cregan PB (2007) A soybean transcript map: Gene distribution, haplotype and SNP analysis. Genetics 176: 685–696.

Copenhaver GP, Browne WE, Preuss D (1998) Assaying genome-wide recombination and centromere functions with *Arabidopsis* tetrads. Proc Natl Acad Sci USA 95(1): 247–252.

Finn RD, Mistry J, Schuster-Böckler B, Griffiths-Jones S, Hollich V, Lassmann T, Moxon S, Marshall M, Khanna A, Durbin R, Eddy SR, Sonnhammer ELL, Bateman A (2006) Pfam: clans, web tools and services. Nucl Acids Res 34: D247–D251.

Haas BJ, Delcher AL, Mount SM, Wortman JR, Smith RK Jr, Hannick LI, Maiti R, Ronning CM, Rusch DB, Town CD, Salzberg SL, White O (2003) Improving the *Arabidopsis* genome annotation using maximal transcript alignment assemblies. Nucl Acids Res 31(19): 5654–5666.

Jackson S, Stacey G, Shoemaker R, Schmutz J, Grimwood J, Rokshar D (2006) Toward a reference sequence of the soybean genome: A multi-agency effort. Crop Sci 46: S55–S61.

Kent WJ (2002) BLAT—The BLAST-Like Alignment Tool. Genome Res 12 (4): 656–664.

Kurtz S. Phillippy A, Delcher AL, Smoot M, Shumway M, Antonescu C, Salzberg SL (2004) Versatile and open software for comparing large genomes. Genome Biol 5: R12.

Lin J-Y, Jacobus BH, SanMiguel P, Walling JG, Yuan Y, Shoemaker RC, Young ND, Jackson SA (2005) Pericentromeric regions of soybean (*Glycine max* L. Merr.) chromosomes consist of retroelements and tandemly repeated DNA and are structurally and evolutionarily labile. Genetics 170(3): 1221–1230.

Mi H, Guo N, Kejariwal A, Thomas PD (2007) PANTHER version 6: protein sequence and function evolution data with expanded representation of biological pathways. Nucl Acids Res 35: D247–D252.

Nelson, W. and Soderlund, C (2009) Integrating sequence with FPC fingerprint maps. Nucl. Acid. Res. 37:e36.

Pfeil BE, Schlueter JA Shoemaker RC, Doyle JJ (2005) Placing paleopolyploidy in relation to taxon divergence: A phylogenetic analysis in legumes using 39 gene families. Syst Biol 54: 441–454.

Salamov AA, Solovyev VV (2000) Ab initio gene finding in Drosophila genomic DNA. Genom Res 10(4): 516–522.

Schuler GD (1997) Sequence mapping by electronic PCR. Genome Res 7: 541–550.

Schlueter JA, Dixon P, Granger C, Grant D, Clark L, Doyle JJ, Shoemaker RC (2004) Mining EST databases to resolve evolutionary events in major crop species. Genome 47: 868–876.

Shoemaker R, Schlueter J, Doyle JJ (2006) Paleopolyploidy and gene duplication in soybean and other legumes. Curr Opin Plant Biol 9: 104–109.

Shoemaker RC, Grant D, Olson T, Warren WC, Wing R, Yu Y, Kim H, Cregan P, Joseph B, Futrell-Griggs M., et al. (2008) Microsatellite discovery from BAC end sequences and genetic mapping to anchor the soybean physical and genetic maps. Genome 51: 294–302.

Soderlund C, Humphray S, Dunham A, French L (2000) Contigs built with fingerprints, markers, and FPC V4.7. Genome Res 10: 1772–1787.

Tatusov RL, Fedorova ND, Jackson JD, Jacobs AR, Kiryutin B, Koonin EV, Krylov DM, Mazumder R, Mekhedov SL, Nikolskaya AN, Rao BS, Smirnov S, Sverdlov AV, Vasudevan S, Wolf YI, Yin JJ, Natale DA (2003) The COG database: an updated version includes eukaryotes. BMC Bioinformat 4: 41.

Tuskan G, DiFazio S, Jansson S, Bohlmann J, Grigoriev I et. al (2006) The genome of black cottonwood, *Populus trichocarpa* (Torr. & Gray). Science 313(5793): 1596.

Vahedian M, Shi L, Zhu T, Okimoto R, Danna K, Keim P (1995) Genomic organization and evolution of the soybean SB92 satellite sequence. Plant Mol Bio 29(4): 857–862.

Venter JC, Adams M, Myers E, Li P, Mural R et al. (2001) The sequence of the human genome. Science 291: 1304–1351.

Yeh RF, Lim LP, Burge CB (2001) Computational inference of homologous gene structures in the human genome. Genome Res 11(5): 803–816.

# Soybean Comparative Genomics

Jianxin Ma,[1] Steven Cannon,[2] Scott A Jackson[1] and Randy C. Shoemaker[2*]

## ABSTRACT

The soybean (*Glycine max* L. Merr.) has developed into a reference species complete with a full set of genomic resources including an assembled whole-genome sequence. In addition, the soybean has one of the densest molecular maps of any crop species. Molecular markers include RFLPs, SSRs and most recently, SNPs. Several bacterial artificial chromosome libraries have been produced and physical maps have been assembled in genotypes representing both Northern and Southern germplasm. The GeneChip by Affymetrix is being used to identify additional markers and measure global gene expression under a wide range of conditions. High-throughput sequencing also has been applied to facilitate transcript profiling, identify coding regions of unannotated genes and to identify SNPs for marker development. A very large expressed sequence tag (EST) collection has been developed. Synonymous substitutions within homoeologous pairs of ESTs have been evaluated to demonstrate that the soybean genome has undergone at least two rounds of large-scale duplication events. Transposable elements (TEs), sources of misannotation of genic sequences in plant genomes, and major players in genome evolution, are being methodically identified and analyzed. A comprehensive TE database is currently being developed. The USDA public databases, SoyBase and The Soybean Breeders Toolbox, provide access to map-based and sequence-based data, as well as germplasm information. More recently, the databases are providing genome browser interfaces to access the whole-genome sequence assembly and annotation.

**Keywords:** genome resources; transposable elements; comparative genomics; genome evolution; gene family

[1]Department of Agronomy, Purdue University, West Lafayette, IN 47907, USA.
[2]USDA-ARS Corn Insects and Crop Genetics Research Unit, Ames, IA 50011, USA.
*Corresponding author: *Randy.Shoemaker@ars.usda.gov*

## 11.1 Introduction

Soybean (*Glycine max* L. Merr.) is a member of the legumes, the third largest family of flowering plants. The legumes are critical in global agriculture, providing the majority of plant protein, and more than a quarter of the world's food and animal feed (Graham and Vance 2003). The legumes are highly diverse, with more than 20,000 species and 700 genera, three commonly recognized subfamilies, and origin around 60 million years ago (MYA) (Doyle and Luckow 2003; Cronk et al. 2006). The family is unusual in its ability to fix nitrogen, through association with *Rhizobia* bacteria. The legumes are also of interest for the diverse biochemistries enabled by their greater access to nitrogen, and for their structural genomic evolution. The legumes show large variation in chromosome numbers and genome sizes, driven partly by episodes of polyploidization in the family (reviewed in Shoemaker et al. 2006), and partly by different rates of amplification and elimination of repetitive DNA, as has been observed in other plant species (Kumar and Bennetzen 1999; Devos et al. 2002; Ma et al. 2004). On the other hand, comparative genetic mapping and recently expanding comparative genomic sequence analyses from multiple legume species have demonstrated extensive macro- and micro-colinearity of genes (Choi et al. 2004; Cannon et al. 2006; Schlueter et al. 2008).

Over the last several years, genome sequencing efforts have been under way in the model legumes *Medicago truncatula* (barrel medic) and *Lotus japonicus* (birdsfoot trefoil). These two model plants were chosen for sequencing partially because of their diploid genomes with relatively small genome size (~470 Mb). In contrast, soybean has a much larger (~1,100 Mb) and more complex genome, which has evidently undergone two rounds of polyploidization within the last 60 million years (Shoemaker et al. 1996; Schlueter et al. 2004). Despite its complexity, as one of the most economically important legume crops, soybean was chosen for sequencing by the whole-genome-shotgun (WGS) sequencing approach in early 2006. At the time of writing, all three of these legume genomes are rapidly moving towards completion. In 2009, the full WGS sequence of the soybean genome was made available. Soon to be made available are the assembled euchromatic regions of the genomes of *Medicago truncatula* and *Lotus japonicus* (both of which were clone-by-clone sequencing projects). Additionally, the genome of *Phaseolus vulgaris* (common bean) will be characterized with extensive genetic and physical maps. These available and emerging genomic resources will provide unprecedented opportunities for researchers to initiate and conduct both basic and applied studies.

## 11.2 Comparative Genome Sequencing

The ability to translate information between species in the legume family depends significantly on features of legume evolutionary history. The legumes appear to have undergone a burst of speciation and diversification relatively early in the origin of the family (Lavin et al. 2005). As a consequence, many species within the Papilionoid subfamily are widely separated from one another (Fig. 11-1). For example, *Lupinus* (lupin), *Arachis* (peanut), *Medicago*, *Lotus*, and *Glycine*, all within the Papilionoid subfamily, are each

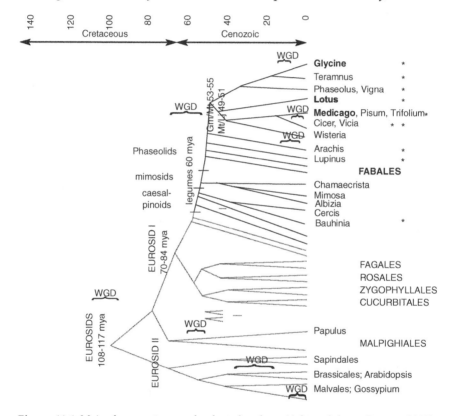

**Figure 11-1** Major legume taxa and selected orders. (Adapted from Cannon 2008). Orders for the legumes (Fabales) and several outgroup taxa are listed on the right. Genera with genome sequencing projects are in bold, and those with EST sequencing projects have asterisks. Estimated timeframes for genome duplications (WGD) are shown with braces. Note early radiation in the legumes, indicated by long branch terminal lengths for many lineages (Lavin et al. 2005).

separated from one another by approximately 45–55 Mya (Hu et al. 2000; Lavin et al. 2005). Thus, the genome sequences from *Medicago, Lotus,* and *Glycine* together span roughly 50 Mya. These species represent three diverse clades in the legumes: the hologalegina (galegoid) clade, comprised of herbaceous, cool-season legumes (with *Medicago* and *Lotus*); and the milletioid clade (with *Glycine* and common bean) (Cronk et al. 2006). More basally separated species in the papilionoid subfamily have received relatively little sequencing effort, although some expressed sequence tag (EST) sequencing has been carried out in *Lupinus* and *Arachis* (peanut). As of mid-2008, EST sequencing projects have been conducted in the following species (largest numbers of ESTs first, from JVCI TIGR Transcript Assemblies; Childs et al. 2007): *Glycine max, Medicago truncatula, Lotus japonicus* (syn. *L. corniculatus*), *Phaseolus vulgaris* (common bean), *Glycine soja* (putative progenitor and interfertile with *Glycine max*), *Phaseolus coccineus* (scarlet runner bean), *Medicago sativa* (alfalfa), *Pisum sativum* (pea), *Arachis hypogaea* (peanut), and *Lupinus albus* (white lupine).

Together, the various sequencing projects begin to flesh out much of the transcript and genomic space across the legumes. As transcriptome sequences are generated in multiple related species, it should, in principle, be possible to trace patterns of gene duplications and losses. Similarly, given multiple complete genomes, it should be possible to infer evolutionary histories and patterns of genomic rearrangement and dynamics.

Several focused comparative genomic sequencing projects are also underway. NSF project 0321664, led by Roger Innes, has sequenced from corresponding one megabase regions (and their homoeolog) from two *Glycine max* lines (cv. Williams and PI96982), and from a diploid *Glycine tomentella* and a recently tetraploid *Glycine tomentella*, from *Teramnus labialis* (sister to the *Glycine* genus), and from *Phaseolus vulgaris, Medicago truncatula,* and *Lotus japonicus* (Innes et al. 2008). The selected region is rich in disease resistance genes, making it an interesting model of evolution of a complex gene family, in the presence or absence of genome duplications, over a wide time frame.

Relatively little genomic effort has been paid outside the papilionoid subfamily. Plants in the mimosid and caesalpinoid subfamilies, however, play critical roles in tropical ecosystems, and exhibit enormous diversity, so should be worthy of additional study. *Acacia* trees (in Mimosoideae) are dominant members of tropical deciduous forests, and trees in the caesalpinoid subfamily comprise the largest number of trees in many tropical rainforests (ter Steege et al. 1993; Pons et al. 2006). Several representatives from these clades are, however, beginning to receive greater attention. For example, targeted sequencing across a wide range of species from all three subfamilies has been used to trace the evolution of diverse floral structures in the legumes (Citerne et al. 2003).

In another example of work outside the papilionoids, deep transcript sequencing is underway in the caesalpinoid species *Chamaecrista fasciculata* (partridge pea), through a project led by Susan Singer (NSF 0746571). *C. fasiculata*, although not widely studied, is interesting evolutionarily (as a nodulating representative of the subfamily dominated by non-nodulators), developmentally (for flowers with near radial symmetry, in contrast to pea-like flowers typical in the papilionoids), and ecologically (for very diverse ecotypes, habitats, and phenotypes) (Fenster and Galloway 1999; Tucker 2003). *C. fasciculata* also has several characteristics that would make it a good model. It is a physically small, selfing, diploid, short-lifecycle annual plant. It has a range of populations and ecotypes across North America, from dry short-grass prairies to northeastern wetlands (Fenster and Galloway 1999; Etterson, 2004).

New short-read sequencing technologies, capable of producing vast quantities of sequence, have the potential to rapidly change the comparative sequence research landscape in the legumes. At the time of writing, some ongoing projects include genome resequencing, digital gene expression, and whole transcript profiling. Projects are being carried out both in legumes with near-complete genome sequence (soybean, *Medicago*, and *Lotus*), and in less well-developed research systems.

An example of a reseqencing project is one in *Glycine soja*, where the objective was to identify single nucleotide polymorphisms (SNPs) with respect to *Glycine max* cv. Williams (D. Hyten, S. Cannon, G. May, R. Shoemaker, and P. Cregan, Pers. Comm.). This project used one Illumina Genome Analyzer Flow Cell, on size-selected genomic *G. soja* DNA, to generate approximately half a million reads (all 36 nt long). SNPs were identified in reads mapped to draft *Glycine max* assemblies, and then used in a GoldenGate assay to map selected SNPs.

An example of a transcript profiling project is one in which five 454 plates were used to sequence from pathogen-infected soybean hypocotyls, from resistant and susceptible strains (S. Cannon, D. Sandhu, B. Roe, and M. Bhattacharyya, Pers. Comm.). This generated sequences and counts from both soybean and from the pathogen, producing more than 100 thousand reads per run, with average sequence lengths of 230 nt, or a total of approximately 130 million bases for the experiment. Aligning the reads to both genomes (soybean and *Phytophthora*) enabled this project to collapse transcripts into probable gene sequences and produce transcript counts; and, as a side-benefit, to aid in annotation of both genome sequences.

A final example is a project that is using a combination of 454 and Solexa sequencing from a wide range of tissues in the legume *C. fasciculata*, mentioned earlier, to simultaneously generate transcript sequences and counts (S. Singer, J. Doyle, and G. May, Pers. Comm.). Relatively inexpensive sequencing could open the door to explorations of many other plants of interest in the very diverse legume family.

## 11.3 Genome Evolution

### 11.3.1 Genome Duplications

Several lines of evidence suggest that the functionally diploid $2n = 40$ *Glycine* species are ancestral polyploids. It has not yet been determined whether this event was auto- or allopolyploidy (Doyle et al. 2003; Walling et al. 2006). Early evidence for the polyploid nature of the soybean genome was cytogenetic (Palmer 1976). Early structural genomic analysis of the leghemoglobin gene family showed that *Glycine* species have two near-identical copies of a four-gene tandem array found in *Phaseolus* (Lee and Verma 1984). More recently, fluorescent in situ hybridization (FISH) on whole chromosomes has also indicated polypoloidy in soybean's history (Walling et al. 2006).

Mapping evidence using restriction fragment length polymorphisms (RFLPs) has suggested that more than 90% of the non-repetitive fraction of the genome is present in two or more copies (Shoemaker et al. 1996). Additionally, nested duplications suggest that some of the homoeologous chromosomes are themselves products of an earlier duplication (Shoemaker et al. 1996). The prevalence of duplicated segments present in three or more linkage groups also supports this hypothesis, although a high rate of segmental duplications cannot be ruled out (Shoemaker et al. 1996). Paleopolyploidy and gene duplication in soybean has been recently reviewed by Shoemaker et al. (2006).

Analysis of the coalescence times of 256 duplicated gene pairs in soybean has shown a distinctive peak at ca. 15 MYA (Schlueter et al. 2004). This study also identified a second distinctive peak in coalescence times consistent with findings based on the RFLP map that at least two rounds of large-scale gene duplications (perhaps by polyploidy) have occurred. The average coalescence time of this later peak was estimated to be 42 MYA (Schlueter et al. 2004).

A detailed sequence analysis of homoeologous regions by Schlueter et al. (2006, 2007a) identified a high degree of sequence conservation within coding regions of duplicated genes. Often, the sequence conservation extended outside of the coding regions. Occasional gene loss or addition, and gene rearrangements were observed. The sequence identity among duplicated regions raised the question of the success of a shotgun sequence strategy for the whole genome of soybean. An analysis of multiple homoeologus bacterial artificial chromosome (BAC) clones indicated that the soybean genome possessed two organizational models; one of which followed that of cotton, and the other more similar to that of maize (Schlueter et al. 2007b). The latter study also demonstrated that even though a high degree of identity exists between duplicated regions, proper assembly of a shotgun sequence, distinguishing both homoeologs, is possible.

Comparative genomics analyses have revealed that different plant species often use homologous genes for very similar functions (reviewed in Bennetzen et al. 2000); it is thus reasonable to believe that duplicated genes within one organism may still maintain the same function. However, it is not uncommon for duplicated genes to acquire a new function, or to lose a function. The organization of duplicated regions can be altered through gene addition, gene loss, rearrangements, or insertion of transposable elements into genes. This can cause structural and functional fractionation of synteny among related genomes. Such fractionation was observed in analyses between soybean, *Medicago* and *Arabidopsis* (Schlueter et al. 2008), between maize and sorghum (Ilic et al. 2003; Lai et al. 2004; Ma et al. 2005), and between many other grass species investigated (reviewed in Salse and Feuillet 2007).

## 11.3.2 Dynamics of Transposable Elements

Transposable elements (TEs) are the most abundant DNA components of all characterized genomes of higher eukaryotes. Based on molecular mechanisms responsible for TE transposition, the eukaryotic TEs can be classified into two classes: class 2 or DNA elements, and class 1 or retroelements. DNA elements, are further classified into at least 10 superfamilies based on their structural features and transposase similarity, whereas retroelements, mainly based on their structures, are traditionally separated into two subfamilies, the long terminal repeat (LTR)-retrotransposons and the non-LTR elements (Wicker et al. 2007). TE content has been shown to vary dramatically among species. For instance, the most abundant TEs in the vertebrate genomes are non-LTR retrotransposons, while LTR-retrotransposons make up the largest fraction of repetitive DNA in all large plant genomes investigated, such as maize, wheat and barley.

In addition to polyploidization, rapid amplification of LTR-retrotransposons has been suggested to be the major force for genome expansion in plants. In large-genome species such as maize, barley and wheat, LTR-retrotransposons make up more than 70–80% of the DNA complement of the genomes, and many of these elements were amplified within the past several million years (SanMiguel et al 1996; Vicient et al. 1999; Wicker et al. 2001). A recent analysis shows that *Oryza australiensis*, a wild relative of the rice *O. sativa*, has accumulated about a million copies of LTR-retrotransposons belonging to three families during the last three million years, leading to a rapid two-fold increase of its genome size without the process of polyploidization (Piegu et al. 2006). Dramatic variation of copy numbers and chromosomal distribution of LTR-retrotransposons has also been observed between two subspecies of rice, *indica* and *japonica*, (Ma et al. 2004), between maize inbred lines (Wang et al. 2006), and even between different generations of a same variety (Hirochika et al. 1996).

On the other hand, recent studies suggest that unequal homologous recombination and illegitimate recombination can generate abundant small deletions, as major mechanisms for counteracting genome expansion caused by aggressive accumulation of LTR-retrotransposons (Devos et al. 2002; Ma et al. 2004). It was estimated that at least 190 Mb of retrotransposon DNA has been removed from the rice genome in the last 5 million years, leaving a genome that currently only contains less than 100 Mb of detectable retroelements or fragments (Ma et al. 2004). In addition, intergenic sequences, and TEs and other non-TE components, have been found to be diverging much faster than genes (Ma and Bennetzen 2004). The rapid amplification and elimination of TEs, plus the higher rates of intergenic sequence divergence largely account for the dramatic size variation and general lack of intergenic homology between orthologous regions of closely related species, such as maize and sorghum (Ilic et al. 2003; Lai et al. 2004), or between duplicated regions within the same organisms, such as two homoeologous chromosomal sets of maize (Illic et al. 2003; Ma et al. 2005; Bruggmann et al. 2006).

TEs have been found to be the major culprits for misannotation of genes in the higher plant genomes. In rice, nearly half of the predicted genes initially reported by the International Rice Genome Sequencing Project (IRGSP) were later found to be TEs or TE remnants (Bennetzen et al. 2004). Recent studies indicate that some TEs, particularly two types of recently discovered TEs, *Helitrons* and *Pack-MULEs*, frequently harbor gene fragments (Jiang et al. 2004; Morgante et al. 2005), further challenging the current imperfect gene prediction programs for gene annotation. Given that TE contents vary greatly across species, it is necessary to identify TEs in any given organism that is sequenced, prior to undertaking a detailed annotation of the genome.

To date, only one TE database has been reported from any legumes. This database was constructed by mining ~32 Mb of transformation-competent bacterial artificial chromosome (TAC) sequences (Holligan et al. 2006). Recently, our laboratories have been developing a semi-automated approach, a combination of structure-based analysis and homology-based comparison, to identify TEs in the soybean genome using the assembled draft sequence. More than 15,000 LTR-retrotransposons have been identified, making up ~35% of the soybean genome (J. Du, S. Cannon, S. Jackson, R. Shoemaker and J. Ma, Pers. Comm.). These elements are grouped into approximately 500 families, including SIRE1, an endogenous retrovirus family previously described (Laten et al. 2003). In addition, a large collection of DNA transposons, including *MULE, CACTA, PIF/Pong*, and *Helitron* elements, has been obtained. This comprehensive TE database, upon completion, will facilitate the annotation of soybean genes and lay a foundation for further study of the organization, structure and evolution of the soybean genome.

## 11.3.3 Gene Family Evolution

Early indications from three legume genome sequencing projects are that legumes show some distinct differences compared with other plant genomes sequenced and described to-date (*Arabidopsis*, poplar, rice, grape). Satisfyingly, many of the differences are related to features of legume biology: nodulation and related signaling, nitrogen-handling and transport, disease response, and production of novel secondary metabolites. Description of all differences is beyond the scope of this chapter, but several examples will be illustrative.

*Lotus japonicus* has 75 auxin response transcription factors (ARFs), which is more than three times the number found in *Arabidopsis* (Sato et al. 2008). Soybean also has elevated numbers of ARF transcription factors. A representative legume ARF used as a TBLASTN query against the 7x soybean assembly identifies 62 matches (E-value < 1e-6). Some of the additional ARFs may play roles in auxin signaling, which plays an active and complex role in nodule initiation and development (reviewed in Eckardt 2006).

The calmodulin-binding transcription activator (CAMTA) transcription factors are also more than twice as abundant in *Lotus* as in *Arabidopsis* (Sato et al. 2008). This transcription factor regulates calcium signaling, which has been found to be critical to nodule development (Mitra et al. 2004). Other proteins in the calcium signaling pathway also show differences in the legumes, and not just numerically. Local duplication and subsequent structural modification of nodule-active Calmodulin-like proteins is observed in *Medicago* (Liu et al. 2006).

Another gene family showing differences in the legumes is the cysteine cluster protein (CCP) family. These short peptides, with similarities to antimicrobial defensins are described in both plants and animals. More than 300 CCPs in *Arabidopsis* are expressed in seeds and flowers. In *Medicago truncatula*, CCPs are also expressed in flowers and seeds, but several hundred CCPs in *Medicago truncatula* are expressed exclusively in nodules (Graham et al. 2004, 2008). Intriguingly, few if any CCPs have been found in soybean. The nodule-active CCPs seem to be remarkably labile, evolving rapidly in genomic clusters, under strong positive selection (Samac and Graham 2007).

## 11.4 Genome Based Molecular Marker Discovery

Because of their ease of use and relative higher rates of polymorphism, simple sequence repeats (SSRs) are a marker type of choice over hybridization-based RFLPs. Conversion of RFLP markers to SSR markers was carried out by Song et al. (2004). In this approach RFLP markers were used to identify BACs. The BACs were then skim-sequenced to identify microsatellites. The SSRs that were subsequently mapped were closely targeted to the original RFLP locus.

Existing EST data coupled with sequencing technologies resulted in the first transcript map for soybean. The transcript map was generated by mapping 1,141 SNPs generated from an equal number of EST sequences. The SNPs were identified by designing primers from the EST sequences, and then using them to resequence genomic regions corresponding to the EST sequence of six genotypes (Choi et al. 2007).

SSRs are commonly detected by a polymerase chain reaction (PCR) amplification that encompasses the repeat sequence followed by gel electrophoresis of the amplification product. SSR repeat sequences are widely distributed throughout the genomes of plants and animals (Burow and Blake 1998) and are useful as molecular markers in a large number of species of plants and animals. Their usefulness comes from their high frequency of length polymorphisms among different genotypes.

In soybean, SSRs were first described more than 15 years ago (Akkaya et al. 1992). They were first used for genetic mapping (Morgante et al. 1994; Akkaya et al. 1995) and then for assessment of variation among *G. max* and *G. soja* accessions (Powell et al. 1996). SSRs have been essential for the development of an integrated genetic linkage map (Cregan et al. 1999; Song et al. 2004). In addition to map construction, SSRs have been used to align the molecular-marker linkage map to the cytogenetic map (Zou et al. 2003). Microsatellites are particularly useful in an ancient paleopolyploid, such as soybean, because of their ability to detect single locus.

Genetic maps have proven to be invaluable for map-based cloning, marker-assisted selection, and studies of genome evolution. As genomic resources continue to be developed, genetic maps will become even more useful. The integration of a physical map into a genetic map is an essential first step to translating genomic resources into genetic advances. To accomplish the integration, existing mapped markers are associated with sequences from BACs comprising the physical map, or DNA sequences generated from BACs comprising the physical map are searched for markers and the markers are subsequently mapped (Shoemaker et al. 2008). The latter approach was used by Shultz et al. (2006 and 2007) and Shoemaker et al. (2008) to genetically anchor the physical maps of the cultivars Forrest and Williams 82, respectively, to the soybean genetic map. The ability of SSRs to detect single loci, even in a paleopolyploid genome such as soybean would seem to imply that a single SSR should be associated with a single locus on a map. However, for both the "Forrest" and the "Williams 82" physical maps it was reported that many DNA markers seemed to "anchor" several distinct contigs (Shultz et al. 2006; Shoemaker et al. 2008), a situation that may be due to some combination of misassociation of markers to BACs, improper inclusion of some BACs in fingerprint contigs (FPCs), BAC library contaminations, or other laboratory tracking errors.

## 11.5 National and International Initiatives: Countries/ Laboratories and Chromosomes

In 2007, an international consortium (International Soybean Genomics Consortium, *http://genome.purdue.edu/isgc/*) was formed to help in coordinating the soybean genomic works worldwide. The first meeting was held in Japan and the consortium now includes members from Brazil, China, Japan, South Korea and the US. This consortium was formed shortly after the Department of Energy embarked on a whole-genome shotgun sequencing project of the "Williams 82" genotype. The goals of the group were: 1) to ensure access to genomic information for researchers worldwide, 2) to establish standards for data sharing and eventual annotation, and 3) as far as possible, reduce duplication of efforts between the member countries. The consortium also functions as a forum for data and research objective sharing between member countries.

There are many national genomics initiatives; these will only skim the surface without in-depth detail. In the US, the sequencing of the Williams 82 genotype by the DOE was completed in 2009. This will serve as a reference for many of the other genome projects. Japan, for instance, is sequencing to a much lower depth of coverage a local paddy-grown variety, "Enrei". They plan to use the "Williams 82" reference sequence to anchor their shotgun sequences in order to look at differences in genome structure and derive genetic mapping markers. South Korea has a sequencing project that was initially focused on specific disease-resistant loci, but has expanded, with advances in sequencing technologies, to include a whole-genome shotgun of a local variety. In combination with the Japan effort, this will provide valuable information at the genome level of diversity between three important varieties from US, Japan and Korea.

In Brazil, recent funding has allowed researchers to begin to leverage the reference genome sequence to understand the genetic architecture underlying biotic and abiotic resistance loci important for cultivation. Part of this is the development of BAC libraries from Brazilian genotypes. These libraries will be an excellent resource for cloning resistance genes.

China is the center of origin for cultivated soybean and thus holds one of the most important germplasm collections for understanding the domestication and evolution of soybean. In that regard, exploration and application of genomics tools for core germplasm collection and exploitation of genetic diversity for germplasm enhancement have been one of their major research focuses.

## 11.6 Community Genomic Resources

The recognition of an organism as a genomic entity among the giants of genomic research (rice, *Arabidopsis*, maize, etc.) is often a gradual process and occurs by incremental advances and tool development. Although the development of soybean genomic resources has been an ongoing process for many years, movement of soybean into the "fast lane" didn't occur until the establishment of an elected Soybean Genetics Executive Committee and the culmination of a National Science Foundation-sponsored Workshop to assess genomic resources, identify needs, and develop a research strategy. This first meeting was held in the fall of 2003. The plan for soybean genomics was published shortly afterwards (Stacey et al. 2004). The plan developed by the research community laid out the weaknesses hindering soybean genomic advances and laid out the path and priorities needed to address those hindrances. Since then, soybean has developed into a credible genomic model. Two additional strategic planning meetings have been held. The latest, in May 2007, lays out updated priorities and can be found at *www.Soybase.org*.

The soybean genome itself is considered a genomic resource. It serves as a model for the study of polyploid evolution. As a paleopolyploid that has undergone at least two discernable rounds of large-scale duplication, soybean provides a unique resource for the study of genome evolution and the evolution of duplicated genomes. Biological resources for soybean are extensive. The USDA maintains a very large germplasm collection of accessions acquired worldwide. The USDA also maintains a Genetic Type collection of well-characterized mutants. A large EST collection has been developed that encompasses over 80 different cDNA libraries from various organs, stages of development, and stressed conditions (Shoemaker et al. 2002). Several BAC libraries have been made publicly available, along with thousands of molecular markers (Choi et al. 2007). Whole-genome expression arrays are available, as are transformation systems and gene knockout systems. An overview of these resources can be found in Stacey et al. (2004).

A number of soybean genomic databases have been created and are accessible by the public. A genome browser, GBrowse (*http://soybeangenome.siu.edu/*) hosted by Southern Illinois University provides a customized database for presenting an integrated view of soybean genomic features. These include, but are not limited to, the cv. "Forrest" physical map and a variety of soybean sequence types such as BAC end sequence, ESTs, and genomic sequence. The USDA-ARS at Ames, Iowa hosts several genomic databases including the original SoyBase (*http://www.soybase.org*) and a breeder-oriented Soybean Breeder's Toolbox (SBT) (*http://www.soybeanbreederstoolbox.org/*). SoyBase is a map-based database containing numerous classes of data (markers, maps, locus, QTL, pathology, etc.). It has converted from its original AceDB version to a relational structure. At SoyBase additional resources have been developed and are available, such as the Affymetrix GeneChip annotation. The Soybean Breeder's Toolbox is a relational database that contains a large subset of data originally in

SoyBase but displayed in a manner more useful to plant breeders. SBT also has map comparative abilities provided by CMap. The "Williams 82" physical map is accessible through SoyBase or SBT and can be found at *http://soybeanphysicalmap.org/*. A broader, more encompassing database that contains information on a number of legumes is the Legume Information System (LIS) hosted at Santa Fe, NM, by the National Center for Genome Resources. LIS can be accessed at *http://comparative-legumes.org/*. The whole-genome shotgun sequence of soybean has provided a tremendous resource for genomic research and is available at a site hosted by the Department of Energy's Joint Genome Institute at *http://www.phytozome.net*.

## 11.7 Conclusion

In just the past few years we have witnessed tremendous progress in soybean genomics and an explosive expansion of new resources. We have seen the development of high-density genetic maps, construction of physical and transcript maps, EST sequencing and analysis, development of high-density cDNA and oligo arrays, and sequencing and comparison of homoeologous segments. These resources and the resultant studies have shed much light on the structure, organization and evolution of the soybean genome.

The whole-genome sequence of the soybean is now assembled. The availability of a whole-genome assembly places the legume research community in a pivotal position, with a unique opportunity to participate in a paradigm shift in genomic/genetic research. With the availability of the whole-genome sequence of the soybean genome, large-scale genomic sequences from two other reference legume species, *M. truncatula* and *L. japonicus*, plus the emerging genomic data from other legumes, multi-species genome-wide comparisons can be achieved. These approaches will allow researchers to decipher the evolutionary history and genomic complexity of legumes. We will be able to further explore genomic approaches to the elucidation of key genes or functional components that control complex agronomical and physiological traits.

The new generation sequencing technology and sequencing strategies have shown significant impact on all aspects of life sciences. The new strategies are already proving valuable for the resequencing of genomes. This not only provides information on the evolution of the genome, but provides data on genetic diversity. Specifically, resequencing is proving useful in identification of SNPs, useful for map saturation and haplotype identification. An unexpected benefit of high-throughput sequencing strategies is in the identification of differentially expressed genes and the identification of previously unannotated genes.

The burgeoning amount of scientific information is satisfying from an academic and an intellectual perspective. However, knowledge of an organism's genetic make-up seems sterile and barren unless it can be translated into safer and more secure food and feed production. The ultimate goal of genetic/genomic advances is to also provide genomics resources and tools for breeding and germplasm enhancement. No longer will it be sufficient to simply add to the knowledge base with little thought to practical application. Many of the tools and resources that have been developed can be more directly applied to that end. Redirecting and refocusing our scientific questions and experimental approaches with translational objectives in mind will be somewhat more difficult. Perhaps that is where the real paradigm shift should occur.

# References

Akkaya MS, Bhagwat AA, Cregan PB (1992) Length polymorphism of simple sequence repeat DNA in soybean. Genetics 132: 1131–1139.

Akkaya MS, Shoemaker RC, Specht JE, Bhagwat AA, Cregan PB (1995) Integration of simple sequence repeat DNA markers into a soybean linkage map. Crop Sci 35: 1439–1445.

Bennetzen JL (2000) Comparative sequence analysis of plant nuclear genomes: Microcolinearity and its many exceptions. Plant Cell 12: 1021–1029.

Bennetzen JL, Coleman C, Liu R, Ma J, Ramakrishna W (2004) Consistent over-estimation of gene number in complex plant genomes. Curr Opin Plant Biol 7: 732–736.

Bruggmann R, Bharti AK, Gundlach H, Lai J, Young S, Pontaroli AC, Wei F, Haberer G, Fuks G, Du C, Raymond C, Estep MC, Liu R, Bennetzen JL, Chan AP, Rabinowicz PD, Quackenbush J, Barbazuk WB, Wing RA, Birren B, Nusbaum C, Rounsley S, Mayer KF, Messing J (2006) Uneven chromosome contraction and expansion in the maize genome. Genome Res 16: 1241–1251.

Burow MD, Blake TK (1998) Molecular tools for the study of complex traits. In: AH Paterson (ed) Molecular Dissection of Complex Traits. CRC Press, Washington, DC, USA, pp 13–29.

Cannon SB (2008) Legume comparative genomics. In: G Stacey (eds) Genetics and Genomics of Soybean. Springer, New York, USA, pp 35–54.

Cannon SB, Sterck L, Rombauts S, Sato S, Cheung F, Gouzy J, Wang X, Mudge J, Vasdewani J, Schiex T, Spannagl M, Monaghan E, Nicholson C, Humphray SJ, Schoof H, Mayer KF, Rogers J, Quétier F, Oldroyd GE, Debellé F, Cook DR, Retzel EF, Roe BA, Town CD, Tabata S, Van de Peer Y, Young ND (2006) Legume genome evolution viewed through the *Medicago truncatula* and *Lotus japonicus* genomes. Proc Natl Acad Sci USA 103: 14959–14964.

Childs KL, Hamilton JP, Zhu W, Ly E, Cheung F, Wu H, Rabinowicz PD, Town CD, Buell CR, Chan AP (2007) The TIGR Plant Transcript Assemblies database. Nucl Acids Res 35: D846–D851.

Choi HK, Mun JH, Kim DJ, Zhu H, Baek JM, Mudge J, Roe B, Ellis N, Doyle J, Kiss GB (2004) Estimating genome conservation between crop and model legume species. Proc Natl Acad Sci USA 101: 15289–15294.

Choi IY, Hyten DL, Matukumalli LK, Song Q, Chaky JM, Quigley CV, Chase K, Lark KG, Reiter RS, Yoon MS, Hwang EY, Yi SI, Young ND, Shoemaker RC, van Tassell CP, Specht JE, Cregan PB (2007) A soybean transcript map: Gene distribution, haplotype and single-nucleotide polymorphism analysis. Genetics 176: 685–696.

Citerne HL, Luo D, Pennington RT, Coen E, Cronk QCB (2003) A Phylogenomic Investigation of CYCLOIDEA-Like TCP Genes in the Leguminosae. Plant Physiol 131: 1042–1053.

Cregan P, Jarvik T, Bush AL, Shoemaker RC, Lark KG, Kahler AL (1999) An integrated genetic linkage map of the soybean genome. Crop Sci 39: 1464–1490.

Cronk Q, Ojeda I, Pennington RT (2006) Legume comparative genomics: progress in phylogenetics and phylogenomics. Curr Opin Plant Biol 9: 99–103.

Devos KM, Brown JK, Bennetzen JL (2002) Genome size reduction through illegitimate recombination counteracts genome expansion in Arabidopsis. Genome Res 12: 1075–1079.

Doyle JJ, Luckow MA (2003) The rest of the iceberg. Legume diversity and evolution in a phylogenetic context. Plant Physiol 131: 900–910.

Doyle JJ, Doyle JL, Harbison C (2003) Chloroplast-expressed glutamine synthetase in *Glycine* and related Leguminosae: Phylogeny, gene duplication, and ancient polyploidy. Syst Bot 28: 567–577.

Eckardt NA (2006) The role of flavonoids in root nodule development and auxin transport in *Medicago truncatula*. Plant Cell 18: 1539–1540.

Etterson JR (2004) Evolutionary potential of *Chamaecrista fasciculata* in relation to climate change. II. Genetic architecture of three populations reciprocally planted along an environmental gradient in the great plains. Evolution 58: 1459–1471.

Fenster CF, Galloway LF (1999) Inbreeding and outbreeding depression in natural populations of *Chamaecrista fasciculata* (Fabaceae). Conserv Biol 14: 1406–1412.

Graham PH, Vance CP (2003) Legumes: Importance and constraints to greater use. Plant Physiol 131: 872–877.

Graham MA, Silverstein KA, Cannon SB, VandenBosch KA (2004) Computational identification and characterization of novel genes from legumes. Plant Physiol 135: 1179–1197.

Graham MA, Silverstein KAT, VandenBosch KA (2008) Defensin-like Genes: Genomic Perspectives on a Diverse Superfamily in Plants. Crop Sci 48: S3–S11.

Hirochika H, Sugimoto K, Otsuki Y, Tsugawa H, Kanda M (1996) Retrotransposons of rice involved in mutations induced by tissue culture. Proc Natl Acad Sci USA 93: 7783–7788.

Holligan D, Zhang X, Jiang N, Pritham EJ, Wessler SR (2006) The transposable element landscape of the model legume *Lotus japonicus*. Genetics 174: 2215–2228.

Hu JM, Lavin M, Wojciechowski M, Sanderson MJ (2000) Phylogenetic systematics of the tribe Millettieae (Leguminosae) based on trnK/matK sequences, and its implications for the evolutionary patterns in Papilionoideae. Am J Bot 87: 418–430.

Ilic K, SanMiguel PJ, Bennetzen JL (2003) A complex history of rearrangement in an orthologous region of the maize, sorghum, and rice genomes. Proc Natl Acad Sci USA 100: 12265–12270.

Innes RW, Ameline-Torregrosa C, Ashfield T, Cannon E, Cannon SB, Chacko B, Chen NW, Couloux A, Dalwani A, Denny R, Deshpande S, Egan AN, Glover N, Hans CS, Howell S, Ilut D, Jackson S, Lai H, Mammadov J, Del Campo SM, Metcalf M, Nguyen A, O'Bleness M, Pfeil BE, Podicheti R, Ratnaparkhe MB, Samain S, Sanders I, Ségurens B, Sévignac M, Sherman-Broyles S, Thareau V, Tucker DM, Walling J, Wawrzynski A, Yi J, Doyle JJ, Geffroy V, Roe BA, Maroof MA, Young ND (2008) Differential Accumulation of Retroelements and Diversification of NB-LRR Disease Resistance Genes in Duplicated Regions following Polyploidy in the Ancestor of Soybean. Plant Physiol 148: 1740–1759.

Jiang N, Bao Z, Zhang X, Eddy SR, Wessler SR (2004) Pack-MULE transposable elements mediate gene evolution in plants. Nature 431: 569–573.

Kumar A, Bennetzen JL (1999) Plant retrotransposons. Annu Rev Genet 33: 479–532.

Lai J, Ma J, Swigonová Z, Ramakrishna W, Linton E, Llaca V, Tanyolac B, Park YJ, Jeong OY, Bennetzen JL, Messing J (2004) Gene loss and movement in the maize genome. Genome Res 14: 1924–1931.

Laten HM, Havecker ER, Farmer LM, Voytas DF (2003) SIRE1, an endogenous retrovirus family from *Glycine max*, is highly homogeneous and evolutionarily young. Mol Biol Evol 20: 1222–1230.

Lavin M, Herendeen PS, Wojciechowski MF (2005) Evolutionary rates analysis of Leguminosae implicates a rapid diversification of lineages during the tertiary. Syst Biol 54: 530–549.

Lee JS, Verma DPS (1984) Structure and chromosomal arrangement of leghemoglobin genes in kidney bean suggest divergence in soybean leghemoglobin gene loci following tetraploidization. EMBO J 3: 2745–2752.

Liu J, Miller SS, Graham M, Bucciarelli B, Catalano CM, Sherrier DJ, Samac DA, Ivashuta S, Fedorova M, Matsumoto P, Gantt S, Vance CP (2006) Recruitment of novel calcium-binding proteins for root nodule symbiosis in *Medicago truncatula*. Plant Physiol 141: 167–177.

Ma J, Bennetzen JL (2004) Rapid recent growth and divergence of rice nuclear genomes. Proc Natl Acad Sci USA 101: 12404–12410.

Ma J, Devos KM, Bennetzen JL (2004) Analyses of LTR-retrotransposon structures reveal recent and rapid genomic DNA loss in rice. Genome Res 14:860–869.

Ma J, SanMiguel P, Lai J, Messing J, Bennetzen JL (2005) DNA rearrangement in orthologous orp regions of the maize, rice and sorghum genomes. Genetics 170: 1209–1220.

Mitra RM, Gleason CA, Edwards A, Hadfield J, Downie JA, Oldroyd GE, Long SR (2004) A Ca2+/calmodulin-dependent protein kinase required for symbiotic nodule development: gene identification by transcript-based cloning. Proc Natl Acad Sci USA 101: 4701–4705.

Morgante M, Rafalski A, Biddle P, Tingey S, Olivieri AM (1994) Genetic mapping and variability of seven soybean simple sequence repeat loci. Genome 37: 763–769.

Morgante M, Brunner S, Pea G, Fengler K, Zuccolo A, Rafalski A (2005) Gene duplication and exon shuffling by helitron-like transposons generate intraspecies diversity in maize. Nat Genet 37: 997–1002.

Palmer RG (1976) Chromosome transmission and morphology of three primary trisomics in soybeans (*Glycine max*). Can J Genet Cytol 18: 131–140.

Piegu B, Guyot R, Picault N, Roulin A, Sanyal A, Kim HR, Collura K, Brar DS, Jackson SA, Wing RA, Panaud O (2006) Doubling genome size without polyploidization: Dynamics of retrotransposition-driven genomic expansions in *Oryza australiensis*, a wild relative of rice. Genome Res 16: 1262–1269.

Pons TL, Perreijn K, van Kessel C, Werger MJA (2007) Symbiotic nitrogen fixation in a tropical rainforest: 15N natural abundance measurements supported by experimental isotopic enrichment. New Phytol 173: 154–167.

Powell W, Morgante M, Doyle JJ, McNicol JW, Tingey SV, Rafalski AJ (1996) Genepool variation in genus *Glycine* subgenus *soja* revealed by polymorphic nuclear and chloroplast microsatellites. Genetics 144: 793–803.

Salse J, Feuillet C (2007) Comparative genomics of cereals. In: R Varshney, R Tuberosa (eds) Genomics-Assisted Crop Improvement. Springer, The Netherlands, pp 177–205.

Samac DA, Graham MA (2007) Recent advances in legume-microbe interactions: Recognition, defense response, and symbiosis from a genomic perspective. Plant Physiol 144: 582–587.

SanMiguel P, Tikhonov A, Jin Y-K, Motchoulskaia N, Zakharov D, Melake-Berhan A, Springer PS, Edwards KJ, Lee M, Avramova Z, Bennetzen JL (1996) Nested retrotransposons in the intergenic regions of the maize genome. Science 274: 765–768.

Sato S, Nakamura Y, Kaneko T, Asamizu E, Kato T, Nakao M, Sasamoto S, Watanabe A, Ono A, Kawashima K, Fujishiro T, Katoh M, Kohara M, Kishida Y, Minami C, Nakayama S, Nakazaki N, Shimizu Y, Shinpo S, Takahashi C, Wada T, Yamada M, Ohmido N, Hayashi M, Fukui K, Baba T, Nakamichi T, Mori H, Tabata S (2008) Genome structure of the legume, *Lotus japonicus*. DNA Res 15(4): 227–239.

Schlueter JA, Dixon P, Granger C, Grant D, Clark L, Doyle JJ, Shoemaker RC (2004) Mining EST databases to resolve evolutionary events in major crop species. Genome 47: 868–876.

Schlueter JA, Scheffler BE, Schlueter SD, Shoemaker RC (2006) Sequence conservation of homeologous BACs and expression of homeologous genes in soybean (*Glycine max* L. Merr). Genetics 174: 1017–1028.

Schlueter JA, Sanders IF, Deshpande S, Yi J, Siegfried M, Roe BE, Schlueter SD, Scheffler BE, Shoemaker RC (2007a) The FAD2 family of soybean: Insights into the structural and functional divergence of a paleopolyploid genome. Plant Genom 1: 14–26.

Schlueter J, Lin J-Y, Schlueter S, Vasylenko-Sanders I, Deshpandem S, Yi J, O'Bleness M, Roe B, Nelson R, Scheffler B, Jackson S, Shoemaker RC (2007b) Gene duplication and paleopolyploidy in soybean and the implications for whole genome sequencing. BMC Genom 8: 330.

Schlueter JA, Scheffler B, Jackson S, Shoemaker RC (2008) Fractionation of synteny in a genomic region containing tandemly duplicated genes across *Glycine max*, *Medicago truncatula* and *Arabidopsis thaliana*. J Hered 99: 390–395.

Shoemaker R, Schlueter J, Doyle JJ (2006) Paleopolyploidy and gene duplication in soybean and other legumes. Curr Opin Plant Biol 9: 104–109.

Shoemaker R, Grant D, Olson T, Warren WC, Wing R, Yu Y, Kim H-R, Cregan P, Joseph B, Futrell-Griggs M, Nelson W, Davito J, Walker J, Wallis J, Kremitski C, Scheer D, Clifton S, Graves T, Nguyen H, Wu X, Luo M, Dvorak J, Nelson R, Cannon S, Tomkins J, Schmutz J, Stacey G, Jackson S (2008) Microsatellite discovery from BAC end sequences and genetic mapping to anchor the soybean physical and genetic maps. Genome 51: 294–302.

Shoemaker RC, Keim P, Vodkin L, Retzel E, Clifton SW, Waterston R, Smoller D, Coryell V, Khanna A, Erpelding J, Gai X, Brendel V, Raph-Schmidt C, Shoop EG, Vielweber CJ, Schmatz M, Pape D, Bowers Y, Theising B, Martin J, Dante M, Wylie T, Granger C (2002) A compilation of soybean ESTs: Generation and analysis. Genome 45: 329–338.

Shultz JL, Kazi S, Bashir R, Afzal JA, Lightfoot DA (2007) The development of BAC-end sequence-based microsatellite markers and placement in the physical and genetic maps of soybean. Theor Appl Genet 114: 1081–1090.

Shultz JL, Kurunam D, Shopinski K, Iqbal MJ, Kazi S, Zobrist K, Bashir R, Yaegashi S, Lavu N, Afzal AJ, Yesudas CR, Kassem MA, Wu C, Zhang HB, Town CD, Meksem K, Lightfoot DA (2006) The soybean genome database (SoyGD): A browser for display of duplicated polyploid regions and sequence tagged sites on the integrated physical and genetic maps of *Glycine max*. Nucl Acid Res 34: D758–D765.

Song QJ, Marek LF, Shoemaker RC, Lark KG, Concibido VC, Delannay X, Specht JE, Cregan PB (2004) A new integrated genetic linkage map of the soybean. Theor Appl Genet 109: 122–128.

Stacey G, Vodkin L, Parrott WA, Shoemaker RC (2004) National Science Foundation-sponsored workshop report. Draft plan for soybean genomics. Plant Physiol 135: 59–70.

ter Steege H, Jetten VG, Polak M, Werger MJA (1993) Tropical rain forest types and soils in a watershed area in Guyana. J Veget Sci 4: 705–716.

Thijs L, Pons TL, Perreijn K, van Kessel C, Werger MJA (2007) Symbiotic nitrogen fixation in a tropical rainforest: N natural abundance measurements supported by experimental isotopic enrichment. New Phytol 173: 154–167.

Tucker SC (2003) Floral development in legumes. Plant Physiol 131: 911–926.

Vicient CM, Suoniemi A, Anamthawat-Jonsson K, Tanskanen J, Beharav A, Nevo E, Schulman AH (1999) Retrotransposon BARE-1 and its role in genome evolution in the genus *Hordeum*. Plant Cell 11: 1769–1784.

Walling J, Shoemaker R, Young N, Mudge J, Jackson SA (2006) Chromosome level homology in paleopolyploid soybean (*Glycine max*) revealed through integration of genetic and chromosome maps. Genetics 172: 1893–1900.

Wang Q, Dooner HK (2006) Remarkable variation in maize genome structure inferred from haplotype diversity at the *bz* locus. Proc Natl Acad Sci USA 103: 17644–17649.

Wicker T, Sabot F, Hua-Van A, Bennetzen JL, Capy P, Chalhoub B, Flavell A, Leroy P, Morgante M, Panaud O, Paux E, SanMiguel P, Schulman AH (2007) A unified classification system for eukaryotic transposable elements. Nat Rev Genet 8: 973–982.

Wicker T, Stein N, Albar L, Feuillet C, Schlagenhauf E, Keller B (2001) Analysis of a contiguous 211 kb sequence in diploid wheat (*Triticum monococcum* L.) reveals multiple mechanisms of genome evolution. Plant J 26: 307–316.

Zou JJ, Singh RJ, Lee J, Xu SJ, Cregan PB, Hymowitz T (2003) Assignment of molecular linkage groups to soybean chromosomes by primary trisomics. Theor Appl Genet 107: 745–750.

# 12

# Role of Bioinformatics as a Tool

*Julie M. Livingstone,*[1,2] *Kei Chin C. Cheng*[2] *and Martina V. Strömvik*[1,2*]

## ABSTRACT

Bioinformatics, informatics applied to biological data, has rapidly become an integral part of biological experiments. Bioinformatics techniques and tools are used for experiment design, data storage, analyses and web publication. Understanding and being able to use such tools and resources is an essential asset to genome and molecular biologists. This chapter discusses the different types of genomics data and experiments currently being analyzed in the soybean community, such as expressed sequence tags and microarrays. General and soybean specific databases such as GenBank and SoyXpress, respectively, as well as tools that are available for aiding in analysis and experimental design are discussed.

**Keywords:** bioinformatics; database; software; expressed sequence tags; contig; microarray

## 12.1 Introduction

The role of bioinformatics as a tool in molecular biology has quickly expanded hand in hand with the evolution of genomics and with expanding experimental designs and data sets. Informatics techniques have proven to be crucial when working with data produced by high-throughput experiments. The usual role of bioinformatics is gathering, processing and

[1]McGill Centre for Bioinformatics, McGill University, 3775 University Street, Montreal, QC H3A 2B4, Canada.
[2]Department of Plant Science, McGill University, 21111 Lakeshore Road, Sainte Anne de Bellevue, QC H9X 3V9, Canada.
*Corresponding author: *martina.stromvik@mcgill.ca*

organizing information in order to answer biological questions. Bioinformatics is also a powerful tool in the initial stages of designing an experiment. Figure 12-1 shows an example of the flow of data and information on the bioinformatics side of a sequencing project.

This chapter will discuss common types of data that are currently available to be used in bioinformatic analysis, specific databases that exist and numerous tools that can be utilized focusing on how these tools and information can aid the soybean community.

## 12.2 Types of data

### 12.2.1 Expressed Sequence Tags

An expressed sequence tag (EST) is a short sequence (tag) of a transcribed gene. A collection of ESTs are sequenced from cDNA libraries, which are constructed from mRNA extracted from different tissue and organ systems at various developmental stages. The EST sequence is obtained from the raw chromatograms generated by the DNA sequencer and subsequently processed into high-quality sequences for publication by deleting the cloning vector sequences, polyA-tails and short repeat sequences (Fig. 12-1). There are already many ESTs available in public databases, such as NCBI's GenBank (Benson et al. 2008). Grouped together these ESTs can represent the transcriptome (a collection of all transcribed genes) of a specific tissue or whole organism. Currently there are over a million soybean ESTs available in the public NCBI database including approximately 330,000 EST sequences that were generated by the Public EST Project for Soybean (Shoemaker et al. 2002) and the Functional Genomics Program for Soybean (Vodkin et al. 2004).

Before the release of the draft soybean genome, researchers relied on EST information to make predictions about the gene inventory in soybean. Some examples are explored below.

In one experiment, the identification of legume specific genes was performed using assembled EST sequences, called tentative contigs (TCs) from *M. truncatula, L. japonicus* and *G. max/soja* (Quackenbush et al. 2000). These were then compared against other plant sequences such as rice, *Arabidopsis*, tomato, barley, cotton, grape and others by BLAST analysis (Graham et al. 2004). Legume specific TCs were identified and InterProScan, (Quevillon et al. 2005) a program that combines different pattern recognition methods and numerous motif databases, was used to discover common motifs and predict protein function. Three interesting families (F-box related proteins, Pro-rich proteins and Cys cluster proteins) were identified in these legume specific TCs (Graham et al. 2004).

Another study used TIGR TCs to investigate possible gene duplications and evolutionary events in different plant species including *Glycine max*

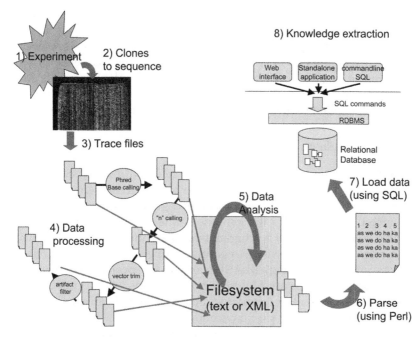

**Figure 12-1** A typical sequence processing and data web publishing project. Biologists perform an experiment (1), which generates clones that are to be sequenced (2). Chromatograms or trace files readable to computer software only, is the output from the sequencer (3). The trace files are base called and trimmed in a processing pipeline to yield readable sequences (4). For each step, output files are generated and these are often stored in a strict file system. Further data analysis and sequence annotation (BLAST, GO Gene Ontology etc) is performed on the processed sequences and output files from these analyses are stored in the file system (5). Pertinent information is extracted, parsed from the analysis output files into a more concise format (6). This is often done using custom scripts written in Perl. Finally, many large projects have their own database where the data is web published and made available to the scientific community (7, 8). Relational database management systems such as MySQL and Oracle are the most popular.

*Color image of this figure appears in the color plate section at the end of the book.*

(Schlueter et al. 2004). The putative open reading frame (ORF) for each contig was found using the program getorf (Rice et al. 2000) and a BLAST analysis was performed using the longest ORF TC set against itself to discover gene copies. Two hundred and seventy five paralogs were identified in soybean and evidence was found for two major genome duplications (15 and 44 MYA) (Schlueter et al. 2004).

EST information can even be used to identify different microbes to which plants have been exposed. By creating a local database of soybean EST sequences and querying 10 known virus genomes including bean pod mottle virus (BPMV), soybean mosaic virus (SMV) and cowpea chlorotic mottle virus (CPMV), a list of ESTs containing viral sequences was obtained

(Strömvik et al. 2006). These EST sequences were then assembled into consensus sequences to represent putative viral genomes using PHRAP (*http://www.phrap.org*). These sequences were aligned with their respective known viral genomes to verify their identity and the frequency of virus sequences among the soybean EST sequences was calculated. This study was able to demonstrate that analysis of EST libraries has the potential to show whether plant tissue used to make libraries was infected with viruses.

### 12.2.2 The Soybean Genome Sequence

The JGI (Joint Genome Institute) recently released the soybean genome sequence. This genome data was created using whole-genome shotgun sequencing, in which the whole *Glycine max* genome was cut, cloned and sequenced. Subsequently the genome sequence was assembled using Arachne2 (Jaffe et al. 2003), a program that aligns and joins sequences using parameters to account for large repetitive sequences (*www.phytozome.net/soybean.php*). By performing a BLAST analysis with specific genes of interest, it is also possible to identify highly similar regions, which could contain homologous sequences or duplicated regions. The soybean genome can be explored through a browsing environment, Gbrowse (Stein et al. 2002), which provides a search interface and shows the location of repeat sequences (16-mers), *Glycine* ESTs, and peptides from different species such as *Arabidopsis* and rice upon a specific section of the genome (*www.phytozome.net/soybean.php*).

### 12.2.3 Microarray Data

Global gene expression profiles can be studied with cDNA microarrays containing around 30,000 elements representing soybean genes (Vodkin et al. 2004) or with Affymetrix GeneChip arrays (Affymetrix 2001). The Affymetrix Soybean GeneChip contains 35,611 soybean transcripts (Affymetrix 2004–05). It also contains probes for the transcriptomes of two pathogens: the fungal pathogen *Phytophthora sojae* (represented by 15,421 probe sets) and the cyst nematode *Heterodera glycines* (represented by 7,431 probe sets). An Affymetrix probe set represents a transcript or a gene and consists of 11 oligonucleotide probe pairs, each 25 nucleotides long and spanning regions of each gene. The probe pair contains two probes, a perfect match probe (perfectly matches its target sequence) and a mismatch probe (where the 13th nucleotide is a mismatch), in order to also assess non-specific hybridization (Affymetrix 2001). Including control probes, there are a total of 61,170 probe sets on the soybean Affymetrix GeneChips, consisting of over 1,340,000 probes in total.

The Affymetrix Soybean GeneChip was used in the analysis of gene expression profiles of host and pathogen in nematode-infected soybean (Ithal et al. 2007) and to study the changes in gene expression affected by the Asian soybean rust disease (Panthee et al. 2007). In the first study, the root tissues of soybean cyst nematode (*H. glycines*) infected and uninfected soybean were compared at three time points (2, 5 and 10 days post-infection) (Ithal et al. 2007). GeneChip Operating Software version 1.0 (GCOS v. 1.0) was used for statistical analysis and genes were identified as being differentially expressed. The soybean EST sequences corresponding to the identified genes were used as query seeds in a search for the *Arabidopsis* orthologs in the TAIR database using WU-BLAST2 search. The top hit was used to annotate the soybean genes. Among the 429 differentially expressed genes, 320 of them were assigned putative functions. This is an example of how biological relevance was discovered from a large data set, using previously created bioinformatics software and annotated databases.

In the second study the Affymetrix soybean GeneChip was utilized for gene expression profiling of soybean with Asian soybean rust disease caused by *Phakopsora pachyrhizi* (Panthee et al. 2007). Gene expression data was analyzed using ArrayAssist Software from Stratagene to identify differentially expressed genes. The putative functions of encoded proteins were assigned to the differentially expressed genes using the ExPasy protein database (Gasteiger et al. 2003; *http://us.expasy.org*). Most of the up-regulated genes are identified as being involved in defense and stress-related responses (Panthee et al. 2007).

Recently, the soybean Affymetrix arrays were used to study the potential global transcriptome differences between different soybean genotypes, including GMO and non-GMO genotypes (Cheng et al. 2008). Statistical analysis was performed with the R (Hornik 2008) and BioConductor (Gentleman et al. 2004) packages. It was discovered that gene expression can differ more between conventional soybean cultivars than between a transgenic cultivar and its closest conventional cultivar (Cheng et al. 2008). In order to investigate exactly which metabolic pathways and biological processes have differentially expressed genes in the different cultivars, the SoyXpress database (Cheng and Strömvik 2008) was created. SoyXpress links the genes represented by the probes on the Affymetrix chips with information from other databases such as protein, gene ontology (GO) and metabolic pathway data through sequence similarity identified by BLAST analysis (Altschul et al. 1990).

The numerous studies described above show the versatility and ease that bioinformatics has brought to analysis of large data sets created by microarray data.

## 12.2.4 *Physical and Genetic Maps*

The original physical map of the soybean was developed from bacterial artificial chromosome (BAC) libraries (Wu et al. 2004; Warren and the Soybean Consortium, 2006) and overlaid with the genetic map consisting of microsatellites (SSRs) (Cregan et al. 1999; Song et al. 2004) and restriction fragment length polymorphism (RFLP) (Keim and Shoemaker 1988). Due to the highly duplicated nature of the soybean genome, RFLP markers tend to hybridize to more than one position and can not reliability be used in the creation of the genetic map. As a solution to this problem, single sequence repeat (SSR) markers were developed and mapped to the 20 linkage groups identified (Cregan et al. 1999; Song et al. 2004). The availability of these two integrated maps has aided researchers in many ways such as marker-assisted selection in plant improvement and molecular map-based cloning experiments (Cregan et al. 1999).

The resources created to make the physical map of soybean have also been used to investigate other properties of soybean. Through the combination of bioinformatics and experimental research, it was discovered that the duplicated regions of soybean are highly conserved with about 86–100% sequence similarity and that genes are clustered in approximately 25% of the genome (Mudge et al. 2003). This was accomplished using a set of BAC contigs anchored with RLFP markers and assembled with the programs Image (Sulston et al. 1988) and FPC (Soderlund et al. 1997). Sequence alignments were created and sequences were identified as genic using BLAST analysis against protein and EST databases. In another study, soybean BAC and cDNA sequences were compared to the *Arabidopsis* five chromosomes using a Java program to identify synteny between the two species (Shultz et al. 2007). The maps created can be viewed (*http://soybeangenome.siu.edu*) or downloaded (*http://msa.ars.usda.gov/ public/jray/1page1.html*).

The comparison of BAC and whole-genome shotgun (WGS) sequences from two different cultivars, Williams 82 and Forrest, identified numerous single nucleotide polymorphisms (SNPs), and single nucleotide insertions (SNIs) within gene sequences (Saini et al. 2008). This information can assist in the creation of better BAC clone assembly methods and annotation of the soybean physical map.

With the release of the soybean genome sequence it is now also possible to locate the SSR markers on the 20 chromosomes and linkage groups (Shoemaker et al. 2008; *http://www.soybeanphysicalmap.org*).

## 12.3 Data Repositories

Public databases house information that is readily available for download and use. Listed below are general databases that are useful. The URLs are listed in Table 12-1.

**Table 12-1** General, soybean and legume bioinformatics resources.

| Resource name | URL | Reference |
|---|---|---|
| *General resources* | | |
| National Center for Biotechnology Information (NCBI) | *http://www.ncbi.nlm.nih.gov* | Benson et al. 2008 |
| SwissProt | *http://ca.expasy.org/sprot* | Boeckmann et al. 2003 |
| Gene Ontology | *http://www.geneontology.org* | The Gene Ontology Consortium 2000 |
| Kyoto Encyclopedia of Genes and Genomes (KEGG) | *http://www.genome.jp/kegg* | Kanehisa et al. 2004 |
| The Institute for Genomic Research (TIGR) | *http://www.tigr.org/db.shtml* | Chan et al. 2007 |
| Gene Expression Omnibus (GEO) | *http://www.ncbi.nlm.nih.gov/geo* | Barrett et al. 2006 |
| Phytozome | *http://www.phytozome.net* | |
| *Soybean specific resources* | | |
| SoyXpress | *http://soyxpress.agrenv.mcgill.ca* | Cheng and Stromvik 2008 |
| SoyBase | *http://www.soybase.org* | |
| Soybean Genomics and Microarray Database (SGMD) | *http://psi081.ba.ars.usda.gov/SGMD/* | Alkharouf and Matthews 2004 |
| Soybean Gene Index (previously TIGR GI) | *http://compbio.dfci.harvard.edu/tgi/cgi-bin/tgi/gimain.pl?gudb=soybean* | Quackenbush et al. 2000 |
| Soybean GBrowse Database | *http://soybeangenome.siu.edu* | Shultz et al. 2006 |
| Gene Networks in Seed Development | *http://estdb.biology.ucla.edu/seed* | |
| *Legume data resources containing soybean data* | | |
| Legume Information System (LIS) | *http://www.comparative-legumes.org* | Gonzales et al. 2005 |
| LegumeDB | *http://ccg.murdoch.edu.au/index.php/LegumeDB* | Moolhujizen et al. 2006 |

## 12.3.1 General Databases

### 12.3.1.1 NCBI (*http://www.ncbi.nlm.nih.gov/*)

The NCBI (National Center for Biotechnology Information) public genome database GenBank (Benson et al. 2008) is part of the International Nucleotide Sequence Database Collaboration, where three large databanks work together to catalogue all available sequence information: NCBI in America, EMBL (European Molecular Biological Laboratory) in Europe and the DNA Databank of Japan (DDBJ) in Japan. This collaboration ensures a comprehensive collection of sequence information available worldwide (Benson et al. 2008). Sequences are obtained through individual submissions from laboratories and large-scale sequencing projects. The GenBank database is organized into different databases called divisions, such as dbEST for transcript data and UniGene for gene-oriented clusters of transcript sequences. Other divisions include whole genome sequences, three-dimensional macromolecular structures, taxonomy, single nucleotide polymorphism, chemical molecules and substances, protein domains, microarray data, cancer and disease related chromosomes, and journals. These databases provide comprehensive information about a gene or protein of interest. The BLAST (Basic Local Alignment Search Tool) web tool can be used to compare nucleotide or protein sequences of interest by sequence similarity searches against the data in the divisions, and all information is conveniently linked (Altschul et al. 1990).

### 12.3.1.2 *SwissProt (http://ca.expasy.org/sprot/)*

Swissprot is a protein knowledgebase maintained by The Swiss Institute for Bioinformatics (SIB) and the European Bioinformatics Institute (EBI). It integrates protein sequences with updated biological knowledge and manually curated entries (Boeckmann et al. 2003). The core data consists of amino acid sequence, protein name, taxonomic data and citation information. Each protein entry is provided with high-quality annotation on protein function, enzymatic information (e.g., enzyme commission (EC) number), secondary and quaternary structure. The nomenclature is standardized to facilitate communication across different databases.

### 12.3.1.3 *GO (http://www.geneontology.org/)*

Gene and protein sequences require annotations to describe their functions. Gene Ontology (GO) provides a standard terminology to describe gene products across different databases consistently (Fig. 12-2a). Gene products are classified according to their biological processes, molecular functions and subcellular location (The Gene Ontology Consortium 2000). The GO terms are organized into a tree-like structure called directed acyclic graphs (DAGs) to resemble a hierarchy (Fig. 12-2b). This allows a more specialized child term to have one or more specialized parent terms. All the child terms inherit all the properties of their parent terms. Therefore, when a gene product is annotated with a child term, all the parent terms also apply to that gene product. GO terms are written, maintained and curated by the GO collaborators. They also make associations between GO and other public genomic and proteomic databases such as Swissprot and TIGR, to facilitate uniform queries across them.

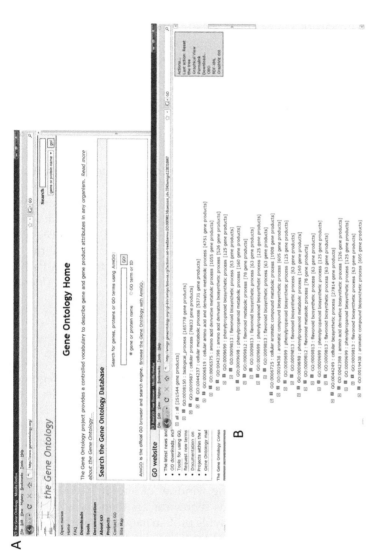

**Figure 12-2** Gene Ontology Home. a) Gene ontology homepage (*http://www.geneontology.org*) displaying the search by "gene or protein name" and "GO term or id" function. b) Example of results showing tree-like structure hierarchy.

*Color image of this figure appears in the color plate section at the end of the book.*

## 12.3.1.4 KEGG (http://www.genome.jp/kegg)

In order to understand the metabolic function of the gene products, the curators of KEGG (Kyoto Encyclopedia of Genes and Genomes) have organized information of metabolic pathways manually entered from published materials (Kanehisa et al. 2006). The pathway database (*http://www.genome.jp/kegg/pathway.html*) integrates current knowledge on molecular interaction networks and biological processes. The reference maps of metabolic networks were generated to show protein interaction such as direct protein-protein interaction, gene expression relation, and enzyme-enzyme relation. All these enzymes are assigned EC numbers.

## 12.3.1.5 TIGR/DFCI (http://compbio.dfci.harvard.edu/tgi/)

The Institute of Genomic Research (TIGR) was formerly the major institute participating in the Human Genome Project. They also collected publicly available EST sequences (including soybean ESTs) to assemble into tentative consensus (TC) sequences, which represent genes (Quackenbush et al. 2000). TIGR developed bioinformatics tools to assemble EST sequences and assign annotations to TCs. Soybean EST and TC information can be retrieved from their Gene Indices web site, which has moved to the Dana-Farber Cancer Institute and the Harvard School of Public Health (*http://compbio.dfci.harvard.edu/tgi/plant.html*).

## 12.3.1.6 The Legume Information System, LIS (http://www.comparative-legumes.org)

The Legume Information System (LIS) is a comparative resource for legumes including soybean, *Medicago truncaula* and *Lotus japonicus* (Gonzales et al. 2005). LIS gathers transcript data from legume plants and *Arabidopsis* from NCBI's High-Throughput Genomic division (Benson et al. 2008). These unfinished sequences were generated from large-scale genomic projects and are undergoing various stages of assembly processing. LIS takes these sequences and their constituent contigs from GenBank, and obtains their consensus sequences using a sliding window of length 10,000 with an overlap of 3,000 bp. These consensus sequences are then analyzed using different sequence databases and provide protein name, protein blocks and motif information. LIS has developed their EST database from NCBI raw EST and cDNA data. The raw EST data is screened for quality and contamination then assembled using PHRAP (*htp://www.phrap.org*). These consensus sequences are annotated with protein names, blocks and motif information. LIS also integrates genetic maps, physical maps and pathway information from collaborate projects such as SoyBase (*http://www.soybase.org/*) and the Southern Illinois University soybean genome project (*http://soybeangenome.siu.edu/*).

### 12.3.1.7 GEO (http://www.ncbi.nlm.nih.gov/geo/)

The Gene Expression Omnibus (GEO) is housed by NCBI and includes data from microarray and other high-throughput experiments (Barrett et al. 2007). Entries in this database must follow the minimum information about a microarray experiment (MIAME) format to ensure that all pertinent biological information is available. Numerous tools allow users to link to related information in other databases such as PubMed, GenBank, and UniGene. Pre-analyzed information centered on experiments (Entrez GEO DataSets) or centered on genes (Entrez GEO Profiles) can be easily found and used with the in house search interface (Barrett et al. 2007).

## 12.3.2 Dedicated Repositories

In addition to publishing their data in central, public databases like GenBank, most scientific communities have their own knowledge bases where they can release genome and related data in custom format. Databases that contain only soybean information are listed and described below.

### 12.3.2.1 SoyBase (http://soybase.org/index.php)

SoyBase and the Soybean Breeder's Toolbox is a website that is dedicated to soybean information (Fig. 12-3). It includes tools to investigate the genetic map, physical map and sequence map of the soybean. There are also many BLAST and search pages to gather data, as well as links to other soybean related resources.

### 12.3.2.2 SoyGD (http://soybeangenome.siu.edu/)

The Soybean Genome Database (SoyGD) browser provides genomic information in relation to the genetic map, which represent linkage groups (chromosomes) based on loci and markers (Shultz et al. 2006). This browser combines Perl MySQL programming with GBrowse (Stein et al. 2002) to allow users to visualize the genetic and physical maps of soybean. The search interface enables users to zoom in on one specific region of interest based on region or landmark such as SSR marker.

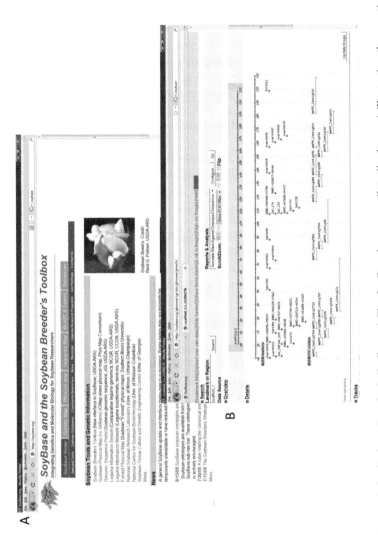

**Figure 12-3** SoyBase and the Soybean Breeder's Toolbox. a) The SoyBase homepage (*http://soybase.org*) illustrating the genetic map, physical map, sequence map and BLAST & Search tabs. b) Picture of SoyBase soybean browser, a Sequence Map tool, displaying the markers and FPC contigs located on Scaffold 1.

*Color image of this figure appears in the color plate section at the end of the book.*

## 12.3.2.3 Gene Networks in Seed Development (http:/estdb.biology.ucla.edu/seed/)

This website stores information used to explore what genes are required to make a soybean seed. The data housed on this site was gathered from soybean and *Arabidopsis* Affymetrix GeneChips, Laser Capture Microdissection (LCM), and 454 high-throughput sequencing technologies and is a collaboration between the Goldberg laboratory at UCLA and Harada laboratory at UCD. Tools are in place to browse and analyze soybean microarray data based on different developmental stages and tissues.

## 12.3.2.4 SGMD (http://psi081.ba.ars.usda.gov/SGMD/Default.htm)

The Soybean Genomics and Microarray Database (SGMD) stores EST and microarray data to explore the interaction of soybean with the major pest, soybean cyst nematode (SCN). The database stores over 50 million rows of DNA microarray data, conformed to the MIAME guidelines, and around 20,000 ESTs (Alkharouf and Matthews 2004). Relevant EST information stored in the database includes cloning information, GenBank accession number, BLAST results and links to PubMed. The web interfaces are embedded with analytical tools, for example analysis of variance (ANOVA), *t*-tests and K-means clustering to show the result and its significance and reproducibility of measurement. The SGMD web interface provides on-the-fly statistical analysis to compare cDNA microarray data.

## 12.3.2.5 Phytozome (http://www.phytozome.net/soybean)

Phytozome is a tool for the comparison of green plants including *Arabidopsis thaliana, Populus trichocarpa, Vitis vinifera, Sorghum bicolor,* and *Oryza sativa*. Users can perform BLAST analysis against the whole-genome sequences of these species or the node consensus (e.g., grass or angiosperm node). There is a keyword search, which will search all available species information. Each gene has been annotated with PFAM, KOG, KEGG, PANTHER RefSeq, UniProt, TAIR, and JGI assignments (http://www.phytozome.net/Phytozome_info.php). The Phytozome currently houses the soybean genome sequence and contains tools for closer investigation. Data can also be easily downloaded using the BioMart tool.

## 12.3.2.6 SoyXpress (http://soyxpress.agrenv.mcgill.ca/)

SoyXpress links gene expression data obtained from Affymetrix chips with related information, such as transcriptome data (annotated EST sequence information), GO (Gene Ontology) terms and KEGG (Kyoto Encyclopedia of

Genes and Genomes) metabolic pathways (Fig. 12-4). SoyXpress contains many different search interfaces, which enables users to search by EST ID, GenBank accession number, Affymetrix probe ID, SwissProt protein ID/ name, EC enzyme number or GO term/number (Cheng and Stromvik 2008). All retrieved information is linked to the original database from which it was retrieved.

## 12.4 Bioinformatics Tools

### 12.4.1 Sequence Alignment Tools

Basic local alignment search tool (BLAST) is a sequence analysis tool that uses a rapid heuristic algorithm to align sequences faster than dynamic methods (Altschul et al. 1990). BLAST has been adapted so that it can align nucleotide sequences, protein sequences, translated sequences and even whole genomes (McGinnis and Madden 2004). BLAST allows users to compare a query sequence against a database of sequences in order to discover sequences that are similar above a certain threshold. This tool is important for many different procedures such as searching for similar sequence in other species, trying to discover homologues or mapping sequences to known genomes.

Multiple sequence alignments can be used to characterize protein families, detect homology between sequences, help predict the secondary or tertiary structures of new sequences and develop oligonucleotide primers for PCR (Thompson et al. 1994). Sequences can be aligned globally (across the whole length of the sequences) or locally (only in certain regions) depending on the purpose of the alignment (Chenna et al. 2003). The Clustal series is an example of a global multiple alignment tool that includes ClustalW (Thompson et al. 1994), a command line tool and ClustalX (Thompson et al. 1997), which has a user interface. Alignments are created using dynamic programming to assign scores for matches and mismatches using different matrices (ex BLOSUM62 or PAM250) and penalties for insertions and deletions until an optimal alignment is created. ClustalW uses biological information to improve alignment techniques, decreasing gap opening penalties when there is a loop or coil region and increasing penalties when there are numerous gaps within eight residues of each other (Thompson et al. 1994).

### 12.4.2 EST Data Processing

Assembly software, such as PHRAP (*www.phrap.org/*), is used to align and assemble EST or genome sequences into longer consensus sequences (contigs), which represents a gene. PHRAP uses a modified version of the

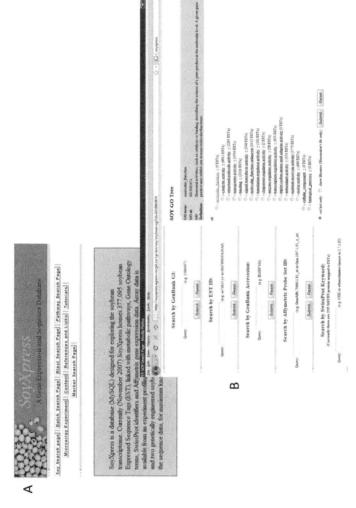

**Figure 12-4** SoyXpress A gene expression and sequence database. a) The SoyXpress homepage (*http://soyxpress.agrenv.mcgill.ca*) illustrating the tools available such as the Soy Search page, Batch Search Page, Blast Search page and Pathway Search page b) Snapshot of the Soy Search Page displaying the different terms available for search (GenBank ID, EST ID, GenBank Accession, Affymetrix Probe Set ID and SwissProt Keyword) and the searchable Soy GO tree.

*Color image of this figure appears in the color plate section at the end of the book.*

Smith-Waterman-Gotoh alignment algorithm to assemble contigs (de la Bastide and McCombie 2007). PHRAP uses user-supplied quality information (the PHRED scores generated by the base calling program) about each EST sequence to improve assembly. PHRAP removes the 5' and 3' poor regions of each EST based on the quality score assigned to each base. Overlaps between EST sequences are then computed and consensus sequences are created. PHRAP is run with specific arguments such as minmatch, which indicates the length of word that must match 100% and minscore, which indicates the minimum alignment score (*www.phrap.org*). PHRAP provides a quality value for each base in the consensus sequence which represents the confidence that the base is correct, based on whether the sequence is covered by both DNA strands, or if the EST was sequenced with more than one type of chemistry (de la Bastide and McCombie 2007).

Another EST assembling program is CAP3 which is very similar to PHRAP, as it trims the low quality 5' and 3' ends of the EST before assembly and provides a quality value for each base in the consensus sequence (Huang and Madan 1999). One special feature is that it allows the use of forward-reverse constraints which occur when a clone is sequenced both from the forward and reverse direction (i.e. opposite strands). Results produced by these two programs are comparable. PHRAP tends to create longer contigs, whereas CAP3 produces fewer errors in consensus sequences (Huang and Madan 2007).

### 12.4.3 *Microarray Data Processing*

A single hybridization experiment produces massive amounts of data. Handling, processing and analyzing the data poses a challenging task, and therefore the application of bioinformatics to microarray data analysis is essential. Before analysis, data can to be stored and organized in a database, which will serve as a repository. It can incorporate statistic methods and algorithms to allow the detection of unintended effects on gene expression. It is possible to combine transcript data with information on genetics, homology, functions, metabolic regulations and toxicology.

Biological interpretation of microarray experiments is needed to provide biological knowledge and facilitate communication among different laboratories and across platforms. The list of differentially expressed genes resulting from microarray analysis is usually translated into functional annotations by searching through literature and multiple public databases gene-by-gene manually (Draghici et al. 2003). However, this is a tedious and slow process. Several tools for automatically assigning functional annotations (such as GO terms) to microarray experiments have been developed, for example GOStat (Beissbarth and Speed 2004) and FatiGO

(Al-Shahrour et al. 2007). Both programs make use of the annotations and structure of the GO database to identify the function of a large dataset of genes. They will automatically retrieve the GO terms from a list of identifiers and determine which annotations are over-represented by comparing it to a control set that is made up of a complete list of annotated genes.

### 12.4.4 Programming Languages

Basic knowledge in some computer programming languages is a valuable asset when analyzing large data sets. A scripting language like Perl (*http:// www.perl.org*) is preferred in bioinformatics because it is powerful in pattern finding and data manipulation. Perl works by saving data into variables and arrays, which can be manipulated using different built-in functions. Perl is often used to connect to relational databases (the DBI module) to retrieve information and used to create interactive web pages (the CGI module). Another asset of Perl is that it uses regular expressions, which is a type of syntax, in order to search for different patterns. Example 1 shows a basic script that will take a tab delimited file, search and parse out wanted information and save information to a new file using regular expressions.

BioPerl (*http://www.bioperl.org*) packages, which are Perl functions that have been created specifically for biological data purposes, are also very useful. Some examples include packages to parse BLAST analysis files (Bio::SeachIO) or download files from databases such as GenBank (Bio::DB::GenBank). Example 2 shows a script that will open a file of GenBank accession identifiers and retrieve the sequences associated with these identifiers from NCBI's GenBank.

Another popular bioinformatics statistical language is R (*http://www.r-project.org*). This free software environment uses its own programming language including built-in functions that allow the user to perform basic statistical tests and to create graphics (Hornik 2008). Example 3 is a basic script that can be used to create a bar graph.

### 12.4.5 Promoter Discovery

Computational promoter motif discovery is an important tool needed to complement experimental methods. Because plant promoters are, in general, much less conserved than are animal promoters, specific algorithms geared towards plant promoters are needed. One such program that can be used to predict plant specific promoter regions is Seeder (Fauteux et al. 2008). This algorithm works by searching a set of sequences to find motifs that are over-represented compared to a set of background sequences (Fauteux et al. 2008).

Currently there are backgrounds available for six plant species including *Arabidopsis thaliana, Medicago truncatula, Glycine max, Populus trichocarpa, Vitis vinifera* and *Oryza sativa*. The Seeder Perl distribution is available at *http://www.cpan.org* or as an online tool at *http://seeder.agrenv.mcgill.ca*.

STAMP is a DNA-binding motif tool that will search existing motif databases to find similar motifs based on sequence similarity (Mahony and Benos 2007). This tool uses "position-specific scoring matrices" (PSSM) that represent the frequency or transcription factor preference of binding at each position in a motif to identify unknown motifs by comparison to annotated databases such as TRANSFAC (Matys et al. 2003) or JASPER (Sandelin et al. 2004) or specific plant databases such as PLACE (Higo et al. 1999) or AthaMap (Steffens et al. 2005) and identify similar motifs. This tool enables researchers to identify transcription factors that may interact with newly identified motifs (Mahony and Benos 2007). Important promoter and regulatory element databases include PLACE (plant cis-acting regulatory DNA elements) (Higo et al. 1999) and PlantCare (Lescot et al. 2002), which both include motifs found in plants. However, though these databases are heavily used, funding for database curation does not always allow them to be kept up to date.

## 12.5 Conclusion

Soybean researchers have an array of databases and tools at their disposal to assist in designing and analysis of experiments. This chapter has illustrated many of the databases and tools that can be utilized and provides many examples of how they have been used. In conclusion, bioinformatics is a growing field that is necessary to enable quick and efficient analysis of large datasets and computationally intensive methods. The future for bioinformatics is bright as the field is merging into systems biology and integration of large scale heterologous data sets from fields such as genomics, proteomics and metabolomics.

# References

Affymetrix (2001) Technical notes: GeneChip arrays provide optimal sensitivity and specificity for microarray expression analysis. Documentation from Affymetrix website *https://www.affymetrix.com/support/technical/technotes/25mer_technote.pdf* 1–4 (Accessed 27 Jul 2009).

Affymetrix (2004–2005) Data Sheet: GeneChip Soybean Genome Array. Documentation from Affymetrix website *http://www.affymetrix.com/support/technical/datasheets/soybean_datasheet.pdf* 1–2 (Accessed 27 Jul 2009).

Al-Shahrour F, Minguez P, Tarraga J, Medina I, Alloza E, Montaner D and Dopazo J. (2007) FatiGO: A functional profiling tool for genomic data. Integration of functional annotation, regulatory modifs and interaction data with microarray experiments. Nucl Acids Res 35 (Web Serv Iss): W91–W96.

Alkharouf NW, Matthews BF (2004) SGMD: The soybean genomics and microarray database. Nucl Acids Res 32 (Database Iss): D398–D400.

Altschul SF, Gish W, Miller W, Myers EW, Lipman DJ (1990) Basic local alignment search tool. J Mol Biol 215: 403–410.

Barrett T, Troup D, Wilhite S, Ledoux P, Rudnev,D, Evangelista C, Kim I, Soboleva A, Tomashevsky M, Edgar R (2007) NCBI GEO: Mining tens of millions of expression profiles-database and tools update. Nucl Acids Res 35 (Database Iss): D760–D765.

Beissbarth T, Speed TP (2004) GOstat: Find statistically overrepresented gene ontologies within a group of genes. Bioinformatics 20: 1464–1465.

Benson DA, Karsch-Mizrachi I, Lipman DJ, Ostell J, Wheeler, DL (2008) GenBank. Nucl Acids Res (Database Iss) 36: D25–D30.

Boeckmann B, Bairoch A, Apweiler R, Blatter MC, Estreicher A, Gasteiger E, Martin MJ, Michoud K, O'Donovan C, Phan I, Pilbout S, Schneider M (2003) The SWISS-PROT protein knowledgebase and its supplement TrEMBL in 2003. Nucl Acids Res 31: 365–370.

Chan A, Rabinowicz P, Quackenbush J, Buell C, Town C (2007) Plant database resourced at The Institute for Genomic Research. Meth Mol Biol 406: 113–136.

Cheng KC, Strömvik MV (2008) SoyXpress: A database for exploring the soybean transcriptome. BMC Genom 9: 368.

Cheng KC, Beaulieu J, Iquira E, Belzile FJ, Fortin MG, Strömvik MV (2008) Effect of transgenes on global gene expression in soybean is within the natural range of variation of their conventional counterparts. J Agri Food Chem 56: 3057–3067.

Chenna R, Sugawara H, Koike T, Lopez R, Gibson T, Higgins D, Thompson J (2003) Multiple sequence alignment with the Clustal series of programs. Nucl Acids Res 31: 3497–3500.

Cregan P, Jarvik T, Bush A, Shoemaker R, Lark K, Kahler A, Kaya N, VanToai T, Lohnes D, Chung J, Specht J (1999) An integrated genetic linkage map of the soybean genome. Crop Sci 39: 1464–1490.

de la Bastide M, McCombieWR (2007) Assembly genomic DNA sequences with PHRAP. Current Protocol of Bioinformatics. Chapter11: Unit 11.4.

Draghici S, Khatri P, Martins RP, Ostermeier C, Krawetz SA (2003) Global functional profiling of gene expression. Genomics 81: 98–104.

Fauteux F, Blanchette M, Strömvik M (2008) Seeder: Discriminative seeding DNA motif discovery. Bioinformatics 24: 2030–2037.

Gasteiger E, Gattiker A, Hoogland C, Ivanyi I, Appel RD, Bairoch A (2003) ExPASy : The proteomics server for in-depth protein knowledge and analysis. Nucl Acids Res 31: 3784–3788.

Gentleman RC, Carey VJ, Bates DM, Bolstad B, Dettling M, Dudoit S, Ellis B, Gautier L, Ge Y, Gentry J, Hornik K, Hothorn T, Huber W, Iacus S, Irizarry R, Leisch F, Li C,

Maechler M, Rossini AJ, Sawitzki G, Smith C, Smyth G, Tierney L, Yang JY, Zhang J (2004) Bioconductor: Open software development for computational biology and bioinformatics. Genome Biol 5: 1–16.

Gonzales MD, Archuleta E, Farmer A, Gajendran K, Grant D, Shoemaker R, Beavis WD, Waugh ME (2005) The Legume Information System (LIS): An integrated information resource for comparative legume biology. Nucl Acids Res 33 (Database Iss): D660–D665.

Graham M, Silverstein K, Cannon S, VandenBosch K (2004) Computational identification and characterization of novel genes from legumes. Plant Physiol 135: 1179–1197.

Higo K, Ugawa Y, Iwamoto M, Korenaga T (1999) Plant *cis*-acting regulatory DNA elements (PLACE) database. Nucl Acids Res 27: 297–300.

Hornik K (2008) The R FAQ. Documentation from R-Project website *http://www.r-project.org* (ISBN 3-900051-08-9) (Accessed 23 Nov 2008).

Huang X, Madan A (1999) CAP3: A DNA sequence assembly program. Genome Res 9: 867–877.

Ithal N, Recknor J, Nettleton D, Hearne L, Maier T, Baum TJ, Mitchum MG (2007) Parallel genome-wide expression profiling of host and pathogen during soybean cyst nematode infection of soybean. Mol Plant-Micr Interact 20: 293–305.

Jaffe D, Butler J, Gnerre S, Mauceli E, Lindblad-Toh K, Mesirov J, Zody M, Lander E (2003) Whole-genome sequence assembly for mammalian genomes: Arachne 2. Genome Res 13: 91–96.

Kanehisa M, Goto S, Hattori M, Aoki-Kinoshita KF, Itoh M, Kawashima S, Katayama T, Araki M, Hirakawa M (2006) From genomics to chemical genomics: New developments in KEGG. Nucl Acids Res 34 (Database Iss): D354–D357.

Keim P, Shoemaker R (1988) Construction of a random recombinant DNA library that is primarily single copy sequences. Soybean Genet Newsl 15: 147–148.

Lescot M, Dehais P, Thijs G, Marchal K, Moreau Y, Van de Peer Y, Rouze P, Rombauts S (2002) PlantCARE, a database of plant *cis*-acting regulatory elements and a portal to tools for in silico analysis of promoter sequence. Nucl Acids Res 30: 325–327.

Mahony S, Benos P (2007) STAMP: A web tool for exploring DNA binding motif similarities. Nucl Acids Res (Web Serv Iss) 35: W253–W258.

Matys V, Fricke E, Geffers R, Gossling E, Haubrock M, Hehl R, Hornischer K, Karas D, Kel AE, et al. (2003) TRANSFAC: Transcriptional regulation, from patterns to profiles. Nucl Acids Res 31: 374–378.

McGinnis S, Madden T (2004) BLAST: At the core of a powerful and diverse set of sequence analysis tools. Nucl Acids Res 32 (website Iss): W20–W25.

Mudge J, Huihuang Y, Denny R, Howe D, Danesh D, Marek L, Retzel E, Shoemaker R, Young N (2003) Soybean bacterial artificial chromosome contigs anchored with RFLPs: Insight into genome duplication and gene clustering. Genome 47: 361–372.

Panthee DR, Yuan JS, Wright DI, Marois JJ, Mailhot D, Stewart Jr CN (2007) Gene expression analysis in soybean in response to the causal agent of Asian soybean rust (*Phakopsora pachyrhizi* Sydow) in an early growth stage. Funct Integr Genom 7: 291–303.

Quackenbush J, Lian F, Holt I, Pertea G, Upton J (2000) The TIGR Gene Indics: Reconstruction and representation of expressed gene sequences. Nucl Acids Res 28: 141–145.

Quevillon E, Silventoinen V, Pillai S, Harte Mulder N, Apweiler R, Lopez R (2005) InterProScan: Protein domain identifier. Nucl Acids Res (Web Serv Iss): W116–120.

Rice P, Longden I, Bleasby A (2000) EMBOSS: the European molecular biology open software suite. Trends Genet 16: 276–277.

Saini N, Shultz J, Lightfoot D (2008) Re-annotation of the physical map of *Glycine max* for polyploidy-like regions of BAC and sequence driven whole shotgun read assembly. BMC Genom 8: 323.

Sandelin A, Alkema W, Engstrom P, Wasserman W, Lenhard B (2004) JASPAR: An open-access database for eukaryotic transcription factor binding profiles. Nucl Acids Res 32 (Database Iss): D91–D94.

Schlueter J, Dixon P, Granger C, Grant D, Clark L, Doyle J, Shoemaker R (2004) Mining EST databases to resolve evolutionary events in major crop species. Genome 47: 868–876.

Shoemaker R, Keim P, Vodkin L, Retzel E, Clifton SW, Waterston R, Smoller D, Coryell V, Khanna A, Erpelding J, Gai X, Brendel V, Raph–Schmidt C, Shoop EG, Vielweber CJ, Schmatz M, Pape D, Bowers Y, Theising B, Martin J, Dante M, Wylie T, Granger C (2002) A compilation of soybean ESTs: Generation and analysis. Genome 45: 329–338.

Shoemaker R, Grant D, Olson T, Warren W, Wing R, Yu Y, Kim H, Cregan P, Joseph B, Futrell-Griggs M, Nelson W, Davito J, Walker J, Wallis J, Kremitski C, Scheer D, Clifton S, Grave, T, Nguyen H, Wu X, Luo M, Dvorak J, Nelson R, Cannon S, Tomkins J, Schmutz J, Stacey G, Jackson S (2008) Microsatellite discovery from BAC end sequence and genetic mapping to anchor the soybean physical and genetic maps. Genome 51: 294–302.

Shultz JL, Kurunam D, Shopinski K, Iqbal MJ, Kazi S, Zobrist K, Bashir R, Yaegashi S, Lavu N, Afzal AJ, Yesudas CR, Kassem MA, Wu C, Zhang HB, Town CD, Meksem K, Lightfoot DA (2006) The Soybean Genome Database (SoyGD): A browser for display of duplicated, polyploid, regions and sequence tagged sites on the integrated physical and genetic maps of *Glycine max*. Nucl Acids Res 34 (Database Iss): D758–D765.

Shultz J, Ray J, Lightfoot D (2007) A sequence based synteny map between soybean and *Arabidopsis thaliana*. BMC Genom 8: 8.

Soderlund C, Longden I, Mott, R (1997) FPC: A system for building contigs from restriction fingerprinted clones. Comput Appl Biosci 13: 523–535.

Song Q, Marek L, Shoemaker R, Lark K, Concibido V, Delannay X, Specht J, Cregan P (2004) A new integrated genetic linkage map of the soybean. Theor Appl Genet 109: 122–128.

Steffen N, Galuschka C, Schindler M, Bulow L, Hehl R (2005) AthanMap web tools for database-assisted identification of combinatorial cis-regulatory elements and the display of highly conserved transcription factor binding sites in *Arabidopsis thaliana*. Nucl Acids Res 33 (Web Serv Iss): W397–W402.

Stein L, Mungall C, Shu S, Caudy M, Mangone M, Day A, Nickerson E, Stajich J, Harris T, Arva A, Lewis S (2002) The generic genome browser: A building block for a model organism system database. Genome Res 12: 1599–1610.

Strömvik M, Latour F, Archambault A, Vodkin L (2006) Identification and phylogenetic analysis of sequences of Bean pod mottle virus, Soybean mosaic virus and Cowpea chlorotic mottle virus in expressed sequence tag data from soybean. Can J Plant Pathol 28: 289–301.

Sulston J, Mallett F, Staden R, Durbin R, Horsnell T, Coulson A (1988) Software for genome mapping by fingerprinting techniques. Comput Appl Biosci 4: 125–132.

The Gene Ontology Consortium (2000) Gene Ontology: Tool for the unification of biology. Nat Genet 25: 25–29.

Thompson J, Higgins D, Gibson T (1994) Clustal W: Improving the sensitivity of progressive multiple alignment through sequence weighting, position-specific gap penalties and weight matrix choice. Nucl Acids Res 22: 4673–4680.

Thompson J, Gibson T, Plewniak F, Jeanmougin F, Higgins D (1997) The ClustalX window interface: Flexible strategies from multiple sequence alignment aided by quality analysis tools. Nucl Acids Res 25: 4876–4882.

Vodkin LO, Khanna A, Shealy R, Clough SJ, Gonzalez DO, Philip R, Zabala G, Thibaud-Nissen F, Sidarous M, Stromvik MV, Shoop E, Schmidt C, Retzel E, Erpelding F, Shoemaker RC, Rodiguez-Huete AM, Polacco JC, Coryell T, Kleim P, Gong G, Lui L, Pardinas J, Schweitzer P (2004) Microarrays for global expression constructed with a low redundancy set of 27,500 sequenced cDNAs representing an array of developmental stages and physiological conditions of the soybean plant. BMC Genom 5: 1–18.

Warren W, (2006) A physical map of the "William 82" soybean (*Glycine max*) genome W151. In: Plant & Anim Genom XIV Conf, San Diego, CA, USA: *http://www.intl-pag.org/14/abstracts/PAG14_W151.html*

Wu C, Sun S, Nimmakayala P, Santos F, Meksem K, Springman R, Ding K, Lightfoot D, Zhang H (2004) A BAC- and BIBAC-based physical map of the soybean genome. Genome Res 14: 319–326.

**Exapmle 1**

**BLAST Results.txt**

| | | | | | |
|---|---|---|---|---|---|
| Contig1 | AR282250 | 846 | 415 | 5e-113 | Sequence 1 from patent US 6521433 |
| Contig1 | AR399206 | 846 | 415 | 5e-113 | Sequence 1 from patent US 6617493 |
| Contig1 | AY595413 | 1157 | 407 | 1e-110 | Glycine max chalcone isomerase 1A mRNA, complete cds |
| Contig1 | D63577 | 745 | 286 | 3e-074 | Pueraria lobata mRNA for chalcone flavanone isomerase, complete cds |
| Contig1 | M91080 | 836 | 139 | 5e-030 | Medicago sativa (cultivar Iroquois) chalcone isomerase (MsCHI-2) mRNA, 3'end |
| Contig1 | M91079 | 950 | 139 | 5e-030 | Medicago sativa (cultivar Iroquois) chalcone isomerase (MsCHI-1) mRNA, complete cds |
| Contig1 | AB024988 | 555 | 135 | 7e-029 | Cicer arietinum mRNA for chalcone isomerase, partial cds |
| Contig1 | AB054801 | 681 | 133 | 3e-028 | Lotus japonicus mRNA for putative chalcone isomerase, complete cds, clone:LjCHI1 |
| Contig1 | AP004250 | 79947 | 133 | 3e-028 | Lotus japonicus genomic DNA, chromosome 5, clone:LjT47K21, TM0260, complete sequence |
| Contig1 | AX825011 | 670 | 133 | 3e-028 | Sequence 5 from Patent WO03072790 |
| Contig1 | AB073787 | 678 | 115 | 7e-023 | Lotus japonicus mRNA for chalcone isomerase, complete cds |
| Contig1 | AF308140 | 309 | 98 | 2e-017 | Lotus corniculatus chalcone isomerase (chi2) gene, partial cds |
| Contig1 | AF308141 | 305 | 98 | 2e-017 | Lotus corniculatus chalcone isomerase 3 (chi3) gene, partial cds |
| Contig1 | AF307301 | 200 | 90 | 4e-015 | Lotus corniculatus chalcone isomerase 1 (chi1) mRNA, partial cds |
| Contig2 | AY595419 | 786 | 1503 | 0.0 | Glycine max chalcone isomerase 1B2 mRNA, complete cds |
| Contig2 | AY595414 | 809 | 1130 | 0.0 | Glycine max chalcone isomerase 1B1 mRNA, complete cds |
| Contig2 | D63577 | 745 | 80 | 6e-012 | Pueraria lobata mRNA for chalcone flavanone isomerase, complete cds |
| Contig2 | M91080 | 836 | 64 | 4e-007 | Medicago sativa (cultivar Iroquois) chalcone isomerase (MsCHI-2) mRNA, 3'end |
| Contig2 | M91079 | 950 | 64 | 4e-007 | Medicago sativa (cultivar Iroquois) chalcone isomerase (MsCHI-1) mRNA, complete cds |
| Contig2 | AY595413 | 1157 | 58 | 2e-005 | Glycine max chalcone isomerase 1A mRNA, complete cds |
| Contig2 | AR282250 | 846 | 58 | 2e-005 | Sequence 1 from patent US 6521433 |
| Contig2 | AR399206 | 846 | 58 | 2e-005 | Sequence 1 from patent US 6617493 |
| Contig3 | AY595414 | 809 | 1561 | 0.0 | Glycine max chalcone isomerase 1B1 mRNA, complete cds |
| Contig3 | AY595419 | 786 | 1170 | 0.0 | Glycine max chalcone isomerase 1B2 mRNA, complete cds |
| Contig3 | D63577 | 745 | 82 | 2e-012 | Pueraria lobata mRNA for chalcone flavanone isomerase, complete cds |
| Contig3 | M91080 | 836 | 68 | 3e-008 | Medicago sativa (cultivar Iroquois) chalcone isomerase (MsCHI-2) mRNA, 3'end |
| Contig3 | M91079 | 950 | 68 | 3e-008 | Medicago sativa (cultivar Iroquois) chalcone isomerase (MsCHI-1) mRNA, complete cds |
| Contig3 | AY595413 | 1157 | 60 | 6e-006 | Glycine max chalcone isomerase 1A mRNA, complete cds |
| Contig3 | AR282250 | 846 | 60 | 6e-006 | Sequence 1 from patent US 6521433 |
| Contig3 | AR399206 | 846 | 60 | 6e-006 | Sequence 1 from patent US 6617493 |
| Contig4 | AY595413 | 1157 | 1719 | 0.0 | Glycine max chalcone isomerase 1A mRNA, complete cds |

| | | | | | |
|---|---|---|---|---|---|
| Contig4 | AR282250 | 846 | 1606 | 0.0 | Sequence 1 from patent US 6521433 |
| Contig4 | AR399206 | 846 | 1606 | 0.0 | Sequence 1 from patent US 6617493 |
| Contig4 | D63577 | 745 | 974 | 5e-281 | Pueraria lobata mRNA for chalcone flavanone isomerase, complete cds |
| Contig4 | AB024988 | 555 | 240 | 3e-060 | Cicer arietinum mRNA for chalcone isomerase, partial cds |
| Contig4 | AB054801 | 681 | 205 | 2e-049 | Lotus japonicus mRNA for putative chalcone isomerase, complete cds, clone:LjCHI1 |
| Contig4 | AX825011 | 670 | 205 | 2e-049 | Sequence 5 from Patent WO03072790 |
| Contig4 | M91080 | 836 | 189 | 1e-044 | Medicago sativa (cultivar Iroquois) chalcone isomerase (MsCHI-2) mRNA, 3'end |
| Contig4 | M91079 | 950 | 181 | 3e-042 | Medicago sativa (cultivar Iroquois) chalcone isomerase (MsCHI-1) mRNA, complete cds |
| Contig4 | AP004250 | 79947 | 133 | 5e-028 | Lotus japonicus genomic DNA, chromosome 5, clone:LjT47K21, TM0260, complete sequence |
| Contig4 | AF307301 | 200 | 125 | 1e-025 | Lotus corniculatus chalcone isomerase 1 (chi1) mRNA, partial cds |
| Contig4 | AB011794 | 912 | 117 | 3e-023 | Citrus sinensis mRNA for chalcone isomerase, complete cds |
| Contig4 | AB073787 | 678 | 117 | 3e-023 | Lotus japonicus mRNA for chalcone isomerase, complete cds |
| Contig4 | AF308140 | 309 | 98 | 3e-017 | Lotus corniculatus chalcone isomerase (chi2) gene, partial cds |
| Contig4 | AF308141 | 305 | 98 | 3e-017 | Lotus corniculatus chalcone isomerase 3 (chi3) gene, partial cds |
| Contig4 | AY595414 | 809 | 60 | 7e-006 | Glycine max chalcone isomerase 1B1 mRNA, complete cds |
| Contig4 | AY595419 | 786 | 58 | 3e-005 | Glycine max chalcone isomerase 1B2 mRNA, complete cds |

Example1.pl

```perl
#!/usr/bin/perl

use strict;
use warnings;

#open file with BLAST results and save into an array
open FILE, "BLASTResults.txt" or die;

my @array = <FILE>;
close FILE;

my @wanted;
my $count=0;
```

```
#cycle through each line of file and look to see if it is a hit against Glycine max, if it is place in new array
foreach(@array){
    if ($_ =~ / Glycine max/) {
        $wanted[$count] = $_;
        $count++;
    }
}

#open new file and print only wanted hits
open FILE2, ">> AllWantedHits.txt" or die;
foreach(@wanted) {
    print FILE2 "$_ \n";
}
close FILE2
```

AllWantedHits.txt

| | | | | | |
|---|---|---|---|---|---|
| Contig1 | AY595413 | 1157 | 407 | 1e-110 | Glycine max chalcone isomerase 1A mRNA, complete cds |
| Contig2 | AY595419 | 786 | 1503 | 0.0 | Glycine max chalcone isomerase 1B2 mRNA, complete cds |
| Contig2 | AY595414 | 809 | 1130 | 0.0 | Glycine max chalcone isomerase 1B1 mRNA, complete cds |
| Contig2 | AY595413 | 1157 | 58 | 2e-005 | Glycine max chalcone isomerase 1A mRNA, complete cds |
| Contig3 | AY595414 | 809 | 1561 | 0.0 | Glycine max chalcone isomerase 1B1 mRNA, complete cds |
| Contig3 | AY595419 | 786 | 1170 | 0.0 | Glycine max chalcone isomerase 1B2 mRNA, complete cds |
| Contig3 | AY595413 | 1157 | 60 | 6e-006 | Glycine max chalcone isomerase 1A mRNA, complete cds |
| Contig4 | AY595413 | 1157 | 1719 | 0.0 | Glycine max chalcone isomerase 1A mRNA, complete cds |
| Contig4 | AY595414 | 809 | 60 | 7e-006 | Glycine max chalcone isomerase 1B1 mRNA, complete cds |
| Contig4 | AY595419 | 786 | 58 | 3e-005 | Glycine max chalcone isomerase 1B2 mRNA, complete cds |

**Example 2**

**MySequences.txt**

```
BE346826
BE440777
BF066359
BF070466
BF070519
BF597034
BG046035
BG157194
BG650415
BG881357
BI320956
BI321546
BI423805
BI497866
BI787393
BI944928
BI967307
BI972287
BI974245
BI974353
BM093262
...
```

```perl
#!usr/bin/perl
use strict; #information needed by perl to run
use warnings;
use Bio::DB::GenBank;
#Opens file and saves to an array
open FILE, "MySequences.txt" or die;
my @array = <FILE>;
my @string;
my @save;
my $i = 0;

#a loop that will get the Accession for each line in the file
foreach (@array){
        $save[$i] = $array[$i];

        #created a Bio::DB::Genbank object
        my $gb = new Bio::DB::GenBank;

        #retrives the sequence from GenBank
        my $seq = $gb->get_Seq_by_acc("$array[$i]"); # Accession Number
        $string[$i] = $seq->seq();
        open FILE, ">> Output.txt" or die $!;

        #prints information to file in fasta format
        print FILE "> $save[$i]";
        print FILE "\t";
        print FILE "$string[$i]\n";

        #closes file
        close FILE;

        $i++;
}
```

**Data.txt**

**Example 3 Bar Graph**

| Bar1 | Bar2 | Bar3 |
|------|------|------|
| 1 | 2 | 3 |
| 3 | 4 | 4 |
| 6 | 4 | 6 |

```
#Data can manually be entered as follows
#all data points to be plotted
h<-c(1,2,4,3,5,4,6,4,6)

#put information into 3 X 3 matrix
info<-array(h,c(3,3))

#or read from file Data.txt
data <-read.table("C:\\Data.txt", header=T, sep="\t")

#create bar graph
barplot(as.matrix(info), main="Title", ylab="Y Axis", xlab= "X Axis", beside=TRUE,
col=rainbow(3)) #NumberOfBars

#add legend
legend("topleft", c("Bar1", "Bar2", "Bar3"),cex=0.6, bty="n", fill=rainbow(3))
```

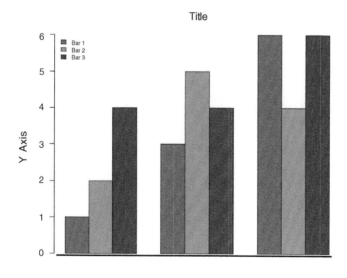

# Soybean Proteomics

*Savithiry S. Natarajan,*[*][1a] *Thomas J. Caperna,*[2] *Wesley M. Garrett*[2] *and Devanand L. Luthria*[3]

## ABSTRACT

Soybeans [*Glycine max* (L.) Merr.] are the most important commercial legume crops grown in the world and are an excellent source of protein, oil, dietary fiber, and phytonutrients. In this chapter, we discuss proteomics tools for soybean protein analysis with emphasis on protein extraction, two-dimensional polyacrylamide gel electrophoresis (2D-PAGE) separation, and matrix-assisted laser desorption/ionization time of flight (MALDI-TOF) coupled with liquid chromatography mass spectrometry (LC-MS) for identification of different types of soybean seed and leaf proteins. Recent advances in mass spectrometry and protein sequence databases enabled identification of complex proteins.

**Keywords:** soybean; proteomics; *G. soja*; *G. max*; 2D-PAGE; MALDI-TOF-MS; LC-MS/MS

## 13.1 Soybean Seed Proteins

Soybean seeds contain 40–50% protein on a dry matter basis. Soybean proteins possess unique physiochemical properties suitable for various

[1]US Department of Agriculture, Agricultural Research Service, Soybean Genomics and Improvement Laboratory, PSI, Beltsville, MD 20705, USA.

[a]Mention of trade name, proprietary product or vendor does not constitute a guarantee or warranty of the product by the US Department of Agriculture or imply its approval to the exclusion of other products or vendors that also may be suitable.

[2] US Department of Agriculture, Agricultural Research Service, Animal Biosciences and Biotechnology Laboratory, Beltsville, MD 20705, USA.

[3]US Department of Agriculture, Agricultural Research Service, Food Composition and Methods Development Laboratory, Beltsville, MD 20705, USA.

*Corresponding author: *savi.natarajan@ars.usda.gov*

human and animal food uses. Soybean seed proteins are used in baby formula, flour, protein supplements, concentrates, and textured fibers. Several reviews (Mityko et al. 1990; Warner and Wemmer 1991) describe soybean seed proteins and the comprehensive studies that have been conducted to identify these proteins and understand their functions (Beachy et al. 1983b; Nielsen et al. 1995). Soybean storage proteins are grouped into two types based on their sedimentation coefficients, β-conglycinin and glycinin (Thanh and Shibasaki 1978). β-conglycinin, a 7S globulin, is a trimeric glycoprotein consisting of three types of non-identical but homologous polypeptide subunits. Glycinin is an 11S globulin and its subunits are composed of acidic and basic polypeptides linked by a disulfide bond. These two major groups account for about 70–80% of the total seed protein. Soybean seeds also contain less abundant proteins including β-amylase, cytochrome c, lectin, lipoxygenase, urease, Kunitz Trypsin Inhibitor (KTI) and the Bowman Birk Inhibitor (BBI) of chymotrypsin and trypsin. Soybeans are also sources of several secondary metabolites including isoflavones, saponins, phytic acid, flatus-producing oligosaccharides, and goitrogens (Liener 1994; Friedman and Brandon 2001). In soybean, storage proteins are deposited in specialized protein storage vacuoles (PSV) and endoplasmic reticulum (ER)-derived protein bodies (Herman and Larkins 1999; Chrispeels and Herman 2000; Kinney et al. 2001).

The β-conglycinin storage protein, a 7S globulin, is a 180 kDa trimeric glycoprotein and encoded by two mRNA groups (Schuler et al. 1982b). The first mRNA group encodes α and α′ β-conglycinin subunits and the second mRNA group encodes the β-subunit of β-conglycinin (Thanh and Shibasaki 1976; Koshiyama 1983). These three subunits of β-conglycinin are encoded by a gene family containing at least 15 members, which are divided into two major groups encoding 2.5-kb and 1.7-kb mRNAs and are clustered in several regions within the soybean genome and assembled in seven different combinations (Koshiyama 1983; Harada et al. 1989).

Glycinin, a hexameric 11S globulin (360 kDa), consists of acidic (A) and basic (B) polypeptides (Nielson 1985; Wright 1987; Nielson et al. 1989; Renkema et al. 2001). Glycinin is encoded by five non-allelic genes, *Gy1*, *Gy2*, *Gy3*, *Gy4* and *Gy5*, which code for five major precursor protein molecules, G1, G2, G3, G4 and G5, respectively (Nielsen et al. 1989; Renkema et al. 2001). Based on the physical properties, these five subunits are classified into two distinct major groups; group I consists of G1 (A1aBx), G2 (A2B1a), and G3 (A1aB1b) proteins, and group II contains G4 (A5A4B3) and G5 (A3B4) subunits. The group I subunits contain more methionine residues than group II, which is an important feature for plant breeders to increase the methionine content in soybean seeds to improve their nutritional quality (Cho et al. 1989; Utsumi et al. 1997; Clarke and Wiseman 2000). There is 82–86% amino acid sequence homology within the subunit groups and

42–45% between the groups (Fukuda et al. 2005; Prak et al. 2005). Each subunit consists of acidic (A) and basic (B) polypeptide components (Staswick et al. 1984; Dickinson et al. 1989; Nielsen et al. 1989). The subunit pairings of glycinin molecules are heterogeneous complexes of A1aBx, A1aB1b, A2B1a, A3B4 and A5A4B3, all of which are covalently linked by disulfide bonds except for the A4 subunit present in the G4 polypeptide (Staswick et al. 1981; Mooney and Thelen 2004). Beilinson et al. (2002) identified two additional genes in the soybean variety, Resnik. Glycinin is responsible for the gel matrix structure that imparts hardness and fracture resistance to tofu (Utsumi 1992; Utsumi et al. 1997). The gel formed by glycinin is turbid, in contrast to the transparent gel formed by β-conglycinin. The subunit compositions of glycinin are considered to play different roles in tofu gel formation (Saio et al. 1969). The presence of more G4 subunits is closely related to gel formation rate and transparency, whereas the presence of more G5 is related to gel hardness (Fukushima 1991). In addition, glycinin lowers cholesterol levels in human serum (Kito et al. 1993).

Other major soybean proteins are the 2S proteins, which are mainly trypsin inhibitors. Two major groups of serine protease inhibitors have been studied extensively in plants such as KTIs and BBIs (Ryan 1990). Soybean lines that lack either KTIs (Orf and Hymowitz 1979) or BBIs (Stahlhut and Hymowitz 1983) have been reported; thus, these proteins apparently are not essential for normal growth and development.

KTIs are the major trypsin inhibitors found in soybean. They are also antinutritional factors and are involved in respiratory hypersensitivity reactions (Besler et al. 2000). Ritt et al. (2004) reported significant differences in the quantity of KTIs between different genotypes of soybeans. Soybean KTIs are encoded by three genes, *KTI1*, *KTI2* and *KTI3*. *KTI1* and *KTI2* genes have nearly identical nucleotide sequences and the *KTI3* gene has diverged (20%) from the other two genes. Jofuku and Goldberg (1989) reported that the KTI3 transcript was detected only in the soybean seed, while KTI1 and KTI2 transcripts are expressed in leaf, root and stem. KTI's, with molecular weights of about 21.5 kDa, are larger than BBIs, the other major group of soybean serine protease inhibitors.

BBIs are cys-rich protease inhibitors with molecular masses of about 8–16 kDa. These protease inhibitors are double-headed, with two reactive sites in a single inhibitor molecule. BBIs are encoded by a family of related genes. BBIs identified in Fabaceae, such as soybean (*G. max*) and lima beans (*Phaseolus lunatus*) are 8 kDa proteins (Qu et al. 2003). Each BBI is comprised of two homologous subdomains consisting of a consensus motif of three β-strands held together by a conserved array of seven disulfide bridges that play a prominent role in the stabilization of the reactive site configuration (Ikenaka and Norioka 1986). The independent reactive sites on the outermost exposed loops of each subdomain adopt a conformation similar to that of a productively bound substrate (Laskowski and Kato 1980). They interact with two (not necessarily identical) molecules of proteinases (Harry and Steiner 1970) without conformational change (Ikenaka and Norioka 1986; Sierra et al. 1999). Interestingly, this type of inhibitor possesses anticarcinogenic activity (Laskowski and Kato 1980; Birk 1993; Kennedy 1993).

Three soybean proteins, Gly m Bd 60 K, Gly m Bd 30 K and Gly m Bd 28 K are the major seed allergens in soybean-sensitive patients (Ogawa et al. 2000). The molecular identities of the genes encoding these protein allergens have been revealed in the recent years. Gly m Bd 60 K allergen proteins consist of α-subunit of β-conglycinin and acidic polypeptides of glycinin G1 and acidic and basic polypeptides of glycinin G2 (Helm et al. 2000). These proteins initiate allergenic reactions in sera of about 25% of soybean sensitive patients (Ogawa et al. 1995). The Gly m Bd 30 K protein was first identified by Kalinski et al. (1992) from fractionated soybean oil body membrane and characterized as a 34-kDa protein. It was later renamed Gly m 1 by Hessing et al. (1996). Kalinski et al. (1990) reported that P34 was encoded by more than one gene. Gly m Bd 28 K is an MP27-MP33 homolog, a minor soybean seed globulin (Tsuji et al. 1995). It is designated as a major allergen and was originally identified as a 28 kDa polypeptide in soybean seed flour (Xiang et al. 2004). The Gly m Bd 28 K allergen is processed into two smaller polypeptides in soybean seeds, an N-terminal polypeptide and a C-terminal polypeptide (Tsuji et al. 2001). Xiang et al. (2004) reported that C-terminal fragments of Gly m Bd 28 K showed slightly stronger binding to IgE as compared to the N-terminal fragments of the protein.

## 13.2 Introduction to Proteomics

Proteomic technologies are powerful tools for examining alterations in protein profiles caused by mutations, the introduction or silencing of genes or responses to various stress stimuli (Görg et al. 2000; Dubey and Grover 2001). Proteomics combines advanced separation techniques such as 2D gel electrophoresis (2D-PAGE), identification techniques such as mass spectrometry (MS) and bioinformatics tools to characterize proteins in complex biological mixtures. Separation of proteins by 2D-PAGE has been practiced and refined over several decades (O'Farrell 1975). In 2D-PAGE, the proteins are separated first by their electrical charge characteristics in one direction and then by their size in the orthogonal direction in a polyacrylamide gel. 2D-PAGE technology utilizing commercially available immobilized pH gradient (IPG) strips has overcome the former limitations of mobile carrier ampholytes with respect to reproducibility, handling, resolution and separation of very acidic and/or basic proteins (Görg et al. 2004). Based on the usage of gel size and pH gradient, 2D-PAGE can resolve more than 5,000 proteins and detect <1 ng of protein per spot. Using 2D analysis, it is possible to identify and characterize proteins that have undergone many forms of post-translational modification including phosphorylation, glycosylation, and limited proteolysis (Görg et al. 2004). There have been a number of further advances in 2D methodologies including improved sample preparation and application to the IPG strips thus allowing more proteins to be arrayed in sub-microgram quantities (Herbert 1999; Görg et al. 2000). The major reasons for inadequate resolution are streaky 2D patterns and background smear which are primarily attributable to poor initial protein solubilization, insufficient solubility of the proteins during first dimension isoelectric focusing or the presence of non-protein contaminants. Therefore, special attention has to be paid to cell lysis conditions, inactivation of protease activities, choice of adequate detergents, chaotropes, and/or amount of reducing agents. Recently reported proteomic studies in plants include studies of chloroplast (Wijk 2000) and mitochondria (Kruft et al. 2001; Millar et al. 2001), *Arabidopsis* seeds (Gallardo et al. 2001), maize root (Chang et al. 2000) and leaves (Porubleva et al. 2001), barrel medic roots (Mathesius et al. 2001, 2002) and medic (Watson et al. 2003), soybean seeds (Mei-Guey et al. 1983; Herman et al. 2003), soybean leaf (Xu et al. 2006), rice (Komatsu et al. 2003) nutrient deficiency (Alves et al. 2006), temperature effect (Sule et al. 2004; Yan et al. 2006), oxidative stress (Wang et al. 2004), herbicide exposure (Castro et al. 2005), salt (Yan et al. 2005), and heavy metal (Labra et al. 2006).

The methods used for protein identification have changed dramatically in the past several years. The traditional technique of Edman degradation followed by identification of progressive N-terminal amino acids using HPLC has been largely replaced by MS (Hunt et al. 1986; Karas and Hillenkamp 1988). Recent advances in ionization technology and the establishment of protein databases have substantially increased the accuracy of protein profile characterization from complex protein mixtures and offer high-throughput analysis. Proteins can be identified by comparison of peptide masses generated following limited proteolysis (peptide mass fingerprinting) with known proteins, and sequence information can be obtained using collision induced dissociation. These techniques require pre-existing databases with annotated protein sequences from the specie of choice or closely related species to obtain statistically accurate protein identifications (see below for discussion of database search engines).

## 13.3 Preparation of Protein for 2D-Analysis

Extraction of soybean seed proteins for 2D-PAGE is challenging. Protein solubilization is sample-dependent and is achieved by optimizing the combination of chaotropic agents, detergents, reducing agents, buffers, enzymes and ampholytes. Generally, extraction of protein begins with cell lysis or disruption of subcellular components and release of biological molecules into an aqueous buffer. Frequently, tissues are first powdered in liquid nitrogen and then solubilized with a strong denaturing agent in order to minimize proteolysis and other modes of protein degradation/ modification. Use of protease inhibitors and removal of interfering substances may be required to prevent protein degradation and improve sample quality (Simpson 2003). Additional purification steps and isolation of subcellular components will allow analysis of less abundant proteins but may result in the unintentional loss of some proteins, which must be considered and may require further concentration of protein prior to 2D analysis (Usuda and Shimogawara 1995; Tsugita et al. 1996; Taylor et al. 2000). Methods of extracting proteins for 2D analysis of plant and seed proteins have been reported for a variety of crops (Görg et al. 1988; Flengsrud and Kobro 1989; Görg et al. 1992; Saravanan and Rose 2004; Carpentier et al. 2005; Lehesranta et al. 2005; Ruebelt et al. 2006; Zarkadas et al. 2007) but, only a limited number of methods have been reported for soybean seed protein analysis (Schuler et al. 1982b; Herman et al. 2003; Natarajan et al. 2005; Xu et al. 2007). In this section, we summarize various approaches used to extract soybean seed proteins from seeds, which have been frozen and powdered in liquid nitrogen. Solubilization reagent combinations include urea and CHAPS, thiourea and urea, phenol extraction followed by urea, thiourea and CHAPS; trichloroacetic acid (TCA)/acetone precipitation/urea solubilization extraction (Hurkman and Tanaka 1986; Berkelman et al. 1998; Cascardo et al. 2001; Herman et al. 2003). Natarajan et al. (2005) compared different solubilization factors, including incubation time, and resolubilization buffers. They found that the optimal method utilized trichloroacetic acid (TCA)/acetone precipitation/followed by solubilization in urea. This method is also efficient in solubilizing soybean leaf proteins (Xu et al. 2006).

## 13.4 Protein Determination

To obtain uniform and readily comparable 2D gels, quantification of the amount of protein loaded is an important preliminary step. Usually protein concentrations are determined by the Bradford method (1976) using a commercial dye reagent (Bio-Rad). Alternatively, we have used a modified Lowry method (1951), which incorporates SDS in the reagent buffer to minimize solubility problems, and an initial precipitation in 10% (w/v) TCA helps to eliminate many potential interfering substances in the extraction buffers (Nerurkar et al. 1981). Variable amounts of protein can be separated by 2D-PAGE depending upon the pH range and size of the IPG strips. Typically, we use 100 μg of soybean seed protein for 2D-PAGE using 13 cm strips (Natarajan et al. 2005).

## 13.5 2D-PAGE

Advances in IPG technology and the development of dedicated high voltage electrophoresis instruments by several manufacturers have improved the reproducibility of the first dimension separation of proteins. In addition, advancements in staining techniques (Colloidal Coomassie Blue, silver, SYPRO and Cy fluorescent dyes), and in analytical software have also enhanced our ability to compare and quantify proteins in 2D-PAGE applications. Commercial IPG strips are readily available in linear and nonlinear gradients, multiple narrow pH ranges and in sizes from 7 to 24 cm (Görg et al. 2004), facilitating separations for different purposes including protein isolation, identification and western blot analysis (Magni et al. 2005). Protein samples are generally added to the dried IPG strips so that rehydration and protein loading are carried out simultaneously at low voltage. For separation at the extremes of the pH ranges some samples are problematic and alternative strategies have been developed (Görg et al. 2004). These include altering reducing conditions and sample application techniques such as applying the protein sample (to pre-equilibrated/hydrated IPG strips) at either end of the strip via a sample cup or paper bridge (Hoving et al. 2002). Although these techniques were found to be essential for first dimension separation of mammalian liver samples in the pH 3–6 and 7–10 range (Caperna and Garrett, unpub. observation), this does not appear to be the case for soybean seed and leaf preparations. We successfully used narrow width gradients in high and low pH ranges and standard loading and running conditions for the separation of soybean seed and leaf proteins (Natarajan et al. 2005; Xu et al. 2006) and identified the separated spots using MALDI-TOF-MS and LC/MS/MS. Several proteins have been identified and the isoelectric point, apparent molecular weight and relative abundance of many of the proteins determined. In addition, we have investigated protein heterogeneity in numerous wild and cultivated soybean genotypes using 2D-PAGE methodologies. We used isoelectric focusing gels in pH 3–10, 4–7 and 6–11 (Fig. 13-1) ranges to separate most of the soybean storage acidic and basic proteins. All the proteins identified by MALDI-TOF-MS, LC-MS and NCBI non-redundant database searches are listed in Table 13-1. The table includes an assigned protein spot number, theoretical isoelectric point (pI) and molecular weight (Mr), protein identity, number of peptides matched, percent sequence coverage, MOWSE score, expect value, and NCBI database accession number of the best match and database that yielded concurrent identification (Natarajan et al. 2006a). In addition, we characterized and identified several soybean allergen and antinutritional proteins (Natarajan et al. 2006b, 2007; Xu et al. 2007). We used pH 3–10 range to separate and identify 71 unique soybean leaf proteins (Fig. 13-2). Using similar approaches, Hajduch and coworkers (Hajduch et al. 2005)

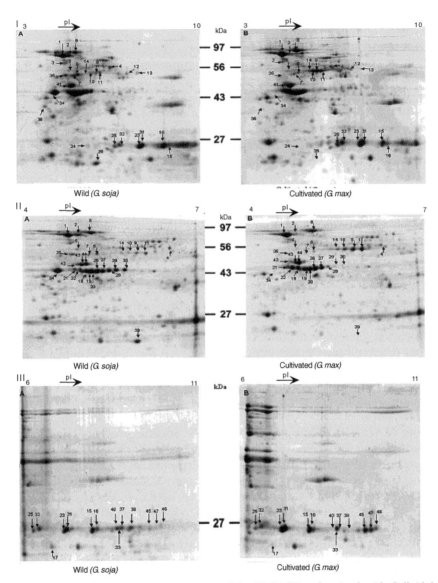

**Figure 13-1** Soybean seed proteins separated by 2D-PAGE and stained with Colloidal Coomassie Blue G-250. Proteins were separated in the first dimension on an IPG strip (I) pH 3.0–10.0, (II) pH 4.0–7.0, (III) pH 6.0–11.0, and in the second dimension on a 12% acrylamide SDS-gel. For more details see Natarajan et al. 2006a.

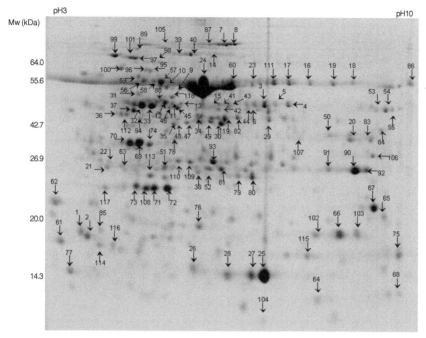

**Figure 13-2** Soybean leaf proteins separated by 2D-PAGE and stained with. Colloidal Coomassie Blue G-250 Proteins were separated in the first dimension on an IPG strip pH 3.0–10.0 and in the second dimension on a 12.5% acrylamide SDS-gel. For more details see Xu et al. 2006.

identified 216 unique non-redundant proteins in the developing soybean seeds. Use of narrow range pH gradients enhances the separation of proteins with similar electrophoretic mobilities; however, it is also clear from our work that fractionation of proteins prior to 2D analysis will be required to probe deeper into the soybean proteome (Mooney and Thelen 2004).

In our laboratory, first dimension separation of soybean proteins is carried out using freshly-prepared rehydration buffer applied to a Bio-Rad Protean IEF system (Hercules, CA, USA) according to manufacturer's recommendations. The voltage settings for isoelectric focusing were 500 V for 1 hour, 1,000 V for 1 hour, and 8,000 V to a total 14.5 kVh. The focused strips were either immediately run on a second dimension SDS gel or stored at –80°C. Second dimensional SDS-PAGE separation can be performed on horizontal or vertical gel systems (Görg et al. 1995); we use a vertical system. Prior to second dimensional electrophoresis, the gel strips are routinely incubated with equilibration buffer 1 (50mM Tris-HCl pH 8.8, 6M urea, 30% glycerol, 2% SDS, 0.002% bromophenol blue, 1% DTT) followed by equilibration buffer 2 (same buffer without DTT, but containing 2.5% iodoacetamide to alkylate the free thiol group of cysteines). The strips are

**Table 13-1** Storage proteins identified by MALDI-TOF-MS and LC-MS/MS analysis in wild and cultivated soybean genotypes.

| Spot ID | Theoretical pI/Mr | Protein Identity | Peptides Matched | Sequence Coverage | Mowes Score | Expect Value | NCBI Accession No. | Wild (G. Soja) | Cultivated (G. Max) | Method/MS |
|---|---|---|---|---|---|---|---|---|---|---|
| 1 | 4.92/63127 | α subunit of β conglycinin | 25 | 39% | 217 | 3.00E-17 | gi\|9967357 | + | + | MALDI-TOF |
| 2 | 4.92/63127 | α subunit of β conglycinin | 25 | 41% | 167 | 4.00E-12 | gi\|9967357 | + | − | MALDI-TOF |
| 3 | 4.92/63184 | α subunit of β conglycinin | 27 | 43% | 250 | 1.50E-20 | gi\|9967357 | + | + | MALDI-TOF |
| 4 | 4.92/63184 | α subunit of β conglycinin | 20 | 34% | 204 | 6.00E-16 | gi\|9967357 | + | + | MALDI-TOF |
| 5 | 5.32/72717 | α subunit of β conglycinin | 9 | 18% | 73 | 7.20E-03 | gi\|15425633 | + | − | MALDI-TOF |
| 6 | 5.32/72717 | α subunit of β conglycinin | 14 | 24% | 112 | 1.10E-03 | gi\|15425633 | + | + | MALDI-TOF |
| 7 | 5.32/72717 | α subunit of β conglycinin | 9 | 23% | 238 |  | gi\|51247837 | + | + | LC-MS/MS |
| 8 | 5.23/65160 | α' subunit of β conglycinin | 20 | 41% | 194 | 5.90E-15 | gi\|9967361 | + | + | MALDI-TOF |
| 9 | 5.67/48358 | β conglycinin β subunit | 26 | 47% | 216 | 5.10E-17 | gi\|63852207 | − | + | MALDI-TOF |
| 10 | 5.67/48358 | β conglycinin β subunit | 18 | 40% | 176 | 5.10E-13 | gi\|63852207 | + | + | MALDI-TOF |
| 11 | 5.67/48358 | β conglycinin β subunit | 27 | 49% | 256 | 5.10E-21 | gi\|63852207 | + | + | MALDI-TOF |
| 12 | 5.67/48358 | β conglycinin β subunit | 22 | 46% | 127 | 4.00E-08 | gi\|63852207 | + | + | MALDI-TOF |
| 13 | 5.67/48358 | β conglycinin β subunit | 11 | 28% | 85 | 6.40E-04 | gi\|63852207 | − | − | MALDI-TOF |
| 14 | 5.67/48358 | β conglycinin β subunit | 23 | 45% | 218 | 3.20E-17 | gi\|63852207 | − | + | MALDI-TOF |
| 15 | 5.89/56299 | Glycinin G1/A1aBx subunit | 7 | 15% | 145 |  | gi\|18635 | + | + | LC-MS/MS |
| 16 | 5.89/55672 | Glycinin G1/A1aBx subunit | 5 | 10% | 157 |  | gi\|18635 | + | + | LC-MS/MS |
| 17 | 6.15/56134 | Glycinin G1/A1aBx subunit | 4 | 9% | 254 |  | gi\|72296 | + | + | LC-MS/MS |
| 18 | 5.46/54927 | Glycinin G2/A2B1 precursor | 9 | 19% | 91 | 1.20E-04 | gi\|1212177 | + | + | MALDI-TOF |
| 19 | 5.46/54927 | Glycinin G2/A2B1 precursor | 9 | 19% | 73 | 7.20E-03 | gi\|1212177 | + | + | MALDI-TOF |
| 20 | 5.46/54927 | Glycinin G2/A2B1 precursor | 8 | 15% | 71 | 1.30E-03 | gi\|1212177 | + | + | MALDI-TOF |
| 21 | 5.46/54927 | Glycinin G2/A2B1 precursor | 9 | 14% | 74 | 6.70E-03 | gi\|1212177 | + | + | MALDI-TOF |
| 22 | 5.46/54927 | Glycinin G2/A2B1 precursor | 12 | 21% | 104 | 6.10E-06 | gi\|1212177 | + | + | MALDI-TOF |
| 23 | 5.78/54047 | Glycinin G2/A2B1 precursor | 6 | 37% | 67 | 3.20E-02 | gi\|169967 | + | + | MALDI-TOF |
| 24 | 5.56/54903 | Glycinin G2/A2B1 precursor | 3 | 9% | 175 |  | gi\|72295 | + | − | LC-MS/MS |
| 25 | 5.56/54903 | Glycinin G2/A2B1 precursor | 9 | 14% | 313 |  | gi\|72295 | + | + | LC-MS/MS |
| 26 | 5.78/54047 | Glycinin subunit G3/A1ab1B | 8 | 19% | 72 | 1.00E-02 | gi\|15988117 | + | + | MALDI-TOF |
| 27 | 5.78/54047 | Glycinin subunit G3/A1ab1B | 8 | 18% | 70 | 1.40E-02 | gi\|15988117 | + | + | MALDI-TOF |
| 28 | 5.78/54047 | Glycinin subunit G3/A1ab1B | 9 | 20% | 74 | 6.40E-03 | gi\|15988117 | + | + | MALDI-TOF |
| 29 | 5.78/54047 | Glycinin subunit G3/A1ab1B | 10 | 20% | 75 | 6.00E-03 | gi\|15988117 | + | − | MALDI-TOF |

| No. | pI/MW | Protein | | % | Score | Expect | Accession | | | Method |
|---|---|---|---|---|---|---|---|---|---|---|
| 30 | 5.78/54047 | Glycinin subunit G3/A1ab1B | 9 | 18% | 96 | 4.20E-05 | gi\|15988117 | + | − | MALDI-TOF |
| 31 | 5.78/54047 | Glycinin subunit G3/A1ab1B | 8 | 18% | 72 | 9.40E-03 | gi\|15988117 | + | + | MALDI-TOF |
| 32 | 5.78/54047 | Glycinin subunit G3/A1ab1B | 9 | 18% | 116 | 3.90E-07 | gi\|15988117 | + | + | MALDI-TOF |
| 33 | 5.73/54835 | Glycinin subunit G3/A1ab1B | 2 | 4% | 148 | | gi\|18639 | − | + | LC-MS/MS |
| 34 | 5.38/64097 | Glycinin G4/A5A4B3 | 13 | 16% | 76 | 4.20E-03 | gi\|99910 | + | − | MALDI-TOF |
| 35 | 5.38/64097 | Glycinin G4/A5A4B3 precursor | 13 | 26% | 96 | 3.50E-05 | gi\|99910 | + | − | MALDI-TOF |
| 36 | 4.46/24349 | Glycinin G4/A5A4B3 precursor | 10 | 38% | 110 | 1.50E-06 | gi\|6015515 | + | − | MALDI-TOF |
| 37 | 6.47/31065 | Glycinin G4/A5A4B3 precursor | 10 | 45% | 142 | 9.50E-10 | gi\|81785 | + | − | MALDI-TOF |
| 38 | 6.47/31065 | Glycinin G4/A5A4B3 precursor | 8 | 40% | 105 | 5.40E-06 | gi\|81785 | + | − | MALDI-TOF |
| 39 | 5.38/64136 | Glycinin G4/A5A4B3 precursor | 6 | 8% | 160 | | gi\|99910 | + | − | LC-MS/MS |
| 40 | 5.38/64136 | Glycinin G4/A5A4B3 precursor | 14 | 14% | 408 | | gi\|99910 | + | − | LC-MS/MS |
| 41 | 5.46/55850 | Glycinin G5/A3B4 subunit | 13 | 26% | 96 | 3.50E-05 | gi\|33357661 | + | + | MALDI-TOF |
| 42 | 5.46/55850 | Glycinin G5/A3B4 subunit | 11 | 18% | 76 | 4.30E-03 | gi\|33357661 | + | + | MALDI-TOF |
| 43 | 5.46/55850 | Glycinin G5/A3B4 subunit | 12 | 22% | 110 | 1.50E-06 | gi\|33357661 | + | + | MALDI-TOF |
| 44 | 5.46/55850 | Glycinin G5/A3B4 subunit | 10 | 21% | 86 | 3.70E-04 | gi\|33357661 | + | + | MALDI-TOF |
| 45 | 5.69/26938 | Glycinin G5/A3B4 subunit | 7 | 30% | 65 | 5.00E-02 | gi\|541941 | + | + | MALDI-TOF |
| 46 | 5.69/26938 | Glycinin G5/A3B4 subunit | 8 | 37% | 69 | 2.50E-02 | gi\|541941 | + | + | MALDI-TOF |
| 47 | 9.64/21482 | Glycinin G5/A3B4 subunit | 3 | 12% | 109 | | gi\|625538 | + | + | LC-MS/MS |

placed onto a 12% polyacrylamide gel (18 × 16 cm) in a tris-glycine buffer system as described by Laemmli (1970) in a Hoefer SE 600 Ruby electrophoresis unit (GE Healthcare, Piscataway, NJ, USA) according to the manufacturer's recommendations. After second dimension electrophoresis, the separated proteins are visualized using various fixations and staining methods.

Many methods have been developed to detect separated proteins in acrylamide gels, and the selection of method is dependent on the experimental purpose. Widely-used methods include staining with anionic dyes such as Coomassie Blue R250/G250 (Neuhoff et al. 1988), fluorescent dyes (Yan et al. 2000a), silver stain (Yan et al. 2000b) and radioactive isotopes, coupled with laser scanning densitometry, fluorescence imaging or autoradiography/fluorography/phosphor-imaging to visualize separated proteins (Görg et al. 2004). In our laboratory, 2D-PAGE gels are visualized by staining with Colloidal Coomassie Blue G-250 as described by Newsholme et al. (2000). This stain has found widespread use for the detection of proteins, due to its low cost, ease of use and compatibility with mass spectrometry (Görg et al. 2004). Stained gels are stored in 20% ammonium sulfate solution and scanned using laser densitometry (PDSI, GE Healthcare). Typically, triplicate samples are used for soybean seed protein extraction and 2D-PAGE.

## 13.6 2D Image Analysis

To evaluate the 2D gel patterns, computerized image analysis system is typically used (Dowsey et al. 2003). The first step in computerized image analysis of 2D protein patterns is capture of the gel images in a digital format. Several devices including modified document scanners, laser densitometers, CCD cameras, and fluorescent and phosphor images are currently available (Gorg et al. 2004). The saved images are then subjected to computer-assisted image analysis. Currently there are several commercially available 2D analysis software including Z3 (Compugen); Melanie III (Genebio and Bio-Rad); PDQuest 2D gel analysis software (Lehesranta et al. 2005; Ruebelt et al. 2006); Phoretix 2D software marketed as Imagemaster 2D (Hajduch et al. 2005), HT analyzer (Genomic solutions), and Phoretix 2D Advance (Mooney and Thelen 2004; Zarkadas et al. 2007). We use Progenesis Samespots (Nonlinear Dynamics Group, UK, *http://www.samespots.com*). The 2D software package analyzes the data using the following steps: 1) Automatic quality analysis of the gel which includes the image format, compression level, saturation, grayscale intensity and bit depth. 2) Image alignment via automatic placement of alignment vectors and/or assistance in placement of manual vectors. 3) Automatic analysis of images which includes spot detection and matching across all gels in the experiment, background subtraction and normalization. Gels can be easily arranged into experimental groups and spots of interest can be ranked by their statistical significance or fold change. Spots of interest can be subjected to further more advanced statistical analysis including an estimate of the false discovery rate (q-value), and an estimate of the spot power can be visualized allowing the researcher to determine if more replicate gels need to be included in the experiment for greater statistical power. The software also provides a graphical view of spot expression profiles, and the user can subject their spots of interest to principal components analysis to determine how well the spots cluster. The principal components analysis is also useful for the determination of data outliers. The Progenesis PG240 module includes an automatic web page builder for ease of visualization of the spot differences, and the tabular data and images can be cut and pasted to the clipboard or exported directly to Microsoft Excel.

## 13.7 In-Gel Digestion of Protein Spots

To identify proteins in 2D gels, spots are excised from the stained gel and washed first with distilled water to remove ammonium sulfate and then with 50% acetonitrile containing 25 mM ammonium bicarbonate to de-stain the gel plug. The gel plug is then dehydrated with 100% acetonitrile, dried under vacuum, and then re-swollen in 20 μl of 10 μg/ml trypsin (modified porcine trypsin, sequencing grade, Promega, Madison, WI) in 25 mM ammonium bicarbonate. Additional ammonium bicarbonate (20 μl) is added to maintain hydration of the gel plug. Digestion is performed overnight at 37°C. The resulting tryptic fragments are extracted in 50% acetonitrile and 5% trifluoroacetic acid with sonication. The extract is dried and dissolved in 50% acetonitrile containing 0.1% trifluoroacetic acid.

## 13.8 Protein Analysis by Mass spectrometry

The ability to analyze and identify proteins by mass spectrometry (MS) required the development of soft ionization techniques to enable these large biopolymers to be vaporized into the gas phase. Currently, the most widely used ionization methods are matrix-assisted laser desorption ionization (MALDI) (Karas and Hillenkamp 1988.) and electrospray ionization (ESI) (Fenn et al. 1989). MALDI ionization involves analysis of a static sample that has been deposited onto a sample plate, while ESI involves a constant flowing liquid, and is thus compatible with simultaneous chromatographic separation techniques. This coupling forms the basis of liquid chromatography-mass spectrometry (LC-MS). To identify proteins, knowledge of their primary sequence is also required, which is obtained using database search engines capable of matching mass spectrometry data to primary sequences in a database. Identification of intact proteins is for all intents and purposes not possible under standard MS analysis, so proteins of interest first have to be digested into smaller fragment peptides; this is usually accomplished using a proteolytic enzyme such as trypsin.

While a MALDI source can be adapted to numerous types of mass spectrometers, this discussion will be limited to time of flight (TOF) instruments. In a TOF MS, ions are separated by their differing velocities as they travel down a flight tube. Drift time down the flight tube is defined by the following equation: $t = s \, (m / \, (2KE) \, z)^{1/2}$ where t = drift time, s = drift distance, m = mass, KE = Kinetic Energy and z = number of charges on the ion (Voyager Biospectrometry Workstation Manual). When a group of ions is subjected to the same electrical field that travels down a flight tube of fixed length, their velocity is proportional to the square root of their mass to charge ratio. In practical terms, lighter ions will travel faster towards a detector than heavier ions. Early TOF instruments suffered from low resolving power, but with the advent of Delayed Extraction (Vestal et al. 1995) and ion mirrors or reflectrons, the current state of the art instruments are capable of much higher resolution. Coupled with the use of external or preferably internal calibration, high mass accuracy and high resolution can be achieved with a MALDI-TOF-MS. These developments have enabled the identification of proteins by a process called Peptide Mass Fingerprinting (PMF). In a typical PMF experiment, acrylamide gel separated proteins are digested with a protease, usually trypsin, and the resulting peptides are co-crystallized onto a sample plate with a UV light absorbing organic acid matrix, typically α-Cyano-4-hydroxycinnamic acid (CHCA). Once the sample plate is introduced into the mass spectrometer, the peptides are ionized and desorbed into the vapor phase via laser energy and accelerated down the flight tube to the detector. Since the lighter objects travel faster than the heavier objects and the relationship between mass and velocity has been determined mathematically, the resulting mass of the analyzed peptides can be calculated very accurately from their flight times. A calibrated list of peptide masses is submitted to a search engine, which uses PMF data to identify proteins from primary sequence databases. Successful identification of unknown proteins by PMF relies upon the accurate assignment of mass to the analyzed peptides, the number of peptides in the spectrum, and as an absolute necessity, the presence of the protein sequence, or a highly homologous sequence from another species in the database.

If a protein cannot be identified by PMF alone, it is possible that a positive identification can be made by fragmenting the individual peptides and analyzing the resulting fragment masses with a database search engine. One of the simplest and relatively least expensive mass spectrometers available for this type of analysis is the quadrupole ion trap mass spectrometer (QIT-MS). As is implied in its name, a QIT-MS has the ability to isolate and trap ionized peptides, and once isolated, to fragment the peptide backbone. Since peptides are soluble in mixed aqueous and organic solvents, they are introduced into the MS via an ESI source, usually coupled to a high-pressure liquid chromatography (HPLC) system. Peptides are separated prior to introduction via any number of chromatographic techniques, reverse phase being one of the most popular. Reverse phase separation takes advantage of the differing hydrophobicities of individual peptides, which are eluted from the column stationary phase with an increasing gradient of an organic solvent such as acetonitrile. A typical duty cycle for a QIT-MS is to perform a survey scan, where a determination of the peptide masses is made (parent ion scan, or survey MS scan), then the three most abundant peptides are individually isolated and fragmented, with subsequent determination of the fragment masses (MS/MS scan). These spectra of fragment masses with their accompanying parent ion masses are submitted to a search engine, which uses the MS/MS data to identify proteins from primary sequence databases. Successful identification of unknown proteins by MS/MS analysis of peptides relies upon matching the observed MS/MS spectra to *in silico* generated theoretical MS/MS spectra of tryptic peptides from the primary sequence databases. As in peptide mass fingerprinting, there is an absolute requirement for the presence of the protein sequence, or a highly homologous sequence, in the database. Unlike peptide mass fingerprinting, MS/MS analysis can theoretically identify proteins from the spectrum of one tryptic peptide, although two or more matched peptides are highly desirable. It should be noted also that neither PMF nor MS/MS analysis of peptides gives protein sequence information. Our current methodology for identifying unknown proteins begins with their initial separation by two dimensional gel electrophoresis, image analysis of the stained gel, spot picking, in-gel tryptic digestion followed by MALDI peptide mass fingerprinting. If the protein spot of interest is identified by MALDI, then the analysis is complete. If however, an interpretable MALDI spectrum is generated that is not identified by PMF, then the remainder of the sample is submitted for MS/MS analysis. If the sample again is not identified, then it is assumed that the protein in question is not present in the database, or possibly that it has undergone extensive post-translational modification.

Once an investigator has acquired PMF or MS/MS data for a protein of interest, positive identification can be achieved by submitting the peak lists to any number of search engines. Protein Prospector (Clauser et al. 1999; *http://prospector.ucsf.edu/*), Profound (*http://prowl.rockefeller.edu/prowl-cgi/profound.exe*) and Aldente (*http://us.expasy.org/tools/aldente/*) are public domain search engines that can identify proteins from PMF data. Sequest (*http://fields.scripps.edu/sequest/*) and Phenyx (*http://phenyx.vital-it.ch/pwi/login/login.jsp*) are search engines, which are designed to identify proteins from tandem MS/MS data. In addition, Mascot by Matrix Science (Perkins et al. 1999; *http://www.matrix-science.com/home.html*) has the capability to identify proteins from FASTA formatted databases for both PMF and MS/MS data. Mascot uses the MOWSE scoring algorithm (Pappin et al. 1993) for protein identification, and includes probability based scoring so identifications are statistically robust. Aside from FASTA formatted protein databases, Mascot is also capable of protein identifications from nucleic acid databases, by performing 6-frame translations of the nucleic acid sequences. The public domain server for Mascot has three popular protein databases, (SwissProt, MSDB, and NCBInr) for PMF and MS/MS ions searches plus three EST databases that are available for MS/MS ion searches. Investigators can license Mascot for use on their own server, thus allowing the use of any FASTA formatted database not made available on the public domain site. We have used both PMF and MS/MS to identify proteins from soybean seeds. Since soybean seed proteins are well characterized and are quite well represented in the public databases, it has been rather simple to perform these identifications from SwissProt and NCBInr (Natarajan et al. 2005, 2006a, b, 2007). On the other hand, the studies we have performed on soybean leaf proteins presented more of a challenge due to the lack of sequence coverage specifically for the soybean leaf proteins (Xu et al. 2006). For these identifications, we had to rely more heavily on MS/MS data, as well as matches to homologous proteins from other plant species, and in the extreme, search MS/MS data against EST databases.

## 13.9 Summary

This chapter describes the current tools utilized for separation and identification of proteins isolated from soybeans. This information is increasingly valuable as new value-added soybeans are continuously introduced in the global market. The development of proteomics tools has enabled the creation of an accurate database of relative quantities of identifiable proteins from various organs, developmental stages, environmental and varietals differences. This database, which is continuously being updated, enables researchers and producers to evaluate the variation in proteins in new soybean hybrids developed through conventional breeding or biotech approaches. The major issues and challenges for preparing samples for protein isolation and characterizations hinge on the wide variety of different kinds of proteins, the wide variety of tissue samples that are used for analysis, and the expense of performing the analysis. The future holds promise for advancements in nuclear magnetic resonance spectroscopy and X-ray crystallography tools, which will provide unique additional three-dimensional structural features about different proteins that are often not feasible with mass spectrometry.

# References

Alves M, Francisco R, Martins I, Ricardo CPP (2006) Analysis of *lupinus albus* leaf apoplastic proteins in response to boron deficiency. Plant Soil 279: 1–11.

Beachy RN, Doyle JJ, Ladin BF, Schuler MA (1983b) Structure and expression of the genes encoding the soybean 7S seed storage proteins. Natl Adv Sci Inst Ser 63: 101–112.

Beilinson V, Chen Z, Shoemaker RC, Fischer RL, Goldberg RB, Nielsen NC (2002) Genomic organization of glycinin genes in soybean. Theor Appl Genet. 104:1132–1140.

Besler M, Helm RM, Ogawa T (2000) Allergen data collection-Update for soybean. In: Internet Symposium on Food Allergens 2: 1–35.

Birk Y (1993) Protease inhibitors of plant origin and role of protease inhibitors in human nutrition. In: W Troll , AR Kennedy (eds) Protease Inhibitors as Cancer Chemo-preventive Agents. Plenum Press, New York, USA, pp 97–106.

Bjellqvist B, Laird N, McDowell M, Olsson I, Westermeier R (1998) In: T Berkelman, T Stenstedt (eds) 2D Electrophoresis using Immobilized pH Gradients—Principles and Methods.Handbooks from Amersham Biosciences, pp 27–55.

Bradford MM (1976) Rapid and sensitive method for the quantitation of microgram quantities of protein utilizing the principle of protein- dye binding. Anal Biochem 72: 248–254.

Carpentier SC, Witters E, Laukens K, Deckers P, Swennen R, Panis B (2005) Preparation of protein extracts from recalcitrant plant tissues: An evaluation of different methods for two-dimensional gel electrophoresis analysis. Proteomics 5: 2497–2507.

Cascardo JCM, Buzeli RAA, Almeida RS, Otoni WC, Fontes EPB (2001) Differential expression of the soybean BiP gene family. Plant Sci 160: 273–281.

Castro AJ, Carapito C, Zorn N, Magné C, Leize E, Dorsselaer AV, Clément C (2005) Proteomic analysis of grapevine (*Vitis vinifera* L.) tissues subjected to herbicide stress. J Exp Bot 56: 2783–2795.

Chang WP, Huang LH, Shen M, Webster C, Burlingame AL, Roberts KM J (2000) Patterns of protein synthesis and tolerance of anoxia in root tips of maize seedlings acclimated to a low-oxygen environment, and identification of proteins by mass spectrometry. Plant Physiol 122: 295–318.

Cho TJ, Davies CS, Nielsen NC (1989) Inheritance and organization of glycinin genes in soybean. Plant Cell 1: 329–337.

Chrispeels MJ, Herman EM (2000) Endoplasmic reticulum derived compartments function in storage and as mediators of vacuolar remodeling via a new type of organelle, precursor protease vesicles. Plant Physiol 123: 1227–1234.

Clarke EJ, Wiseman J (2000) Developments in plant breeding for improved nutritional quality of soybeans. I. Protein and aminoacid content. J Agri Sci 134: 111–124.

Clauser KR, Baker P, Burlingame AL (1999) Role of accurate mass measurement (± 10 ppm) in protein identification strategies employing MS or MS/MS and database searching. Anal Chem 71: 2871–2882.

Dickinson CD, Hussein EHA, Nielsen NC (1989) Role of post-translational cleavage in glycinin assembly. Plant Cell 1: 459–469.

Dowsey AW, Dunn MJ, Yang GZ (2003) The role of bioinformatics in two-dimensional gel electrophoresis. Proteomics 3: 1567–1596.

Dubey H, Grover A (2001) Current initiatives in proteomics research: The plant perspective. Curr Sci 80: 262–269.

Fenn JB, Mann M, Meng CK, Wong SF, Whitehouse CM (1989) Electrospray ionization for mass spectrometry of large biomolecules. Science 246: 64–71.

Flengsrud R, Kobro G (1989) A method for two-dimensional electrophoresis of proteins from green plant tissues. Anal Biochem 177: 33–36.

Friedman M, Brandon DL (2001) Nutritional and health benefits of soy proteins. J Agri Food Chem 49: 1069–1086.

308    *Genetics, Genomics and Breeding of Soybean*

Fukuda T, Maruyams N, Kanazawa A, Abe J, Shimamoto Y, Hiemori M, Tsuji H, Takatoshi T, Utsumi S (2005) Molecular analysis and physicochemical properties of electrophoretic variants of wild soybean *Glycine soja* storage proteins. J Agri Food Chem 53: 3658–3665.

Fukushima D (1991) Structures of plant storage proteins and their functions. Food Rev Int 7: 353–381.

Gallardo K, Job C, Groot SPC, Puype M, Demol H, Vandekerckhove J, Job D (2001) Proteomic analysis of *Arabidopsis thaliana* seed germination and priming. Plant Physiol 126: 835–849.

Görg A, Postel W, Domscheit A, Gunther S (1988) Two-dimensional electrophoresis with immobilized pH gradients of leaf proteins from barley (*Hordeum vulgare*): Method, reproducibility and genetic aspects. Electrophoresis 9: 681–692.

Görg A, Postel W, Weiss W (1992) Detection of polypeptides and amylase isoenzyme modifications related to malting quality during malting process of barley by two-dimensional electrophoresis and isoelectric focusing with immobilized pH gradients. Electrophoresis 13: 759–777.

Görg A, Boguth G, Obermaier C (1995) Two-dimensional polyacrylamide gel electrophoresis with immobilized pH gradients in the first dimension (IPG-Dalt): The state of the art and the controversy of vertical versus horizontal systems. Electrophoresis 16: 1079–1080.

Görg A, Obermaier C, Boguth G, Harder A, Scheibe B, Wildgruber R, Weiss W (2000) The current state of two-dimensional electrophoresis with immobilized pH gradients. Electrophoresis 21: 1037–1053.

Görg A, Weiss W, Dunn MJ (2004) Current two-dimensional electrophoresis technology for proteomics. Proteomics 4: 3665–3685.

Hajduch M, Ganapathy A, Stein JW, Thelen JJ (2005) A systematic proteomic study of seed filling in soybean. Establishment of high-resolution two-dimensional reference maps, expression profiles, and an interactive proteome database. Plant Physiol 137: 1349–1419.

Harada JJ, Barker SJ, Goldberg RB (1989) Soybean β-conglycinin genes are clustered in several DNA regions and are regulated by transcriptional and post transcriptional processes. Plant Cell 1: 415–425.

Harry JB, Steiner RF (1970) A soybean proteinase inhibitor, thermodynamic and kinetic parameters of association with enzymes. Eur J Biochem 16: 174–179.

Helm RM, Cockrell G, Connaughton C, Sampson HA, Bannon G, Beolinson V, Burks AW (2000) A soybean G2 glycinin allergen. Allerg Immunol 123: 213–219.

Herbert B (1999) Advances in protein solubilization for two-dimensional electrophoresis. Electrophoresis 20: 660–663.

Herman EM, Larkins BA (1999) Protein storage bodies. Plant Cell 11: 601–613.

Herman EM, Helm RM, Jung R, Kinney AJ (2003) Genetic modification removes an immunodominant allergen from soybean. Plant Physiol 132: 36–43.

Hessing M, Bleeker H, Tsuji H, Ogawa T, Vlooswijk RAA (1996) Comparison of human IgE- binding soybean allergenic protein Gly m 1 with the antigenicity profiles of calf anti-soya protein sera. Food Agri Immunol 8: 51–58.

Hoving S, Gerrits B, Voshol H, Müller D, Roberts RC, Van Ostrum J (2002) Preparative two dimensional gel electrophoresis at alkaline pH using narrow range immobilized pH gradients. Proteomics 2: 127–34.

Hunt DF, Yates III JR, Shabanowitz J, Winston S, Hauer CR (1986) Protein sequencing by tandem mass spectrometry. Proc Natl Acad Sci USA 83: 6233–6237.

Hurkman WJ, Tanaka CK (1986) Solubilization of plant membrane proteins for analysis by two-dimensional gel electrophoresis. Plant Physiol 81: 802–806.

Ikenaka T, Norioka S (1986) Bowman-Birk family serine proteinase inhibitors. In: AJ Barrett, G Salvenson (eds) Proteinase inhibitors. Elsevier Science, Amsterdam, The Netherlands, pp 361–374.

Jofuku KD, Goldberg RB (1989) Kunitz trypsin inhibitor genes are differentially expressed during the soybean life cycle and in transformed tobacco plants. Plant Cell 1: 1079–1093.

Kalinski A, Weisemann JM, Matthews BF, Herman EM (1990) Molecular cloning of a protein associated with soybean seed oil bodies that are similar to thiol proteases of the papain family. J Biol Chem 265: 13843–13848.

Kalinski A, Melroy DL, Dwivedi RS, Herman EM (1992) Soybean vacuolar protein (P34) related to thiol protease is synthesized as a glycoprotein precursor during seed maturation. J Biol Chem 267: 12068–12076.

Karas M, Hillenkamp F (1988) Laser desorption ionization of proteins with molecular masses exceeding 10,000 daltons. Anal Chem 60: 2299–301.

Kennedy AR (1993) Anticarcinogenic activity of protease inhibitors. In: W Troll, AR Kennedy (eds) Protease Inhibitors as Cancer Chemopreventive Agents. Plenum Press, New York, USA, pp 9–64.

Kinney AJ, Jung R, Herman ER (2001) Co-suppression of the α-subunits of β-conglycinin in transgenic soybean seeds induces the formation of endoplasmic reticulum-derived protein bodies. Plant Cell 13: 1165–1178.

Kito M, Moriyama T, Kimura Y, Kambara H (1993) Changes in plasma lipid levels in young healthy volunteers by adding an extruder-cooked soy protein to conventional meals. Biosci Biotechnol Biochem 57: 354–355.

Komatsu S, Konishi H, Shen S, Yang G (2003) Rice proteomics: A step towards functional analysis of the rice genome. Mol Cell Proteom 2: 2–10.

Koshiyama I (1983) Storage proteins of soybean. In: W Gottschalk , HP Muller (eds) Seed Proteins: Biochemistry, genetics, nutritive value. Martinus Nijhoff/Dr. W. Junk Publishers, Hingham, MA, USA, pp 427–450.

Kruft V, Eubel H, Jansch L, Werhan W, Braun HP (2001) Proteomics approach to identify novel proteins in *Arabidopsis*. Plant Physiol 127: 1694–1710.

Labra M, Gianazza E, Waitt R, Eberini I, Sozzi A, Regondi S, Grassi F, Agradi E (2006) *Zea mays* L. protein changes in response to potassium dichromate treatments. Chemosphere 62: 1234–1244.

Laemmli UK (1970) Cleavage of structural proteins during assembly of the head of bacteriophage T4. Nature 227: 680–685.

Laskowski M Jr, Kato I (1980) Protein inhibitors of proteinases. Annu Rev Biochem 49: 593–626.

Lehesranta SJ, Davies HV, Shepherd LVT, Nunan N, McNicol JW, Auriola S, Koistinen KM, Suomalainen S, Kokko HI, Kärenlampi SO (2005) Comparison of tuber proteomes of potato varieties, landraces, and genetically modified lines. Plant Physiol 138: 1690–1699.

Liener IE (1994) Implication of anti-nutritional components in soybean foods. Crit Rev Food Sci Nutri 34: 31–67.

Lowry OH, Ronserough NJ, Farr AL, Randell RJ (1951) Protein measurement with the folin phenol reagent. J Biol Chem 193: 265–275.

Magni C, Ballabio C, Restani P, Sironi E, Scarafoni A, Poiesi C, Duranti M (2005) Two-dimensional electrophoresis and western-blotting analyses with anti Ara h 3 basic subunit IgG evidence the cross-reacting polypeptides of *Arachis hypogaea*, *Glycine max*, and *Lupinus albus* seed proteomes. J Agri Food Chem 53: 2275–2281.

Mathesius U, Keijzers G, Natera SHA, Weinman JJ, Djordjevic MA, Rolfe BG (2001) Establishment of a root proteome reference map for the model legume *Medicago truncatula* using the expressed sequence tag database for peptide mass fingerprinting. Proteomics 1: 1424–1440.

Mathesius U, Imin N, Chen H, Djordjevic MA, Weinman JJ, Natera SHA, Morris AC, Kerim T, Paul S, Menzel C, Weiller GF, Rolfe BG (2002) Evaluation of proteome reference maps for cross-species identification of proteins by peptide mass fingerprinting. Proteomics 2: 1288–1303.

Mei-Guey L, Tyrell R, Bassette R, Reeck GR (1983) Two dimensional electrophorectic analyses of soybean proteins. J Agri Food Chem 31: 963–968.

Millar AH, Sweetlove LJ, Giege P, Leaver CJ (2001) Analysis of the *Arabidopsis* mitochondrial proteome. Plant Physiol 127: 1711–1727.

Mityko J, Batkai J, Hoddos-Kotvics G (1990) Trypsin inhibitor content in different varieties and mutants of soybean. Acta Agron Hung 39: 401–405.

Mooney BP, Thelen JJ (2004) High-throughput peptide mass fingerprinting of soybean seed proteins: automated workflow and utility of UniGene expressed sequence tag databases for protein identification. Phytochemistry 65: 1733–1744.

Natarajan SS, Chenping X, Caperna TJ, Garrett WM (2005) Comparison of protein solubilization methods suitable for proteomic analysis of soybean seed proteins. Anal Biochem 342: 214–220.

Natarajan SS, Xu C, Bae H, Caperna TJ, Garrett WM (2006a) Characterization of storage proteins in wild (*Glycine soja*) and cultivated (*Glycine max*) soybean seeds using proteomics analysis. J Agri Food Chem 54: 3114–3120.

Natarajan SS, Xu C, Bae H, Caperna TJ, Garrett WM (2006b) Proteomic analysis of allergen and anti-nutritional proteins in wild and cultivated soybean. J Plant Biochem Biotechnol 15: 103–108.

Natarajan SS, Xu C, Bae H, Bailey BA (2007) Proteomic and genomic characterization of Kunitz trypsin inhibitors in wild and cultivated soybean genotypes. J Plant Physiol 164: 756–763.

Nerurkar LS, Marino PA, Adams DO (1981) Quantification of selected intracellular and secreted hydrolases of macrophages. In: JA Bellanti , A Ghaffer (eds) Manual of Macrophage Methodology. Marcel Dekker, New York, USA, pp 229–247.

Neuhoff V, Arold N, Taube N, Ehrhardt W (1988) Improved staining of proteins in polyacrylamide gels including isoelectric focusing gels with clear background at nanogram sensitivity using Coomassie Brilliant Blue G-250 and R-250. Electrophoresis 9: 255–262.

Newsholme SJ, Maleeft BF, Steiner S, Anderson NL, Schwartz LW (2000) Two-dimensional electrophoresis of liver proteins: Characterization of a drug-induced hepatomegaly in rats. Electrophoresis 21: 2122–2128.

Nielsen NC (1985) The structure and complexity of the 11S polypeptides in soybeans. J Am Oil Chem Soc 62: 1680–1686.

Nielsen NC, Dickinson CD, Cho TJ, Thanh VH, Scallon BJ, Fischer RL, Sims TL, Drews GN, Goldberg RB (1989) Characterization of the glycinin gene family in soybean. Plant Cell 1: 313–328.

Nielsen NC, Jung R, Nam YM, Beaman TW, Oliveira LO, Bassuner R (1995) Synthesis and assembly of 11S globulins. J Plant Physiol 145: 641–647.

O'Farrell PH (1975) High resolution two-dimensional electrophoresis of proteins. J Biol Chem 250: 4007–4021.

Ogawa T, Bando N, Tsuji H, Nishikawa K, Kitamura K (1995) Alpha subunit of beta-conglycinin, an allergenic protein recognized by IgE antibodies of soybean-sensitive patients with atopic dermatitis. Biosci Biotechnol Biochem 59: 831–833.

Ogawa T, Samoto M, Takahashi K (2000) Soybean allergens and hypoallergenic soybean products. J Nutr Sci Vitaminol 46: 271–279.

Orf JH, Hymowitz T (1979) Inheritance of the absence of Kunitz trypsin inhibitor in seed protein of soybeans. Crop Sci 19: 107–109.

Pappin DJ, Hojrup P, Bleasby AJ (1993) Rapid identification of proteins by peptide-mass fingerprinting. Curr Biol 1: 327–332.

Perkins DN, Pappin DJC, Creasy DM, Cottrell JS (1999) Probability-based protein identification by searching sequence databases using mass spectrometry data. Electrophoresis 20: 3551–3567.

Porubleva L, Velden KV, Kothari S, Oliver DJ, Chitnis PR (2001) The proteome of maize leaves: Use of gene sequences and expressed sequence tag data for identification of proteins with peptide mass fingerprints. Electrophoresis 22: 1724–1738.

Prak K, Nakatani K, Katsube T, Adachi M, Maruyama N, Utsumi S (2005) Structure-function relationships of soybean proglycinins at subunit levels. J Agri Food Chem 53: 3650–3657.

Qu L-J, Chen J, Liu M, Pan N, Okamoto H, Lin Z, Li C, Li D, Wang J, Zhu G, Zhao X, Chen X, Gu H, Chen Z (2003) Molecular cloning and functional analysis of a novel type of Bowman-Birk inhibitor gene family in rice. Plant Physiol 133: 560–570.

Ryan CA (1990) Protease inhibitors in plants: genes for improving defenses against insects and pathogens. Annu Rev Phytopathol 28: 425–449.

Renkema JMS, Knabben JHM, Vliet T (2001) Gel formation by beta-conglycinin and glycinin and their mixtures. Food Hydrocolloids 15: 407–414.

Ritt AB, Mulinari F, Vasconcelos IM, Carlini CR (2004) Anti-nutritional and/or toxic factor in soybean (*Glycine max*) seeds: Comparison of different cultivars adapted to the southern region of Brazil. J Sci Food Agri 84: 263–270.

Ruebelt MC, Lipp M, Reynolds TL, Schmuke JJ, Astwood JD, DellaPenna D, Engel KH, Jany KD (2006) Application of two-dimensional gel electrophoresis to interrogate alterations in the proteome of genetically modified crops. 3. Assessing unintended effects. J Agri Food Chem 54: 2169–2177.

Saio K, Kamiya M, Watanabe T (1969) Food processing characteristics of soybean 11S and 7S proteins, part I. Effect of difference of protein components among soybean varieties on formation of tofu-gel. Agri Biol Chem 33: 1304–1308.

Saravanan RS, Rose KC (2004) A critical evaluation of sample extraction techniques for enhanced proteomics analysis of recalcitrant plant tissues. Proteomics 4: 2522–2532.

Schuler MA, Ladin BF, Fryer G, Pollaco JG, Beachy RN (1982b) Structural sequences are conserved in the genes coding for the α′, α and β subunits of the soybean seed storage protein and other seed proteins. Nucl Acids Res 10: 8245–8260.

Sierra I, Li de La Quillien L, Flecker P, Gueduen J, Brunie S (1999) Dimeric crystal structure of a Bowman-Birk protease inhibitor from pea seeds. J Mol Biol 285: 1195–1207.

Simpson RJ (2003) Preparative two-dimensional gel electrophoresis with immobilized pH gradients. In: J Inglis , J Argentine , M Zierler (eds) Proteins and Proteomics: A Laboratory Manual. Oxford Univ Press, New York, USA, pp 143–218.

Staswick PE, Hermodson MA, Nielsen NC (1981) Identification of the acidic and basic subunit complexes of glycinin. J Biol Chem 256: 8752–8755.

Staswick PE, Hermodson MA, Nielsen NC (1984) Identification of the cystines which link the acidic and basic components of the glycinin subunits. J Biol Chem 259: 13431–13435.

Sule A, Vanrobaeys F, Hajós G, Van Beeumen J, Devreese B (2004) Proteomic analysis of small heat shock protein isoforms in barley shoots. Phytochemistry 65: 1853–1863.

Stahlhut RW, Hymowitz T (1983) Variation in the low molecular weight proteinase inhibitors of soybean. Crop Sci 23: 766–769.

Taylor RS, Wu CC, Hays LG, Eng JK, Yates JRI, Howell KE (2000) Proteomics of rat liver Golgi complex: Minor proteins are identified through sequential fractionation. Electrophoresis 21: 3441–3450.

Thanh VH, Shibasaki K (1976) Heterogeneity of beta-conglycinin. Biochem Biophys Acta 181: 404–409.

Thanh VH, Shibasaki K (1978) Major proteins of soybean seeds. Subunit structure of β-conglycinin. J Agri Food Chem 26: 695–698.

Tsugita A, Kamo M, Kawakami T, Ohki Y (1996) Two-dimensional electrophoresis of plant proteins and standardization of gel patterns. Electrophoresis 17: 855–865.

Tsuji HN, Okada R, Yamanishi N, Bando MK, Ogawa T (1995) Measurement of Gly m Bd 30K, a major soybean allergen, in soybean products by a sandwich enzyme-linked immunosorbent assay. Biosci Biotechnol Biochem 59: 150–151.

Tsuji H, Heimori M, Kimoto M, Yamashita H, Kobatake R, Adachi M, Fukuda T, Bando N, Okita M, Utsumi S (2001) Cloning of cDNA encoding a soybean allergen, Gly m Bd 28 K. Biochem Biophy Acta 1518: 178–182.

Usuda H, Shimogawara K (1995) Phosphate deficiency in maize. Changes in the two-dimensional electrophoretic patterns of soluble proteins from second leaf blades associated with induced senescence. Plant Cell Physiol 36: 1149–1155.

Utsumi S (1992) Plant food protein engineering. Adv Food Nutr Res 36: 89–98.

Utsumi S, Matsumura Y, Mori T (1997) Structure-function relationships of soy proteins. In: S Damodaran, A Paraf (eds) Food Proteins and their Applications. Marcel Dekker, New York, USA, pp 257–291.

Vestal ML, Juhasz P, Martin SA (1995) Delayed extraction matrix-assisted laser desorption time-of-flight mass spectrometry. Rapid Commu Mass Spectrom 9: 1044–1050.

Wang K, Yamashita T, Watanabe M, Takahata Y (2004) Genetic characterization of a novel Tib-derived variant of soybean Kunitz trypsin inhibitor detected in wild soybean (*Glycine soja*). Genome 47: 9–14.

Warner MH, Wemmer DE (1991) 1H Assignments and secondary structure determination of the soybean trypsin/chymotrypsin Bowman-Birk inhibitors. Biochemistry 30: 3356–3364.

Watson BS, Asrivatham VS, Wang L, Sumner LW (2003) Mapping the proteome of Barrel Medic. Plant Physiol 131: 1104–1123.

Wijk KJ (2000) Proteomics of the chloroplast: experimentation and prediction. Trends Plant Sci 5: 420–425.

Wright DJ (1987) The seed globulins. In: BJF Hudson (ed) Developments in food proteins 5. Elsevier Appl Sci, London, New York, pp 81–157.

Xiang P, Hass EJ, Zeece MG, Markwell J, Sarath G (2004) C-Terminal 23 kDa polypeptide of soybean Gly m Bd 28 K is a potential allergen. Planta 220: 56–63.

Xu C, Garrett WM, Sullivan J, Caperna TJ, Natarajan S (2006) Separation and identification of soybean leaf proteins by two dimensional gel electrophoresis and mass spectrometry. Phytochemistry 67: 2431–2440.

Xu C, Caperna TJ, Garrett WM, Cregan P, Bae H, Luthria DL, Natarajan S (2007) Proteomic analyses of the distribution of the major seed allergens in wild, landrace, ancestral and modern soybean genotypes. J Sci Food Agri 87: 2511–2518.

Yan JX, Harry RA, Spibey C, Dunn MJ (2000a) Post-electrophoresis staining of proteins by two dimensional gel electrophoresis using SYPRO dyes. Electrophoresis 21: 3657–3665.

Yan JX, Walt R, Berkelman T, Harry RA, Westbrook JA, Wheller CH, Dunn MJ (2000b) A modified silver staining protocol for visualization of proteins compatible with matrix-assisted laser desorption/ionization and electrospray ionization-mass-spectrometry. Electrophoresis 21: 3666–3672.

Yan S, Tang Z, Su W, Sun W (2005) Proteomic analysis of salt stress-responsive proteins in rice root. Proteomics 5: 235–244.

Yan S, Zhang Q, Tang Z, Su W, Sun W (2006) Comparative proteomic analysis provides new insights into chilling stress responses in rice. Mol Cell Proteom 5: 484–496.

Zarkadas CG, Gagnon C, Gleddie S, Khanizadeh S, Cober ER, Guillemette RJD (2007) Assessment of the protein quality of fourteen soybeans [*Glycine max* (L.) Merr.] cultivars using amino acid analysis and two-dimensional electrophoresis. Food Res Int 40: 129–146.

# Metabolomics Approach in Soybean

Takuji Nakamura,[1] Keiki Okazaki,[2] Noureddine Benkeblia,[3] Jun Wasaki,[4] Toshihiro Watanabe,[5] Hideyuki Matsuura,[5] Hirofumi Uchimiya,[6] Setsuko Komatsu[1] and Takuro Shinano[2*]

## ABSTRACT

Profiling plant metabolites is able to serve a new technology for understanding metabolic pathway and it may also help picking up the difference beneath plant phenotypes. CE-MS is one of the suitable tools to analyze a large number of plant metabolites, while GC-MS is also a very powerful especially when focusing on primary metabolites, and LC-MS(/MS) is suitable for the detection of secondary metabolites. Several applications of metabolomics approach by using these analytical tools with fractionation of cellular organelles and extraction procedures are demonstrated in this chapter.

**Keywords:** metabolites; flavonoids; compartmentation; mass spectrometry; gas chromatography

## 14.1 Introduction

To integrate the information from genome to phenome, metabolomic approaches may have an intermediary bridge-building role (Fiehn et al.

[1]Soybean Physiology Research Team, National Institute of Crop Science, Tsukuba, Japan.
[2]Rhizosphere Environment Research Team, National Agricultural Research Center for Hokkaido Region, Sapporo, Japan.
[3]Department of Life Sciences, The University of West Indies, Jamaica.
[4]Graduate School of Biosphere Science, Hiroshima University, Higashi Hiroshima, Japan.
[5]Graduate School of Agriculture, Hokkaido University, Sapporo, Japan.
[6]Institute of Molecular and Cellular Biosciences, The University of Tokyo, Tokyo, Japan.
*Corresponding author: *shinano@affrc.go.jp*

2001). The metabolomics approach offers not only comprehensive analysis of a large number of metabolites, but also targets several compounds simultaneously. Metabolome has been developed to investigate comprehensive analysis of metabolites, and the terms are designated mainly by how they focus on the targeting (Nielsen and Oliver 2005). Metabolomics study has progressed especially using gas chromatography mass spectrometry (GC-MS) because of enormous efforts in developing methodological standards and updating a large number of metabolites in plants (e.g., Fiehn et al. 2000; Roessner et al. 2000). Other methodological applications have also been introduced, including fourier transform ion cyclotron resonance mass spectrometry (FT-ICR-MS), liquid chromatography mass spectrometry (LC-MS), capillary electrophoresis mass spectrometry (CE-MS), and nuclear magnetic resonance (NMR). There are several leading references on metabolome analysis itself such as reviewed by Villas-Bôas et al. (2007). Recent progress in plant metabolomics is mainly focusing on specific model plants such as *Arabidopsis thaliana*. This is not only because of the accessibility to its genomics data, but also large efforts have been made to develop tools to integrate transcriptomics and metabolomics data such as KaPPA view (*http://kpv.kazusa.or.jp/kappa-view/*) and Batch Learning (BL)-SOM (Hirai et al. 2004, 2005). However, the advantage of using metabolomics is the possibility to demonstrate the metabolic flow or its physiological traits in non-model plants such as tea (Pongsuwan et al. 2008) and spinach (Okazaki et al. 2008). In this chapter, our recent progress in soybean metabolomics is introduced. Importantly, soybean belongs to *Fabaceae*, and contains a large number of secondary metabolites such as flavonoids, isoflavonoids, etc. Though metabolomics concentrates on comprehensive analysis of overall metabolites in an organism, one major problem is that > 1,000,000 compounds are considered to exist in the plant kingdom (Wink 1988; Weckwerth 2003; Oksman-Caldentey and Inzé 2004), and even in one plant species 5,000-25,000 different compounds may exist (Trethewey 2004), and their quantitative variation is large (Windsor et al. 2005). The number seems to increase with improved analytical methods (Last et al. 2007), and still there are no reliable procedures to cover all plant compounds.

Thus, it is important to select the most suitable (quantitative and/or qualitative) method for extraction and further instrumental analysis. Recently published texts cover these topics (e.g., Villas-Bôas et al. 2007); however, reviews on plant metabolomics are focused especially on low molecular weight compounds (e.g., Sumner et al. 2003; Weckwerth 2003; Hagel and Facchini 2008). Furthermore, the importance of comprehensive analysis is growing, such as in the area of lipids that are proposed to work as signal transduction processes (Bamba et al. 2005).

Flavonoids are considered one of the large and diverse plant secondary metabolites. About 7,000 flavonoids analogs (Harborne and Baxter 1999; Harborne et al. 1999; Andersen and Markham 2006), 25,000 terpenoids, 12,000 alkaloids, and 8,000 phenyl propanoids and related phenolic compounds are reported (Croteau et al. 2000). The *Fabaceae* is a plant family known to contain several characteristic secondary compounds such as isoflavonoids, and the dynamics of primary compounds in *Fabaceae* have unique patterns as compared with *Gramineae*, now known as *Poaceae*. We are especially interested in the dynamics of primary metabolites in soybean leaves to demonstrate the physiological change in metabolic pathway(s) after photosynthesis under several environmental conditions. We have compared the dynamics of primary photosynthate behavior between soybean and rice, and found the carbon fluxes of photorespiratory and starch synthesis are high in soybean leaf (Nakamura et al. 1997; Okazaki et al. 2005). Earlier experiments were mainly based on the utilization of paper chromatography, giving low resolution and lower numbers of target compounds. To overcome this problem, and to understand the plant metabolic differences among $C_3$ plants' photosynthesis-related processes, we introduced a metabolomics approach.

It is known that nitrogen (N) use efficiency (NUE, the amount of dry mass on the basis of N in the plant) of soybean is lower than that of rice (Osaki et al. 1992; Nakamura et al. 1997). Generally, this lower NUE of soybean has been thought to be due to the consumption of large amounts of energy for N fixation and the biosynthesis of protein in seeds (Sinclair and de Witt 1975; Yamaguchi 1978; Ryle et al. 1979; Finke et al. 1982). However, the similarity of NUE, even between nodulating and non-nodulating soybean, indicated that the energy requirement was not high enough to explain the low NUE of soybean (Osaki et al. 1992). Moreover, Shinano et al. (1993) reported that the growth efficiency of soybean pods was higher than that of rice ears, which indicated that respiratory loss of carbohydrate in soybean pods was not the reason for its lower NUE. Furthermore, Shinano et al. (1994, 1995) showed that the growth efficiency of soybean shoot was significantly lower than that of rice; and regardless of light and dark conditions that soybean leaves released a larger amount of [14]C compared to rice when [14]$CO_2$ was assimilated photosynthetically. They confirmed that soybean shoots had a higher respiratory rate than rice and that assimilated [14]C was largely distributed to organic acids and amino acids. Conclusively, it is suggested that metabolic regulation of soybean differs from rice. In our research, we examined diurnal changes in metabolites in rice and soybean leaves, using metabolomics techniques to clearly demonstrate differences in metabolic regulation between rice and soybean.

## 14.2 Metabolites Analysis by CE-MS

In the analysis of the metabolites in the leaf of rice and soybean, the difference of the primary metabolites were investigated because of results that showed that $^{14}C$ distribution into organic and amino acids in the leaf of soybean was higher than those in the leaf of rice (Shinano et al. 1994, 1995). CE-MS was used because it is a strong tool to detect charged compounds such as carboxylic acids, amino acids and nucleotides (Soga and Heiger 2000; Soga et al. 2002a, b). Therefore, we measured the diurnal changes in ca. 90 different metabolites in rice leaf and soybean following the methods of Takahashi et al. (2006) as shown in Figure 14-1.

To detect the differences in metabolic profiling, the metabolite data were processed with principal component analysis (PCA); the first and third components revealed the differences (Fig. 14-2). The first component revealed the greatest difference in metabolic profiling between rice and soybean and accounted for 42.4% of total variance. To further investigate the contributors to the principal components, the metabolic loadings in the first component were compared (Table 14-1). The significant metabolites for the first component were alanine, aspartate, glutamate, ornithine, 2-ketoglutarate, citrate, iso-citrate and malate. The diurnal change in these metabolites is shown in Figure 14-3. 2-Ketoglutarate, citrate and malate had positive

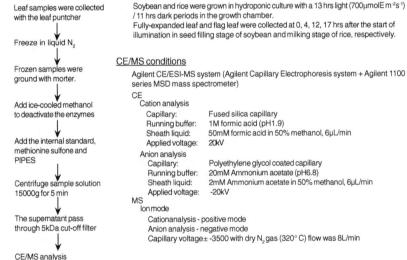

**Materials and methods**

Experimental procedure based on Takahashi et al. (2006)

Extractions

Leaf samples were collected with the leaf puntcher
↓
Freeze in liquid N₂
↓
Frozen samples were ground with morter.
↓
Add ice-cooled methanol to deactivate the enzymes
↓
Add the internal standard, methionine sulfone and PIPES
↓
Centrifuge sample solution 15000g for 5 min
↓
The supernatant pass through 5kDa cut-off filter
↓
CE/MS analysis

Growth conditions

Soybean and rice were grown in hydroponic culture with a 13 hrs light (700μmolE m⁻²s⁻¹) / 11 hrs dark periods in the growth chamber.
Fully-expanded leaf and flag leaf were collected at 0, 4, 12, 17 hrs after the start of illumination in seed filling stage of soybean and milking stage of rice, respectively.

CE/MS conditions

Agilent CE/ESI-MS system (Agilent Capillary Electrophoresis system + Agilent 1100 series MSD mass spectrometer)

CE
Cation analysis
  Capillary:          Fused silica capillary
  Running buffer:     1M formic acid (pH1.9)
  Sheath liquid:      50mM formic acid in 50% methanol, 6μL/min
  Applied voltage:    20kV
Anion analysis
  Capillary:          Polyethylene glycol coated capillary
  Running buffer:     20mM Ammonium acetate (pH6.8)
  Sheath liquid:      2mM Ammonium acetate in 50% methanol, 6μL/min
  Applied voltage:    -20kV
MS
  Ion mode
  Cation analysis - positive mode
  Anion analysis - negative mode
  Capillary voltage± -3500 with dry N₂ gas (320° C) flow was 8L/min

**Figure 14-1** Scheme of the experimental procedure for metabolome analysis of soybean and rice leaves using CE-MS.

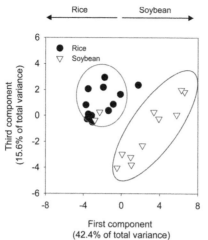

**Figure 14-2** Sample scores for the first (Factor 1), and third principal components (Factor 3) provided by PCA analysis for identified metabolites in leaf extracts of soybean and rice.

correlations with the first component of PCA. These metabolites are part of the TCA cycle and provide carbon to amino acids in soybean leaves (Stitt et al. 1994; Morcuende et al. 1998). Higher amounts of alanine and glutamate accumulated in the leaf of rice and had negative correlations with the first component while γ-aminobutyric acid (GABA) was highly accumulated in soybean leaf (data not shown). Thus, it is suggested that GABA is involved in N metabolism and possibly in the storage and/or transport of N, and that the GABA shunt pathway is important in generating C/N fluxes that enter the TCA cycle in soybean leaves (Bouché and Fromm 2004). Those data show large differences in metabolic regulation between these $C_3$ plants, rice and soybean. However, one limiting factor in evaluating the metabolites' distribution pattern in the plant is the plant organelles. It is quite important to consider the flux of pathways when distinguishing distribution of metabolites in different organelles. Unfortunately, almost all trials of plant metabolomics are at the plant organ level.

**Table 14-1** PCA loadings of metabolites with first principal component in leaves of soybean and rice.

| Metabolites | 1st component |
|---|---|
| Ala | −0.334 |
| Asp | −0.209 |
| Glu | −0.548 |
| Ornithine | 0.117 |
| 2–OG | 0.150 |
| CA | 0.185 |
| ICA | −0.160 |
| MA | 0.647 |

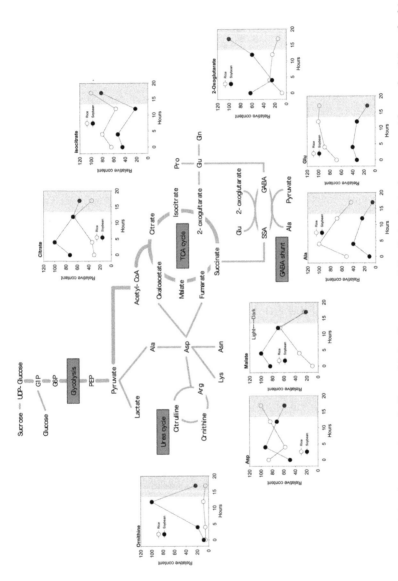

**Figure 14-3** Mapping of the changes of metabolites with loading score of the first principal component beyond ± 0.1 on the biosynthetic pathway. Hours refer to the time after the start of illumination. Relative content indicates metabolite content relative to its maximum content. Gray zone in the graphs indicates the dark period.

## 14.3 Compartmentation of Metabolites in Leaf

Compartmentation of metabolites is an important component of metabolic regulation, because many metabolic pathways are highly compartmented between different subcellular organelles (Browsher and Tobin 2001). Although the role of individual organelles has been the subject of many studies in plant metabolism, cellular compartmentation is less often considered (Browsher and Tobin 2001; Miller et al. 2001), because the vital properties and organized structures of the cell result from not only the compartments but also from the connections between the molecular constituents of which the cells are composed (Kitano 2002; Trewavas 2005). Thus, understanding plant metabolism at the compartmental level requires first the qualitative and quantitative knowledge of the metabolomes, and second, their distributions in the different cell compartments involved in major or minor anabolic or catabolic activities. This compartmentation is the principal distinguishing characteristic of plant metabolism (Goodacre et al. 2004; Fridman and Pichersky 2005). The chloroplast, cytosol and vacuole are the main compartments of the photosynthetic and other metabolic activities in plant leaves; knowledge of the metabolic networks and their regulation should be considered based upon their experimentally determined compartmental distributions. Indeed, understanding of plant metabolism at cellular and compartmental levels requires much data on the different metabolites and their distributions in cellular compartments. Therefore, prior to analysis, it is essential to use tissue fractionation that ensures that metabolites are neither lost nor altered. Nevertheless, measurement of metabolites is difficult because (i) the dynamics of metabolites makes harvesting protocols and experimental design problematic, and (ii) their chemistry makes their identification more complex (Kell 2004). The major problem of applying these compartmentation protocols to metabolomics approaches is that very little attention has been paid to stopping enzymatic reactions in the cell. One of the most efficient approaches to studying metabolites in leaves is to use a non-aqueous fractionation (NAQF) procedure (Gerhardt and Heldt 1984; Stitt et al. 1989). This method gave satisfactory results for different plant tissues; however, the type of cells must be considered because of the presence of some metabolites, e.g. starch granules, which could damage the organelles during the extraction process. Based on these techniques, we developed an extraction method for reliable organelle separation of soybean leaves (Fig. 14-4).

By using this separation technique, we analyzed soybean leaf metabolites with GC-MS and estimated the distribution pattern in vacuole, cytosol and chloroplast fractions. Density gradient centrifugation showed a consecutive spectrum of fractionation, so a very high degree of accuracy of metabolite separation is not expected. However, it is important to analyze the detailed

## Cell Tissue Fractionation
(modified based on Gerhardt and Heldt 1984)

**Sampling**
Leaves are separated from the plant by cutting
the petiole just under the blade

↓

**Leaves are plunged quicly in liquid N₂**
(Keep plant material with care to avoid any stress or
wounding)

↓

**Remove the middle rib**

↓

**Grind leaf tissue**
with mortor or balls grounder

↓

**Freeze-Dry[a]**
0.03 torr, -50 °C, 4 days

↓

**Store the dried material until use**
-35 °C, in a dessicator with $P_2O_5$ as dessicant

Following operations are carried out at
4 °C except otherwise stated

↓

**200 to 300 mg of dry material**

{ Add 20 ml of a mixture' of carbone
tetrachlorure / n-heptane [$CCl_4/C_7H_{16}$] d
= 1.28 g/cm³, dried over molecular
sieve beads.

↓

**Sonicate for 90 seconds**
5 sec then 15 sec pause, during pause, samples are
cooled using dry ice + heptane to avoid over heating

↓

**Filter through a layer of quartz wool**
To remove any remaining coarse material

↓

**Add 2 equal volumes of n-Heptane to the
filtrate, mix and centrifuge at 3,000 g for
2 min**

---

**↦ Discard supernatant**

**Sediment (Pellet)**

{ Add 3 ml of a mixture' of carbone
tetrachlorure / n-heptane [$CCl_4/C_7H_{16}$] d
= 1.28 g/cm³, dried over molecular
sieve beads.

↓

**Sample ( ca. 3 ml)**

Take an
aliquot of ⟨ 200 μl for enzyme
marker activities

200 μl for metabolite
analysis

**Remainder ( ca. 2.5 ml)**

↓

**Layer the remainder as folows:**

Remainder
d = 1.28 g/cm³

Exponential
gradient of
$CCl_4$ /
n-Heptane

9 layers of 2 ml
each[c]

d = 1.50 g/cm³

1 ml pure $CCl_4$ (d =
1.59 g/cm³)

↓

**Centrifuge at 25,000 x g for 2.5 h**
Check the equilibrium distibution of the sample
and eventually adjust time of centrifugation

↓

*Figure 14-4 contd....*

---

behavior of metabolite dynamics among organelles. It is clear
that information on qualitative and quantitative metabolites at the organelle-
level will increase more reliable flux data for metabolites. The problem of
rapid change (quantitative and eventually qualitative) of metabolites in the
plant can distort the analysis, thus sample preparation methodology has
been repeatedly emphasized (e.g., Villas-Boâs et al. 2007). Further
improvement will be required, and isotope microscopy (Yurimoto 2006) with

*Figure 14-4 contd....*

**Cellect the fractions**
**(6 to 8 fractions, of 1 to 2 ml each)**

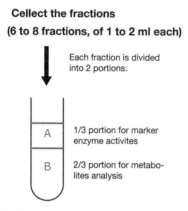

Each fraction is divided
into 2 portions:

A    1/3 portion for marker
      enzyme activites

B    2/3 portion for metabo-
      lites analysis

**Each portion of each fraction (6 to 8**
**portions for both A and B) is treated**
**as follows:**

**To each portion**

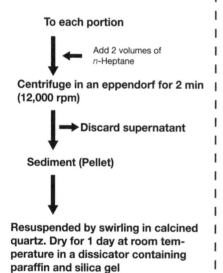

Add 2 volumes of
*n*-Heptane

**Centrifuge in an eppendorf for 2 min**
**(12,000 rpm)**

→ Discard supernatant

**Sediment (Pellet)**

**Resuspended by swirling in calcined**
**quartz. Dry for 1 day at room tem-**
**perature in a dissicator containing**
**paraffin and silica gel**

| **Remarks**

| \* The ratio of the volume of $CCl_4$ to *n*-heptane for
| desired density is estimated as follow:

$$\frac{V_a}{V_b} = \frac{d_m - d_b}{d_a = d_m}$$

| where,
| $d_m$ = density of the mixture
| $d_a$ = density of $CCl_4$ (1.59 g/cm³)
| $d_b$ = density of *n*-heptane (0.684 g/cm³)

| [a] The freeze-drying time could be variable
| depending mainly on the pressure. However, it is
| preferrfed to carry this operation as longer as
| possible (normally for good result, it is recom-
| mended to freeze-dry for 3 to 4 days what ever
| the pressure).

| [b,c] The exponential gradient may vary. It is how-
| ever recommended to make 6 to 8 layers of 1 or
| 2 ml each (not more than 2.2 ml).

**Figure 14-4** Fractionation scheme of soybean leaf into different organelles.

labeled compounds is one possibility. In the case of primary metabolites analysis by GC-MS, it is essential to derivate the compounds before analysis. We used ribitol as an internal standard after lyophilization, then a second dehydration under nitrogen gas, then methoxyamination, and subsequently by using MSTFA (N-methyl-N-(trimethylsilyl)trifluoroacetamide) and TMCS (trichlormethylchlorsilane) to make volatile and thermally stable derivatives for GC-MS applications (Broeckling et al. 2005). In soybean leaf we identified

93 metabolites: 19 organic-acid-related compounds; 21 amino acids, 29 sugars and sugar alcohol-related compounds; 11 fatty acids; 9 phenolic compounds and 4 cyclitols (Benkeblia et al. 2007). The distribution pattern among vacuole, cytosol and chloroplast was estimated in each group compartment (Fig. 14-5).

Several compounds are selectively accumulated, such as glycine and oxalic acid in stroma; alanine and glutamine in cytosol; and malate and citrate in the vacuole. Thus, the combination of NAQF and GC-MS-based metabolomics study is one strategy to study metabolites' flux. However,

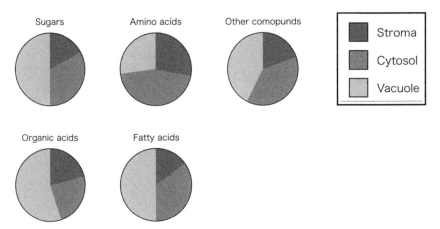

**Figure 14-5** Compartmentation of metabolites into cytosol, chloroplast and vacuole in soybean leaf.

GC-MS-based analysis is not suitable for analysis of high molecular weight compounds such as flavonoids. Soybean belongs to *Leguminosae*, known to contain larger amounts of flavonoids, including iso-flavonoids, and it is important to include these compounds in the metabolomics approach. In our trial, only coumaric acid, ferulic acid and genistin were detected; however, GC-MS is not suitable for these compounds because extraction efficiency should be re-considered.

## 14.4 Detection of Flavonoids by LC-MS(/MS)

High-performance liquid chromatography with tandem mass spectrometry detection (LC-MS(/MS) is better for the detection and quantification of flavonoids. Furthermore, a sophisticated extraction method is also required to separate aglycone and glycoside. Samples for LC-MS(/MS) can be applied directly without the derivatization as required for GC-MS. However, standard analytical schemes for LC-MS(/MS) for plant metabolomics have just reported (Sawada et al. 2009) and further development and improvement of platforms for LC-MS(/MS) is urgently required to include metabolite MS/MS data and reference compounds for identification.

We applied LC-MS(/MS) for quantification of five different flavonoids (quercetin, kaempferol, naringenin, daidzein, and genistein). To extract flavonoids efficiently from the plant samples, acidic conditions are recommended. It is also required to distinguish between the extraction method for LC-MS(/MS) and GC-MS protocol. In this chapter we also included the recent application of UPLC-MS/MS (Ultra Performance LC, Waters Co. Ltd) for the detection of flavonoids of another *Fabaceae*, *Lotus japonicus*, because this application method is readily applicable to soybean.

Flavonoid aglycones and their glycosides were prepared from hydrolyzed and non-hydrolyzed samples, respectively. Flavonoid aglycones extraction was based on the methods described by Burbulis et al. (1996), Moco et al. (2006), and Farag et al. (2007) with some modifications. Shoots and roots of *L. japonicus* were ground to a fine powder in the presence of liquid $N_2$ using multi-beads shocker (Yasuda Kikai Ltd.; Fig. 14-6). Flavonoids were quantified from the standard curves using quercetin, kaempferol, naringenin, daidzein and genistein as standards.

The UPLC column was fitted with an Acquity UPLC BEH-$C_{18}$ (1.7 µm, 50 mm x 2.1 mm) column (Waters). Total run time was 4.8 minutes, and the retention times for each flavonoid were as follows: quercetin –1.98 minutes, kaempferol –2.17 minutes, naringenin –2.05 min, daidzein –1.94 minutes, and genistein –2.10 minutes. The mass spectra were obtained in negative ion electrospray mode. MS/MS parameters for each flavonoid are listed in Table 14-2. Data acquisition and quantification used MassLynx software. Flavonoids detected in samples were identified by comparison of MS/MS spectra and LC retention times with their standards, and the quantity of each flavonoid in the sample was calculated from each standard curve based on the peak area (Fig. 14-7).

Among the strategic responses to phosphorus (P) deficiency, plant accumulation of phenylpropanoids and subsequent flavonoid compounds has been reported (Stewart et al. 2001; Akiyama et al. 2002; Malusà et al. 2006; Weisskopf et al. 2006). Phenylpropanoids are derived from cinnamic acid, which is formed from phenylalanine by phenylalanine-ammonia-lyase, the branch point enzyme between primary and secondary metabolism. Phenylpropanoids can serve as starter units in various polyketide synthase reactions, and the most widely distributed reaction is that catalyzed by chalcone synthase to produce flavonoids (Dixon and Paiva 1995; Paiva 2000). In general, phenylpropanoid accumulation in plants is induced by various biotic and abiotic stresses such as wounding, pathogen attacks, UV radiation, $O_3$ stress, and N or P deficiency (Dixon and Paiva 1995; Booker and Miller 1998). Most stress-induced phenylpropanoids are derived from a C15 flavonoid skeleton; and by hydroxylation, methylation, glycosylation,

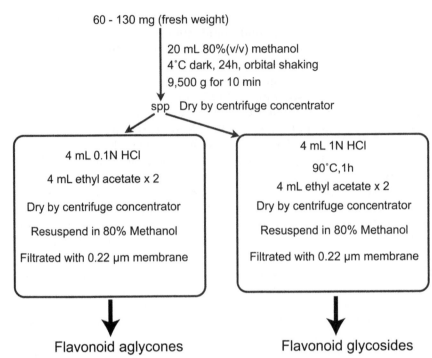

Figure 14-6 Fractionation procedure of flavonoid analysis for UPLC-MS/MS. A Waters Acquity Ultra Performance LC (UPLC) (Waters, MA, USA) coupled with a Quattro Premier XE (Micromass Technologies, Manchester, UK) was used for all analyses.

Table 14-2 Flavonoids specific MS/MS parameters used in UPLC-MS/MS.

| Flavonoids | [M–H]⁻ (m/z) | Transition ion (m/z) | Cone voltage (V) | Collision energy (eV) |
|---|---|---|---|---|
| Quercetin | 300.9 | 106.7 | 53.0 | 32.0 |
| | | 150.7 | | 24.0 |
| | | 178.7 | | 20.0 |
| Kaempferol | 284.97 | 92.6 | 63.0 | 36.0 |
| | | 156.1 | | 32.0 |
| | | 184.7 | | 30.0 |
| Naringenin | 270.97 | 106.6 | 47.0 | 26.0 |
| | | 118.7 | | 28.0 |
| | | 150.6 | | 18.0 |
| Daidzein | 252.97 | 132.2 | 69.0 | 36.0 |
| | | 208.2 | | 28.0 |
| | | 223.5 | | 30.0 |
| Genistein | 268.97 | 132.7 | 59.0 | 32.0 |
| | | 158.6 | | 32.0 |
| | | 180.5 | | 26.0 |

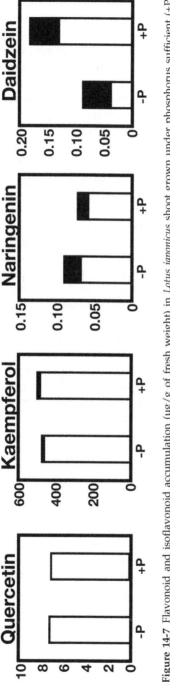

**Figure 14-7** Flavonoid and isoflavonoid accumulation (µg/g of fresh weight) in *Lotus japonicus* shoot grown under phosphorus sufficient (+P) and deficient (–P) phosphorus conditions. Open bars indicates flavonoid glycosides and closed bars indicates flavonoid aglycones. Genistein was not detected.

oxidation, and reduction, hundreds of other flavonoids are synthesized (Dixon and Paiva 1995; Paiva 2000). Flavonoids and isoflavonoids are also involved in many aspects of plant growth and development and act as flower pigments, UV protectants, auxin transporter inhibitors, phytoalexins and signal molecules for plant–microorganism interactions (Harrison and Dixon 1994). Naringenin is a simple flavonoid with various functions in the plant, and is the precursor of several important classes of flavonoids that include kaempferol and quercetin, two widespread flavonols (Paiva 2000). Daidzein and genistein are also derived from naringenin via the isoflavonoid biosynthetic pathway in legumes and act as *nod* gene inducer and precursors of phytoalexins (Paiva 2000). The role of flavonoids in plant–microbe interactions has been well studied. However, involvement of flavonoids in nutritional deficiency has not been studied well. Stewart et al. (2001) reported that N deprivation caused increased flavonols in seedlings of *Arabidopsis thaliana* and *Solanum lycopersicum*. In white lupin root, exudation of genistein and its hydroxylates increased under P deficiency; because genistein exudation peak shortly preceded citrate exudation, it could be deduced that a possible antimicrobial role of genistein prevents microbial degradation of citrate (Weisskopf et al. 2006). In *Phaseolus vulgaris* roots, daidzein accumulated under P-deficiency, whereas naringenin level decreased (Malusà et al. 2006). Accumulation of quercetin, kaempferol and isorhamnetin under P-deficiency was also reported in *Arabidopsis thaliana* and *Solanum lycoperiscum* seedling (Stewart et al. 2001). It was also reported that formononetin-7-O-glucoside, other flavonoid, was significantly higher in alfalfa roots under low-P than high-P treatments (Volpin et al. 1994). In addition, the role of flavonoids in plant–microbe interaction under P-deficiency has been reported. P-deficiency induced C-glycosylflavone, isovitexin 2″-O-β-glucoside, and had a positive effect on colonization of arbuscular mycorrhizal (AM) fungi in melon roots (Akiyama et al. 2002). In contrast, an isoflavonoid phytoalexin, medicarpin-3-O-glucoside, accumulated in alfalfa roots colonized by a pathogenic fungus as well as in AM-treated roots receiving high P (Guenoune et al. 2001).

One of the future perspectives of metabolomics is non-target profiling and the other is flux analysis based on detailed qualitative analysis. *Leguminosae* (including soybean) often have many secondary compounds, and because determination and obtaining standard compounds of each secondary compound are still difficult, application of Flavonoid Viewer (*http://www.metabolome.jp/software/FlavonoidViewer/*) without identification would be useful.

## 14.5 Conclusion

Metabolomics is a very powerful tool to describe the linkage between phenotype and genome. We have applied several analytical technologies on soybean metabolomics; however, it is still insufficient in the meaning of 1) identification, 2) quantification, 3) compartmentation, and 4) high-throughput analysis. It seems that applying several methodologies to one sample is required to describe the metabolic pathway(s) in more detail, while it is not recommended when we are interested in non-target analysis with high-throughput analysis. At the moment, exhaustive analysis of metabolites of soybean is still not available, researchers should select the methodologies depending on the purpose of the experiment.

## References

Akiyama K, Matsuoka H, Hayashi H (2002) Isolation and identification of a phosphate deficifency-induced C-glycosylflavonoid that stimulates arbuscular mycorrhiza formation in melon roots. Mol Plant-Micr Interact 15: 334–340.

Andersen ØM, Markham KR (2006) Flavonoids: Chemistry, Biochemistry and Applications. CRC Press, Tayler & Francis Group, Boca Raton, Florida, USA: ISBN 0849320216.

Bamba T, Fukusaki E, Minakuchi H, Nakazawa Y, Kobayashi A (2005) Separation of polyprenol and dolichol by monolithic silica capillary column chromatography. J Lipid Res 46: 2295–2298.

Benkeblia N, Shinano T, Osaki M (2007) Metabolite profiling and assessment of metabolome compartmentation of soybean leaves using non-aqueous fractionation and GC-MS analysis. Metabolomics 3: 297–305.

Booker F, Miller J (1998) Phenylpropanoid metabolism and phenolic composition of soybean [Glycine max (L.) Merr.] leaves following exposure to ozone. J Exp Bot 49: 1191–1202.

Bouché N, Fromm H (2004) GABA in plants: Just a metabolite. Trends Plant Sci 9: 100–115.

Broeckling CD, Huhman DV, Farag MA, Smith JT, May GD, Mendes P, Dixon RA, Sumner LW (2005) Metabolic profiling of Medicago truncatula cell cultures reveals the effects of biotic and abiotic elicitors on metabolism. J Exp Bot 56: 323–336.

Browsher CG, Tobin AK (2001) Compartmentation of metabolism within mitochondria and plastids. J Exp Bot 52: 513–527.

Burbulis IE, Iacobucci M, Shirley BW (1996) A null mutation in the first enzyme of flavonoid biosynthesis does not affect male fertility in Arabidopsis. Plant Cell 8: 1013–1025.

Croteau R, Kutchan TM, Lewis NG (2000) Natural products (secondary metabolites). In: BB Buchanan, W Gruissem, RL Jones (eds) Biochemistry and Molecular Biology of Plants. Am Soc Plant Physiol, Rockwell, MA, USA, pp 1250–1318.

Dixon RA, Paiva NL (1995) Stress-induced phenylpropanoid metabolism. Plant Cell 7: 1085–1097.

Farag MA, Huhman DV, Lei Z, Sumner LW (2007) Metabolic profiling and systematic identification of flavonoids and isoflavonoids in roots and cell suspension cultures of *Medicago truncatula* using HPLC-UV-EST-MS and GC-MS. Phytochemistry 68: 342–354.

Fiehn O, Kopka J, Dörmann P, Altmann T, Trethewey RN, Willmitzer L (2000) Metabolite profiling for plant functional genomics. Nat Biotechol 18: 1157–1161.

Fiehn O, Kloska S, Altmann T (2001) Integrated studies on plant biology using multiparallel techniques. Curr Opin Biotechnol 12: 82–86.

Finke RL, Harper JE, Hageman RH (1982) Efficiency of nitrogen assimilation by N2-fixing and nitrate-grown soybean plants (*Glycine max* [L.] Merr.). Plant Physiol 70: 1178–1184.

Fridman E, Pichersky E (2005) Metabolomics, genomics, proteomics, and the identification of enzymes and their substrates and products. Curr Opin Plant Biol 8: 242–248.

Gerhardt R, Heldt HW (1984) Measurement of subcellular metabolite levels in leaves by fractionation of freeze-stopped material in nonaqueous media. Plant Physiol 75: 542–547.

Goodacre R, Vaidyanathan S, Dunn WB, Harrigan GG, Kell DB (2004) Metabolimics by numbers: Acquiring and understanding global metabolite data. Trends Biotechnol 22: 245–252.

Guenoune D, Galili S, Phillips DA, Volpin H, Chet I, Okon Y, Kapulnik Y (2001) The defense response elicited by the pathogen *Rhizoctonia solani* is suppressed by colonization of the AM-fungus *Glomus intraradices*. Plant Sci 160: 925–932.

Harborne JB, Baxter H (1999) The Handbook of Natural Flavonoids (2 vols). John Wiley, Chichester, UK: ISBN 0-471-95893-X.

Harborne JB, Baxter H, Moss GP (1999) Phytochemical Dictionary: A Handbook of Bioactive Compounds form Plants. 2nd edn. Taylor & Francis, London, UK: ISBN 0748406204.

Hagel JM, Facchini PJ (2008) Plant metabolomics: Analytical platforms and integration with functional genomics. Phytochem Rev 7: 479–497.

Harrison MJ, Dixon RA (1994) Spatial patterns of expression of flavonoid/isoflavonoid pathway genes during interaction between roots of *Medicago truncatula* and the mycorrhizal fungus *Glomus versiforme*. Plant J 6: 9–20.

Hirai, MY, Yano M, Goodenowe DB, Kanaya S, Kimura T, Awazuhara M, Arita M, Fujiwara T, Saito K (2004) Integration of transcriptomics and metabolomics for understanding of global responses to nutritional stresses in *Arabidopsis thaliana*. Proc Natl Acad Sci USA 101: 10205–10210.

Hirai, MY, Klein M, Fujikawa Y, Yano M, Goodenowe DB, Yamazaki Y, Kanaya S, Nakamura Y, Kitayama M, Suzuki H, Sakurai N, Shibata D, Tokuhisa J, Reichelt M, Gershenzon J, Papenbrock J, Saito K (2005) Elucidation of gene-to-gene and metabolite-to-gene networks in arabidopsis by integration of metabolomics and transcriptomics. J Biol Chem 280: 25590–25595.

Kitano H (2002) Systems biology: A brief overview. Science 295: 1662–1664.

Kell DB (2004) Metabolomics and systems biology: Making sense to the soup. Curr Opin Microbiol 7: 296–307.

Last RL, Jones AD, Shachar-Hill Y (2007) Towards the plat metabolome and beyond. Nat Rev Mol Cell Biol 8: 167–174.

Malusà E, Russo MA, Mozzetti C, Belligno A (2006) Modification of secondary metabolism and flavonoid biosynthesis under phosphate deficiency in bean (*Phaseolus vulgaris* L.) roots. J Plant Nutr 29: 245–258.

Miller AJ, Cookson SJ, Smith SJ, Wells DM (2001) The use of microelectrodes to investigate compartmentation and the transport of metabolized inorganic ions in plants. J Exp Bot 52: 541–549.

Moco S, Bino RJ, Vorst O, Verhoeven HA, de Groot J, van Beek TA, Vervoort J, Ric de Vos CH (2006) A liquid chromatography-mass spectrophotometry-based metabolome database for tomato. Plant Physiol 141: 1205–1218.

Morcuende R, Krapp A, Hurry V, Stitt M (1998) Sucrose-feeding leads to increased rates of nitrate assimilation, increased rates of alpha-oxoglutarate synthesis, and increased synthesis of a wide spectrum of amino acids in tobacco leaves. Planta 206: 394–409.

Nakamura T, Osaki M, Shinano T, Tadano T (1997) Difference in system of current photosynthesized carbon distribution to carbon and nitrogen compounds between rice and soybean. Soil Sci Plant Nutr 43: 777–788.

Nielsen J, Oliver S (2005) The next wave in metabolome analysis. Trends Biotechnol 23: 544–546.

Okazaki K, Shinano T, Osaki M (2005) Difference in carbon distribution of initial photoassimilates between soybean and rice as revealed by 20s pulse-300s chase experiments. Soil Sci Plant Nutr 51: 835–840.

Okazaki K, Oka N, Shinano T, Osaki M, Takebe M (2008) Differences in the metabolite profiles of spinach (*Spinacia oleracea* L.) leaf in different concentrations of nitrate in the culture solution. Plant Cell Physiol 42: 170–177.

Oksman-Caldentey K-M, Inzé D (2004) Plant cell factories in the post-genomic era: new ways to produce designer secondary metabolites. Trends Plant Sci 9: 433–440.

Osaki M, Shinano T, Tadano T (1992) Carbon-nitrogen interaction in field crop production. Soil Sci Plant Nutr 38: 553–564.

Paiva NL (2000) An introduction to the biosynthesis of chemicals used in plant-microbe communication. J Plant Growth Regul 19: 131–143.

Pongsuwan W, Bamba T, Yonetani T, Kobayashi A, Fukusaki E (2008) Quality prediction of Japanese green tea using pyrolyzer coupled GC/MS based metabolic fingerprinting. J Agri Food Chem 56: 744–750.

Roessner U, Wagner C, Kopka J, Trethewey RN, Willmitzer L (2000) Simultaneous analysis of metabolites in potato tuber by gas chromatography-mass spectrophotometry. Plant J 23: 131–142.

Ryle GJA, Powell CE, Gordon AJ (1979) The respiratory costs of nitrogen fixation in soyabean, cowpea, and white clover. II. Comparison of the cost o nitrogen fixation and the utilization of combined nitrogen. J Exp Bot 30: 145–153.

Sawada Y, Akiyama K, Sakata A, Kuwahara A, Otsuki H, Sakurai T, Saito K, Yokota Hirai M (2009) Widely targeted metabolomics based on large-scale MS/MS data for elucidating metabolite accumulation patterns in plants. Plant Cell Physiol 50: 37–47.

Shinano T, Osaki M, Komatsu K, Tadano, T (1993) Comparison of production efficiency of the harvesting organs among field crops. I. Growth efficiency of the harvesting organs. Soil Sci Plant Nutr 39: 269–280.

Shinano T, Osaki M, Tadano T (1994) [14]C-allocaiton of [14]C-compounds introduced to a leaf to carbon and nitrogen components in rice and soybean during ripening. Soil Sci Plant Nutr 40: 199–209.

Shinano T, Osaki M, Tadano T (1995) Comparison of growth efficiency between rice and soybean at the vegetative growth stage. Soil Sci Plant Nutr 41: 471–480.

Sinclair TR, de Wit CT (1975) Photosynthesis and nitrogen requirements for seed production by various crops. Science 189: 565–567.

Soga T, Heiger DN (2000) Amino acid analysis by capillary electrophoresis electrospray ionization mass spectrometry. Anal Chem 72: 1236–1241.

Soga T, Ueno Y, Naraoka H, Ohashi Y, Tomita M, Nishioka T (2002a) Simultaneous determination of anionic intermediates for *Bacillus subtilis* metabolic pathways by capillary electrophoresis electrospray ionization mass spectrometry. Anal Chem 74: 2233–2239.

Soga T, Ueno Y, Naraoka H, Matsuda K, Tomita M, Nishioka T (2002b) Pressure-assisted capillary electrophoresis electrospray ionization mass spectrometry for analysis of multivalent anions. Anal Chem 74: 6224–6229.

Stewart AJ, Chapman W, Jenkins GI, Graham T, Martin T, Crozier A (2001) The effect of nitrogen and phosphorus deficiency on flavonol accumulation in plant tissue. Plant Cell Environ 24: 1189–1197.

Stitt M, Schulze D (1994) Does Rubisco control the rate of photosynthesis and plant growth? An exercise in molecular ecophysiology. Plant Cell Environ 17: 465–487.

Stitt M, Lilley RMC, Gerhardt R, Heldt H (1989) Determination of metabolite level in specific cells and subcellular compartments of plant leaves. Meth Enzymol 174: 518–552.

Sumner LW, Mendes P, Dixon RA (2003) Plant metabolomics: Large-scale phytochemistry in the functional genomics era. Phytochemistry 62: 817–836.

Takahashi H, Hayashi M, Goto F, Sato S, Soga T, Nishioka T, Tomita M, Kawai-Yamada M, Uchimiya H (2006) Evaluation of metabolic alteration in transgenic rice overexpressing dihydroflavonol-4-reductase. Anal Bot 98: 819–825.

Trethewey RN (2004) Metabolite profiling as an aid to metabolic engineering in plants. Curr Opin Plant Biol 7: 196–201.

Trewavas A (2005) Green plants as intelligent organisms. Trends Plant Sci 10: 413–419.

Villas-Boâs SG, Roessner U, Hansen MAE, Smedsgaard J, Mielsen J (2007) Metabolome Analysis: An Introduction. John Wiley, Hoboken, NJ, USA: ISBN-13: 978-0-471-74344-6.

Volpin H, Phillips DA, Okon Y, Kapulnik Y (1995) Suppression of an isoflavonoid phytoalexin defense response in mycorrhizal alfalfa roots. Plant Physiol 108: 1449–1454.

Weckwerth W (2003) Metabolomics in systems biology. Annu Rev Plant Biol 54: 669–689.

Weisskopf L, Abou-Mansour E, Fromin N, Tomasi N, Santelia D, Edelkott I, Neumann G, Aragno M, Tabacchi R, Martionia E (2006) White lupin has developed a complex strategy to limit microbial degradation of the secreted citrate required for phosphate nutrition. Plant Cell Environ 29: 919–927.

Windsor AJ, Reichelt M, Figuth A, Svatos A, Kroymann J, Klibenstein DJ, Gershenzon J, Mitchell-Olds T (2005) Geographic and evolutionary diversification of glucosinolates among near relatives of *Arabidopsis thaliana* (Brassicaceae). Phytochemistry 66: 1321–1333.

Wink M (1988) Plant breeding: Importance of plant secondary metabolites for protection against pathogens and herbivores. Theor Appl Genet 75: 225–233.

Yamaguchi J (1978) Respiration and the growth efficiency in relation to crop productivity. J Fac Agri Hokkaido Univ 180: 59–129.

Yurimoto H (2006) Isotope microscope-Imaging SIMS with high sensitive 2D-ion-detection. Microscopy 41: 134–137.

# Soybean Future Prospects

*Ed Ready*

## ABSTRACT

The United Soybean Board and other representatives of the soybean value chain recently conducted an exercise to consider and describe different scenarios that could potentially impact the soybean industry in the United States. The result of that exercise was, "Soy2020: Our Vision, Our Destination". A description of the scenarios along with their impact on the US soybean industry is described here.

**Keywords:** production; competitors; technology; investments; future

## 15.1 Introduction

Someone has said that we should be careful about making predictions, especially about the future. While such a statement may be amusing, it is also a poor way to plan. It is a cliché, but nonetheless true, that the only thing constant is change. This is at least as true in farming as in other areas of life.

Since no one can know the future, the United Soybean Board, in conjunction with many others in the soybean value chain, conducted an exercise that resulted in "Soy2020: Our Vision, Our Destination". The purpose of this exercise was to consider and describe several different scenarios, both positive and negative, that could potentially impact the US soybean industry. While this exercise focused on the effect of these scenarios on US soybean production and US farmers, by necessity the exercise also considered the effect on other soybean producing areas of the world as well.

United Soybean Board, 16305 Swingley Ridge Rd., Suite 120, Chesterfield, MO 63017, USA; e-mail: *eready@smithbucklin.com*

Some things can be predicted with a high degree of certainty. For example, in the absence of some cataclysmic occurrence, the world population will continue to increase and will likely exceed 8 billion people by 2020, with 93% of that increase taking place in developing countries. People in developing countries with an increasing economic status will seek to improve their diets by adding more animal protein and other higher quality foods. The world's demand for food and feed will grow. Competition between agriculture and urbanization for prime land and water will continue.

Soy2020 considered four possible scenarios that could exist for the future of soybeans, based on the ability of soy and other crops to maintain a sustainable technology innovation advantage relative to other competing technologies. The scenarios are depicted below.

## 15.2 Full Speed Ahead

The ideal scenario of sustainable soybean technology and innovation advantages, while such technology does not advance for competing crops.

Under this scenario, in 2020, the situation will be:

- Integrated distribution channels exist for both commodity and specialized soybeans.

  As soybean breeders develop varieties of soybeans with different traits (reduced linolenic acid, increased oleic acid, increased digestible sugars, etc.), the market can no longer treat all soybeans as a commodity. Mechanisms will develop globally to handle soybeans with value-added traits efficiently and economically, while still allowing for the efficient marketing of commodity soybeans.

- Investment dollars have increased for development of output technologies.

  Researchers are already developing germplasm with value-added traits such as improved fatty acid balance to reduce or eliminate the need for partial hydrogenation, lower saturated fatty acids to produce healthier oil, and reduced phytate-phosphorus to reduce the environmental impact of phosphorus in the waste of livestock and poultry. This research will continue, and, in fact, varieties with some of these traits are available today. In the future, new lines will be developed with increased digestible sugars to provide more metabolizable energy for feed and with an improved balance of amino acids, also to provide a better feed. Other efforts may succeed in producing high oil varieties to meet the growing need for biofuels, as well as developing other oil and meal traits for specific markets such as aquaculture.

  The nature of scientific advancement is such that germplasm improvements in the US will soon be adopted in other soybean producing countries and vice versa, if there is a benefit in doing so.

- The portfolio of soybean technology innovations is sustainable and balanced.

Many changes in soybean composition are possible either through conventional breeding or genetic transformation. However, for them to be successful, the resultant soybeans must meet the needs of end users while adding sufficient value to the soybeans to justify additional handling costs.

- The animal ag industry continues to be strong and is a primary market for improved soybean meal. The improved meal helps deal with odor problems.

The vast majority of soybean meal (roughly 95%) is used for animal feed. Therefore, for the soybean industry to continue to be successful, the animal agriculture industry in the US must also continue to be successful. For this to occur, some of the "Not in my Backyard (NIMBY)" issues surrounding animal production, like odors, must be addressed effectively. Improved soybean meal can help.

- Consumers benefit from soy's new nutritional and pharmaceutical benefits.

Many compositional changes are or will be possible. It will be necessary that these changes meet the needs of consumers. For example, changes in fatty acid composition can reduce or eliminate trans-fat. Other changes can produce varieties of soybeans that are high in omega-3 fatty acids, which have been shown to have health benefits. To date, concerns about trans-fats and the need for omega-3 fatty acids have been largely a US phenomenon, with some concern expressed in other developed countries, but in the future there may be greater awareness globally.

- Soybeans continue to be the preferred feedstock for biodiesel.
- A dual marketing system has developed to compensate the value chain for specialty soybeans, while also allowing the US to compete with global commodity soybeans.

## 15.3 Shared Success

Sustainable technology innovations exist for soybean and other crops, with numerous benefits of other crops challenging competitiveness.

Under this scenario in 2020, the situation will be:

- Soy and other crops enjoy technology innovations.

Innovations in crop quality will not be limited to soybeans, but other crops will also be improved to meet customer needs better.

- Soy and other crops have integrated distribution channels.
- The largest share of research dollars is directed to improvement of infrastructure and logistics.

Effective movement to the hundreds of millions of tons of commodity crops in the US depends upon an efficient infrastructure, and infrastructure is one area where the US is far ahead of other soybean-producing countries. But infrastructure, whether it is locks and dams, rail beds, or enough barges and rail cars, requires ongoing maintenance and improvement. In addition, logistics, how the infrastructure is used, can benefit from advances in computer and communication technology.

- Higher energy corn and Distillers Dried Grains with solubles (DDGs) with higher protein content compete with soybean meal as a feed source. While not as good a source of protein as soybean meal, supplies of DDGs are increasing as ethanol production increases, and ethanol production improvements have resulted not only in more efficient ethanol production, but in DDGs with higher feed value.
- The rural economy is boosted by increased commodity prices and the expansion of biodiesel and ethanol.
- Consumers and producers have multiple appealing choices.

## 15.4 Offshore Migration

Soybean struggles to compete as there are no soy-specific technology innovation advantages.

Under this scenario, by 2020, the situation will be:

- A significant amount of soybean production has shifted offshore.

  The US can produce soybeans more efficiently than other countries, but continuing to do so depends on continued innovation. Other countries enjoy cheaper land and labor, and South American countries can easily bring more land into production. Since the vast majority of soybean meal is used to feed poultry and livestock, there is an advantage to growing soybeans in the US to meet the US need.

  However, the threat to US livestock represented by animal agriculture opponents, whether it is NIMBY concerns or the concerns of animal rights activists, is a significant threat to poultry and livestock production in the US. A relatively easy response to onerous zoning, regulation, or legislation is for animal agriculture to move offshore.

  If livestock production moves offshore to other soybean-producing countries, those countries will gain a logistical advantage in supplying the livestock market, increasing the benefits of off-shore production as compared with the US.

- Research and capital investments have grown rapidly outside of the US.

Many US-trained plant scientists have returned to their home countries to conduct research, and soybean producing countries like Brazil and China have invested substantial sums to build their research capabilities. Therefore, these countries are well positioned to conduct effective research and to achieve technological advances. Investors and companies have a global perspective, and if they see greater profit potential with acceptable levels of risk in other countries, capital investment will go there rather than the US.

- South American countries recognize the intellectual property rights in a manner similar to North America.

Industry is reluctant to invest or to introduce new technologies into areas where intellectual property is not respected and protected. If South American countries adopt and enforce laws protecting intellectual property, high-tech companies are more likely to conduct business there.

- Other crops compete with soybean oil as a feedstock for biodiesel.
- New technologies have been developed and adopted for corn, reducing the need for soybean in crop rotations.

While soybeans are valuable as a crop, they also offer benefits in crop rotations with corn. Pest control is easier when crops are rotated; soybeans fix nitrogen from the atmosphere, so fertilizer requirements are reduced. However, new technologies such as genetic transformation to introduce pest resistance to corn or improved technology to reduce the need for synthetic fertilizer could make rotation less necessary.

- Producers find profits increasing in crops other than soybeans, and research and processing innovations are not focused on soybeans.

If soybeans become less competitive as a crop, farmers will shift to other crops that have a higher profit potential. For example, in 2007, corn prices rose proportionately more than soybean prices, and farmers shifted several million acres from soybeans to corn.

Likewise, technology companies such as seed companies, crop protection chemical companies, etc. focus their research efforts in areas where they expect the best opportunities. If producers move from soybeans to other crops, technology companies will reduce their investments in soybean technology and emphasize other areas.

Assuming that demand for soybeans continues at its present level or grows, other countries with lower labor and production costs will capitalize on the opportunity to increase their production, exacerbating the shift of soybean production.

## 15.5 Forward to the Past

No sustainable technology advances for soybean or other crops creates a constant state.

Under this scenario, by 2020, the situation will be:

- There would be little change in growth or new technology. Stagnation in technology means "we'll always do what we've always done".
- Production and distribution systems support the traditional commodity world. Corn, soybeans, and other crops are produced domestically for efficiency.

  If new traits are not successful in the market, there will be no need for systems to provide identity preservation, so the production and distribution systems will revert to treating soybeans and other crops as commodities.
- Research investment dollars focus primarily on agronomic traits.

  Until relatively recently, most crop improvement research focused on yield and agronomic traits like pest resistance. If improvements in compositional traits are not sufficiently valued by the market to justify the cost of research, production and identity preservation, then research in value-added traits will die off. If soybeans continue to be a commodity product, rather than value-added, the competitive advantage will be to those countries that are the low cost producers.
- Biodiesel production enhanced the US soybean industry for a while, but newer technologies have capped demand and growth.
- The distribution channel has consolidated, reducing the number of soybean producers and ag input suppliers.

  If the soybean market does not continue to grow and prosper, soybean production will evolve to the least cost producer. There will be pressure to consolidate farm acreage to enjoy the economies of scale. This will result in fewer, larger customers for ag inputs, which will in turn result in consolidation among suppliers.

## 15.6 What Will Happen?

While the "Full Speed Ahead" scenario would provide the best future for soybean farmers and others in the soybean value chain, and the "Offshore Migration" scenario would be the least desirable for the US soybean industry, the purpose of the Soy2020 exercise was to consider strategies to optimize the US soybean value chain regardless of what the future world unfolds. In order to react appropriately to whatever scenario develops, it is important to determine the direction the industry is heading as early as possible, so the Soy2020 participants identified a number of early indicators for each scenario.

Some early indicators of these scenarios include (NOTE that the same indicator may appear in more than one scenario):

## 15.6.1 Full Speed Ahead

- Consumers quickly accept new soybean traits.

  New varieties are introduced that meet consumers needs, and consumers accept and pay for the added value. The public accepts new varieties produced by genetic transformation (biotech).
- Integrated value chains form in the soybean industry.
- Producers adopt new technology rapidly.
- Public and private investment for soybean innovations increases significantly.

  If investors and industry see that the global regulatory environment is favorable for innovation such as improvements through genetic transformation and if the public accepts these improvements, then investment will flow into soybean technology.
- Industry groups are successful at creating and deploying incentives for technology adoption.

  It is a major challenge to move from research/development of varieties with improved traits into production and marketing of those varieties. At the introduction, the market is small; it takes time to build demand; each step in the value chain wishes to capture as much of the added value of the new trait as possible. If the value chain works together to introduce the product, share the value, and build the market, then new technologies will be adopted readily and improved soybean varieties will be grown.
- Improved soybeans with traits meeting consumer needs regularly reach the market.
- Soybeans with specific traits provide sustained profitability throughout the value chain.

  It will be difficult to introduce new traits, no matter how valuable they may be, unless the entire value chain benefits from the added value. There is always a risk in bringing a new product to market, and if the value chain is not compensated for that risk, the value chain will stick with the commodity soybean it knows.
- Intellectual property dispute adjudication and mediation is accelerated.
- Technology to protect trait technology is developed and deployed.

  As an open pollinated crop, soybean seed can be saved for planting (bin-run seed). Unlike hybrid corn, which farmers must purchase anew each year if they are to enjoy the benefits of the hybrid, biotech soybean seed harvested one year can be saved and planted again the following year, although this is not permitted by the purchase contracts for biotech seed.

Technology companies must be confident that they will be paid for the value of their intellectual property or they will not spend the tens of millions of dollars required to develop new traits and to shepherd the new product through the regulatory process. Technology to prevent farmers from saving and planting seed would ensure that the intellectual property is protected.

- Intellectual property rights have strong global protection.

In an increasingly global economy, it is not enough for the intellectual property of technology providers to be protected in the US and other developed countries. It is vital that these rights be protected throughout the world.

- The US livestock industry continues and is successful.

Approximately 95% of soybean meal processed in the US is consumed by livestock and poultry. However, there are pressures on the livestock industry including people who do not want new/expanded livestock operations in their neighborhood, animal rights groups, and environmental groups who want to regulate livestock operations more stringently.

If the livestock industry addresses the concerns and continues to grow and to be successful, then the market for soybeans in the US will do likewise. However, if it becomes too difficult for livestock producers to operate in the US, they can exercise the option of moving offshore. As one USB chairman stated, "Pigs and chickens that speak Portuguese do not eat US soybean meal".

- Global trade agreement(s) are passed favorable to technology and trait propagation.
- Large investments are made in identity-preserved systems, especially for storage and transportation.

For soybeans to move from a commodity to value-added crop(s), it is necessary to develop the infrastructure to handle different types of soybeans, rather than handling them as a commodity. This will include ways to produce the crop, to measure the traits of interest accurately and consistently, and to deliver, store, ship, and process the value-added crop in reliable and efficient ways. To build new systems or to modify existing systems to meet these needs will require substantial investments.

## 15.6.2 Shared Success

All of the above conditions apply, in addition to the following:

- R&D investments are greater for other crops than for soybeans.
- Innovation development for competing crops is significant and rapid.
- Integrated value chains form for other crops.

## 15.6.3 Off-Shore Migration

- Increased consumer and environmental lobbies limit growth in soybean innovation.

  Some soybean trait improvements can be accomplished through conventional plant breeding, but others can be done only through genetic transformation. The US cannot bring meaningful additional acres into production and cannot be the low-cost producer. Thus, if the US cannot continue to innovate because of regulatory costs and constraints or because of consumer concern, there is less and less reason to produce soybeans in the US and more justification for moving production offshore.

- Farmers shift significant acres out of soybeans to produce other crops.
- Farmers increasingly feel that they grow soybeans because they have to, not because they want to.

  Farmers who believe that the only reason they grow soybeans is to have something to rotate with their "real" crop, will plant as few acres as possible, rather than choosing to plant more acres of soybeans.

- US farmers increasingly develop foreign-based partnerships for utilizing land in other countries for soybean production.

  As with other sectors of the market, agriculture is global. Some farmers presently have interests in farming in other countries. If the environment for agriculture in general and soybeans specifically becomes too unfriendly or unprofitable, more farmers may choose to seek foreign-based partnerships so they can grow their crops, including soybeans, elsewhere. This will be to the disadvantage of US soybean production, but shifting resources via foreign-based partnerships will benefit soybean production in other countries.

- The production costs for agriculture offshore continue to decrease.
- Infrastructure investment in other countries increases.

  At present, the US enjoys a huge advantage over South America in its roads, railroads, and navigable rivers. While Brazil can produce vast amounts of soybeans, getting them to the terminal or market is often slow and difficult. If other countries make investments in infrastructure and the US fails to maintain and improve its infrastructure, the US could easily lose its advantage.

- There is less collaboration within the soybean industry, limiting growth in innovation and positive change.

  Developing and introducing soybeans with new traits that meet market needs must benefit the entire value chain to be successful. However, introducing value-added traits into what has been a commodity market requires that the industry work together. The United Soybean Board and others in the value chain have formed a coalition called QUALISOY. Among other things, QUALISOY seeks to assist the process of taking new products from the researcher's field to the market. Without industry-wide cooperation, it will be difficult, if not impossible, to move toward value-added soybeans.

- R&D investments increase more for other crops than for soybeans.
- Development of innovations in other crops is rapid and significant.
- Public and private investment in soybean innovation drops significantly.

Investors and industry seek winners. If soybean acreage is going down and soybeans are seen to be less desirable than alternative crop choices, investors and industry will put their efforts elsewhere. This will be a downward spiral. Less investment will mean less innovation and fewer improved soybean varieties. Less innovation will mean the gap will increase between soybeans and "preferred" crops and even fewer soybeans will be planted and grown.

- Integrated value chains form in other crops.
- There is accelerated adjudication and mediation of intellectual property disputes.

For technology providers to feel comfortable sharing new traits with other countries, it is not enough that there be strong protection of intellectual property. It is also necessary that disputes be adjudicated in a timely fashion. Otherwise, an issue can drag on for years without resolution. During this time, the technology provider loses the value of its property.

- Trait protection technology is developed and deployed.

If technology providers can protect their products by, for example, methods to prevent germination of farmer-saved seed, then they will be less concerned about distributing this technology to other countries where protection of intellectual property is less robust than in the US. It should be noted that such technology can be used selectively, allowing for example, peasant farmers to save and replant their seed, while applying the technology to seed sold to large-scale commercial farmers.

- Protection of intellectual property rights is stronger.
- US agribusinesses maintain strong global intellectual property protection.
- Price basis reflects a lack of nearby soybean demand.

If the US loses its customers, demand for US soybeans will decrease. If, for example, a substantial percentage of livestock production moves offshore , demand in the US for soybeans will decrease. In addition, transportation costs to new markets will be reflected on the basis of soybean pricing, reducing the price the farmer receives.

- The increase in biodiesel production continues without new technology innovation in soybean for food and feed use.
- Processing plants slow or shut down in the US and expand in other countries.

  Fewer markets for US soybeans and fewer farmers choosing to plant soybeans will mean a decreased need for soybean processing in the US. Likewise, if production increases elsewhere and/or if livestock moves offshore , the need for processing will grow in other countries, and industry will expand to meet that demand.
- US funding does not increase significantly for infrastructure and transportation system repair and improvement.

  The US presently has an advantage over other countries in infrastructure, but this is not something that maintains itself. The US could lose its advantage in this area, if the US fails to maintain and improve locks and dams, railroad beds, highways, bridges and so forth.

## 15.6.4 Forward to the Past

- Consumer and environmental lobbying limit growth and increase in innovation.

  As has been seen in Europe, if activist groups are successful at raising the regulatory or political hurdles to bringing technological innovation to market, or if activists are successful at convincing the public that the risks of new products outweigh their benefits, then technology providers will not spend the millions of dollars required to bring new products to market.
- Offshore production costs for agriculture continue to decrease.
- Offshore investment in infrastructure increases.
- Collaboration within the soybean industry decreases, limiting growth in innovation and positive change.

  As described above, moving a new product from research and development into the market is difficult. It is a badly over-used term, but in this case an accurate one to say that moving from a commodity focus to a value-added market is a paradigm shift, and such shifts are difficult to accomplish. If the soybean industry fails to work together, few new products will succeed in the market.
- Consolidation throughout the industry increases rapidly.
- Investment in all crop innovations, including soybeans, drops.
- There is a continued successful livestock industry.

  As previously stated, the vast majority of soybean meal processed in the US goes to feed livestock and poultry. For US soybean production to stay essentially the same, the US livestock industry must continue to thrive.

- Biodiesel production continues to increase, but without new technology innovation in soy for food and feed use.
- Industry groups have little success creating technology adoption incentives in production policy and programs in the farm bill.

  Government policies and programs can help introduce new products or new processes. For example, there are programs to encourage ethanol and biodiesel production. Similar programs could encourage farmers for the first few years to grow soybeans with improved compositional traits. If such programs are not implemented, it may be more difficult to make the transition from commodity to value-added crop.

- Funding is not significantly increased for US infrastructure and transportation repair and improvement.

  Depending upon who is asked, present levels of funding for US infrastructure are either just sufficient to maintain the present level or not enough even for that. In any case, to improve the infrastructure to adapt to value-added crops will require increases in funding. If such funding is not available, it will be difficult or impossible for the present system to handle value-added crops efficiently.

It is clear that the direction of soybean production will be strongly influenced by technological innovation and meeting the needs of the value chain. In some cases, technological development will involve infrastructure and transportation. Such development will allow efficient identity preservation of soybeans with different traits and will allow soybeans to be moved quickly and efficiently to meet market demands. Other technological innovations will involve changing the soybean itself by increasing yield, increasing its ability to resist biological or environmental stresses and by changing the composition to meet the needs of the end user better.

If the US soybean industry can continue to innovate and to improve soybean quality to meet the needs of the value chain, then the Full Speed Ahead or the Shared Success scenario will be the result, and the US soybean industry will grow and stay strong. If technological innovation is stymied by lack of R&D investment or an inhibitory regulatory environment and, as a result, soybeans cannot be improved to meet end users' needs, then soybean production will move offshore or, at best, the soybean industry will stagnate and remain a commodity market.

While some strategies will vary, depending on how the future develops, there are other strategies that will be common to all scenarios. These strategies should be implemented regardless of the scenario that may develop. They fall into three areas: Anticipate, Collaborate, and Act.

Among the strategies is the need to improve soybean quality and to lead global soy improvements by focusing R&D expenditures to improve yield, quality and functionality of soybeans in food, feed, fuel and other end user markets.

Research is needed to produce soybeans efficiently. This includes the need to protect soybeans from biological and environmental stresses, like diseases and drought.

Research is also needed to change the composition of soybeans to meet the needs of end users better. Food companies do not want trans-fat labels on their products. Thus, it is necessary to modify soybean oil so that it has the oxidative stability and functionality required without the need for partial hydrogenation, since trans-fats are an artifact of hydrogenation. This means soybean oil with lower-linolenic acid and higher levels of oleic acid.

Although it has been somewhat overlooked since the trans-fat issue has caught the headlines, reduced saturated fat in oil is also needed.

Soybean processors want to be sure the soybeans they buy contain enough protein to make a high-protein meal. Feed integrators are concerned not only about the total amount of protein, but also about the amino acid balance in the soy meal.

Phytate-phosphorus in soybean meal is not readily digested by swine and poultry. This means that feed rations must be supplemented with inorganic phosphorus to meet the animal's needs. It also means that undigested phytate passes out in the waste stream and into the environment, creating concerns. Reducing the phytate-phosphorus in soybeans would help to address these concerns.

Soybean meal is not only a source of protein for feed rations; it is also a source of calories, or metabolizable energy. If hard-to-digest sugars like raffinose and stachyose can be reduced and replaced with sucrose, the caloric value of soybean meal will be increased, adding value for the feed integrator.

To meet the needs of the end user and to increase soybean yields requires modification of the soybean's genetics. For decades, plant breeders have used conventional breeding techniques to improve soybeans. This process is long and laborious, requiring many years. In some cases, desired improvements simply cannot be achieved by conventional breeding.

Genetic transformation offers another approach to soybean improvement. It allows traits to be introduced that cannot be introduced through conventional breeding, traits like herbicide tolerance or insect resistance. However, the regulatory hurdles are high and public acceptance in some parts of the world has been slow.

Whether one wants to improve soybeans through conventional breeding or transformation, genomics tools are vital. Genetic markers, micro arrays, functional genomics, maps, etc. are all tools breeders and other researchers can use to make the process of soybean improvement easier, faster and more efficient.

The US soybean industry cannot predict exactly what will happen in the future, and the US soybean industry can only influence to a limited extent what takes place elsewhere in the world. . If research and technical innovation are supported and encouraged, if infrastructure and transportation meet the market's needs, and if the regulatory environment is conducive to adoption of new technology, then the future for soybeans in the US is bright and Full Speed Ahead. If research is stifled, infrastructure is allowed to deteriorate, and inhibitory regulations are implemented, then the US can watch its soybean industry contract and move offshore.

At present, global soybean production is not a zero-sum game. As long as the global demand for soybeans grows and the "pie" gets bigger, then soybean producers in all countries can enjoy their "slice". However, if in the future the market matures and demand stabilizes, then for one slice of the pie to grow will mean that another must decrease. Research to increase soybean production and improve soybean quality will help create conditions under which the pie will continue to grow for many years to come.

# Index

# Color Plate Section

## Chapter 7

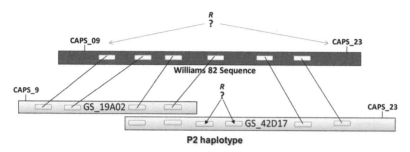

Figure 7-6 Physical map of the *R* region.
In the hypothetical example, the *R* gene to be cloned is present only in parent P2. Blue rectangles represent putative genes. Physical mapping data (Fig. 7.5) showed that the P2 haplotype is bigger than the Williams 82 haplotype. Therefore, a BAC library for the P2 haplotype was constructed and two overlapping BACs, GS_19A02 and GS_42D17, carrying CAPS_09 and CAPS_23, respectively, were isolated. Sequencing of these two BACs revealed that there are two additional open reading frames in the *R* region of P2 haplotype as compared to that in Williams 82.

## Chapter 9

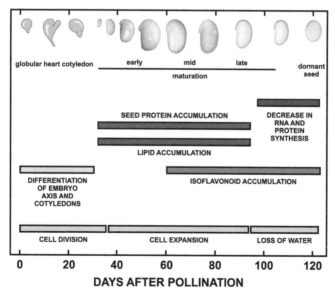

Figure 9-1 A diagram representing physiological and developmental events that occur during soybean seed development (modified from Le et al. 2007).

## Chapter 10

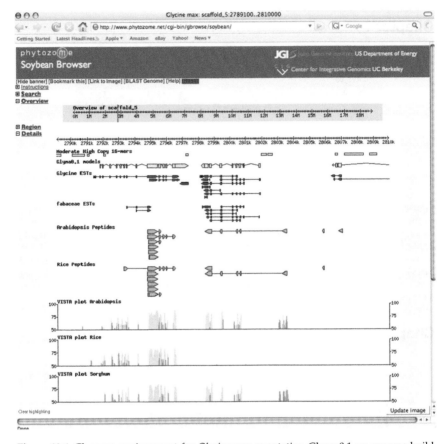

**Figure 10-1** Gbrowse environment for *Glycine max* annotation Glyma0.1 on genome build Glyma0 available at *http://www.phytozome.net/cgi-bin/gbrowse/soybean*

**Figure 10-4** An FPC representation of scaffold 22 that shows a potential scaffold break (inverted segment) between 5,928 kb and 6,712 kb.

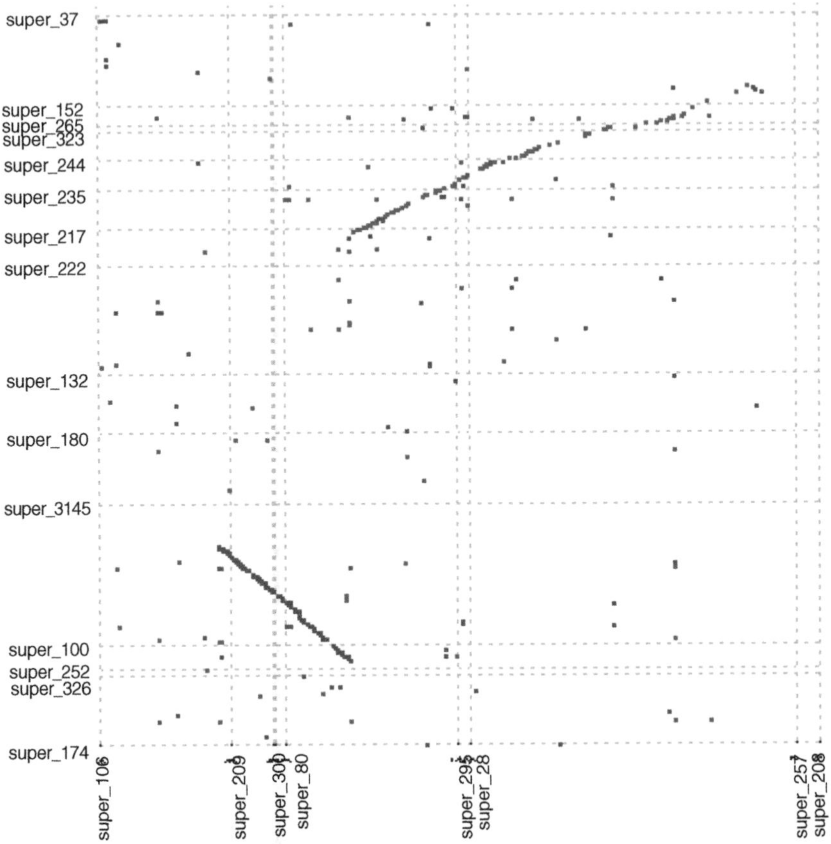

**Figure 10-6** Synteny comparisons between parts of two chromosomes. The horizontal axis is approximately one quarter of chromosome 1 (formerly D1A), and the vertical axis is approximately one quarter of chromosome 2 (formerly D1B). Each dot is the top "promer" match between predicted genes on two chromosomes. Thus, a given gene may have up to one top match per chromosome pair, or up to a total of 20 "top matches." Scaffold boundaries are gray dotted lines.

# Chapter 12

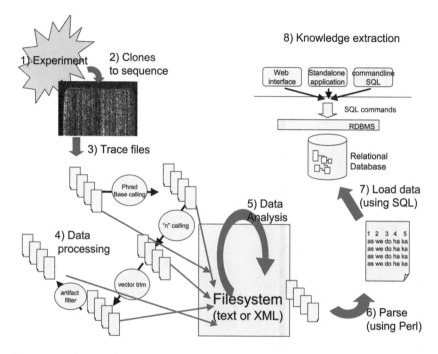

**Figure 12-1** A typical sequence processing and data web publishing project. Biologists perform an experiment (1), which generates clones that are to be sequenced (2). Chromatograms or trace files readable to computer software only, is the output from the sequencer (3). The trace files are base called and trimmed in a processing pipeline to yield readable sequences (4). For each step, output files are generated and these are often stored in a strict file system. Further data analysis and sequence annotation (BLAST, GO Gene Ontology etc) is performed on the processed sequences and output files from these analyses are stored in the file system (5). Pertinent information is extracted, parsed from the analysis output files into a more concise format (6). This is often done using custom scripts written in Perl. Finally, many large projects have their own database where the data is web published and made available to the scientific community (7, 8). Relational database management systems such as MySQL and Oracle are the most popular.

**Figure 12-2** Gene Ontology Home. a) Gene ontology homepage (*http://www.geneontology.org*) displaying the search by "gene or protein name" and "GO term or id" function. b) Example of results showing tree-like structure hierarchy.

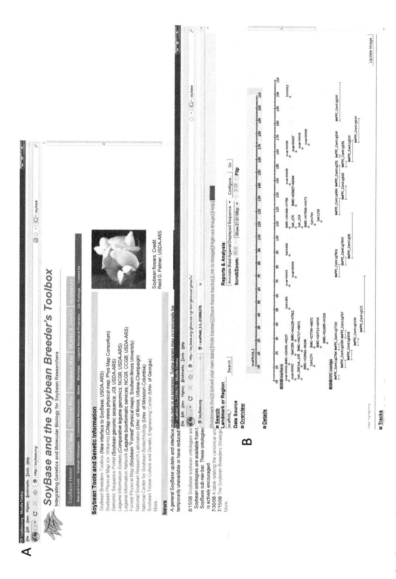

**Figure 12-3** SoyBase and the Soybean Breeder's Toolbox. a) The SoyBase homepage (*http://soybase.org*) illustrating the genetic map, physical map, sequence map and BLAST & Search tabs. b) Picture of SoyBase soybean browser, a Sequence Map tool, displaying the markers and FPC contigs located on Scaffold 1.

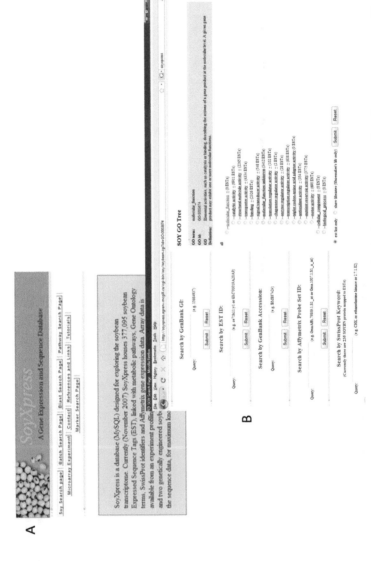

**Figure 12-4** SoyXpress A gene expression and sequence database. a) The SoyXpress homepage (*http://soyxpress.agrenv.mcgill.ca*) illustrating the tools available such as the Soy Search page, Batch Search Page, Blast Search page and Pathway Search page b) Snapshot of the Soy Search Page displaying the different terms available for search (GenBank ID, EST ID, GenBank Accession, Affymetrix Probe Set ID and SwissProt Keyword) and the searchable Soy GO tree.